The Growing Classroom

Garden-Based Science

Roberta Jaffe
Gary Appel

Developed by Life Lab Science Program, Inc.

Illustrator: Kate Murray

Dale Seymour Publications
Parsippany, New Jersey

Life Lab Science Program, Inc. is a private non-profit organization dedicated to the enhancement of garden-based science education. Its focus is to offer teacher inservices nationwide as part of the U.S. Department of Education's National Diffusion Network,* and to develop new curriculum with support from the National Science Foundation and Addison-Wesley Publishing Company.

Life Lab is validated as an exemplary science program by:
U.S. Department of Education
National Science Teachers Association
California State Department of Education
California School Board Association

**Programs in the National Diffusion Network have been developed by local school districts, institutions of higher education, state departments of education, and private non-profit agencies, and have been examined by the U.S. Department of Education to determine that they are appropriate for dissemination. While the Department has determined that the programs are appropriate for dissemination, no Federal endorsement of content or philosophy should be inferred.*

Dale Seymour Publications®
An imprint of Pearson Learning
299 Jefferson Road, P.O. Box 480
Parsippany, New Jersey 07054-0480
www.pearsonlearning.com
1-800-321-3106

PHOTO CREDITS:

Bill Aron Photography: Page 103

The following photographs were provided expressly for the publisher by Tim Davis:

Inside jacket flap; pages 13, 19, 20 (top), 21 (top), 39, 151 (all except top left), 171, 209, 229, 253 (center, bottom), 281, 315 (center, bottom), 345, 369

A previously published edition was made possible as follows:

1st printing: April, 1982, California Department of Education, Title IV-C
2nd printing: November, 1982, California Department of Food and Agriculture, Division of Integrated Pest Management
3rd printing: March, 1985, California Department of Education, Vocational Agriculture Program

ISBN 0-201-21539-X

9 10 11-DR-02 01 00

Contents

Nutrition Units

Contributors

Margaret Cadoux
Lisa Glick
Teresa Buika
Stephen Tracy
Dawn Binder
Kim Fine
Kay Thornley
Thomas Wittman
Christine Aldecoa
Stephen Rutherford
Vivian Gratton
Don Harmon
Marion Dresner

Science Advisor

Stephen Gliessman, Professor of Environmental Studies, University of California, Santa Cruz

Other Supportive People and Organizations

Ann Leavenworth [deceased], former president, California Board of Education

George Buehring, former principal, Green Acres School, Santa Cruz, CA

Neil Schmidt, former superintendent, Live Oak School District, Santa Cruz, CA

Robert Yager, former president, National Science Teachers Association

Lucille and David Packard Foundation

Live Oak School District

Santa Cruz County Office of Education

University of California, Santa Cruz

Foreword

Dr. Roger Johnson, Professor of Curriculum and Instruction and Co-Director of Cooperative Learning Center, University of Minnesota

In 1982, I was part of a National Science Teachers Association (NSTA) team whose mission was to find the most exemplary science programs in schools throughout the United States.

We were searching for programs that matched the criteria for ideal science instruction identified by the National Science Foundation study, Project Synthesis. Nominations came in from all over the country, from schools big and small, rural and urban, rich and poor. The committee read, evaluated, and debated.

Finally eleven exemplary elementary programs were chosen, and two of those were further selected to be "Centers of Excellence"; one of those was a small garden-based project in California called *Life Lab.*

A site visit was arranged and Bob Yager, then President of NSTA, and I arrived not quite knowing what to expect. We were met by an enthusiastic group of Life Lab teachers. We talked to students, teachers, administrators, community members, and volunteers. Students were in the garden, or "living laboratory," setting up experiments, collecting data, and reporting results. Cooperative groups were making compost and transplanting seedlings. We heard students talking about being entomologists as they examined pests in their test plots. We watched as teachers worked alongside their students. This was science in action; it was integrated; and it was working!

At the conclusion of our visit, Bob Yager wrote:

"In many respects, the science program at Green Acres Life Lab best represents the new directions for elementary school science that were synthesized from current data/indicators by the (Project) Synthesis researchers . . . . The program emphasizes appropriate consumer behavior; it involves students in gathering first hand information. It is designed to make a difference in the daily lives of students. The program visualizes the students as central, a focus of study. Personal health is emphasized with a goal of affecting habits and quality of life. The program has a strong environmental component where the interdependence of humans and their environment is central. The emphasis is placed on stewardship and responsible personal behavior and actions . . . . One can only wonder what would happen to school graduates, what a future generation would be like if most had such science experiences as those that exist at Green Acres Life Lab."

I like Life Lab. I am charmed by its humble beginnings in a low-income school's dirt parking lot. Created by teachers for teachers, its activities and processes match what we know about good science education. It is a concrete, active program which promotes cooperation among students. Lessons are built on questions that encourage inquiry rather than answers that halt thinking. The "Living Laboratory" is real. Once you see the excitement of a young child who is harvesting a first carrot or radish, you have to wonder how education ever moved so far from its roots.

How To Use This Book

Organization of Sections

This sourcebook is a curriculum for teachers in developing a garden-based science program. It can be divided into four sections.

BREAKING GROUND, the first section, provides general information about Life Lab and explains why the garden as a "Living Laboratory" provides an important and exciting context for learning science. This section provides valuable information for starting a school garden and incorporating it into the classroom, and gives suggestions for adapting the Life Lab program to your specific needs and resources. Techniques for managing the class in the garden and methods for cultivating community support are explained.

The second section is the SCIENCE curriculum, which includes both outdoor and indoor activities. It is divided into ten science units:

Let's Work Together (Problem Solving and Communication)
We Are All Scientists (Awareness and Discovery)
The Living Earth (Soil)
Growing
Living Laboratory (Outdoor Gardening)
Cycles and Changes
Interdependence
Garden Ecology
Garden Creatures
Climate

The third section is the NUTRITION curriculum, and is divided into three nutrition units:

Food Choices
Nutrients
Consumerism

Other useful information for both the curriculum and gardening is contained in the APPENDIX. Included are blackline masters, equipment designs and planting guides, a scope and sequence chart, a complete materials list for each unit, an English/Spanish vocabulary list, a seed company list, and additional resources.

It is important to plan how you will use this sourcebook. A Scope and Sequence chart recommends grade levels for each activity. (Grade levels are repeated with each unit introduction and on each activity page.) We recommend starting slowly and adding sections in increments. Select units that relate to what the class is doing in the garden. For example, if you're preparing the garden soil, it would be most appropriate to incorporate The Living Earth unit.

Format of Activities

All activities follow a simple format as outlined in the following example.

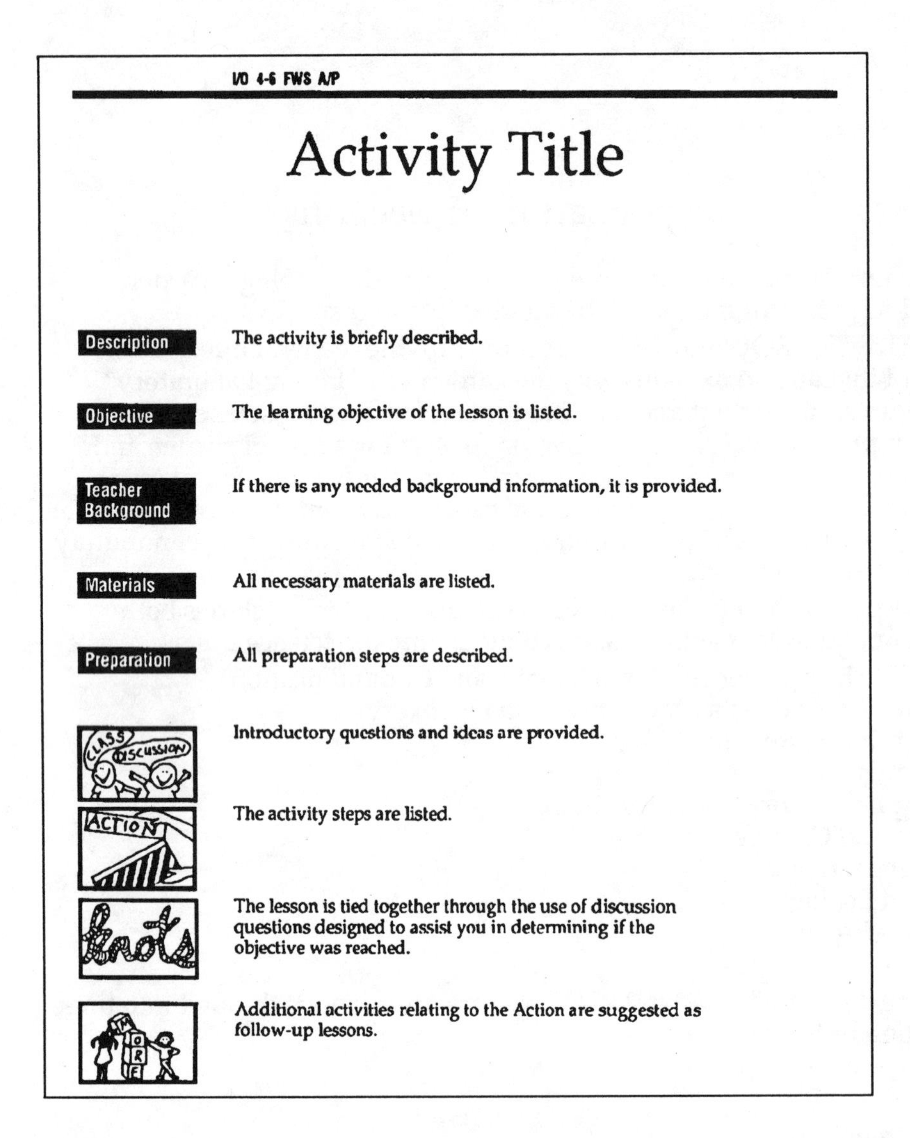

I/O 4-6 FWS A/P

Activity Title

Description — The activity is briefly described.

Objective — The learning objective of the lesson is listed.

Teacher Background — If there is any needed background information, it is provided.

Materials — All necessary materials are listed.

Preparation — All preparation steps are described.

Class Discussion — Introductory questions and ideas are provided.

Action — The activity steps are listed.

Knots — The lesson is tied together through the use of discussion questions designed to assist you in determining if the objective was reached.

More — Additional activities relating to the Action are suggested as follow-up lessons.

The code line at the top of each activity is interpreted as follows:

I	indoor activity
O	activity best done outdoors
4-6	suggested appropriate grade levels
FWS	seasons that are best suited for the activity (Fall, Winter, Spring)—this designation may vary in your particular region
A	project can be completed in a class session or period (approximately 40 minutes)
P	project requires more than one class session

Student Journals

Student journals are referred to in many lessons. Journals can serve as both a place for students to record data and information regarding their experiments, and as a focus for feelings and observations. The more kids explore their world, the more they want to express and communicate the results of their explorations. We encourage you to capitalize on this natural motivation to write and draw.

Taste with Care

Your Living Laboratory will be a sensory delight, and part of the learning and excitement includes the students' eating the fruit and leaves of their efforts. As a word of caution, please be sure that students know the difference between edible and inedible plants; that you and other teachers are aware of any food allergies children may have; that any school district guidelines or restrictions are observed; and that Life Lab participants wash their hands after gardening and before eating.

Dedicated to

All of the teachers, administrators, and communities throughout the United States who have helped to make Life Lab what it is

and

Our Kids (The Walking Laboratories)

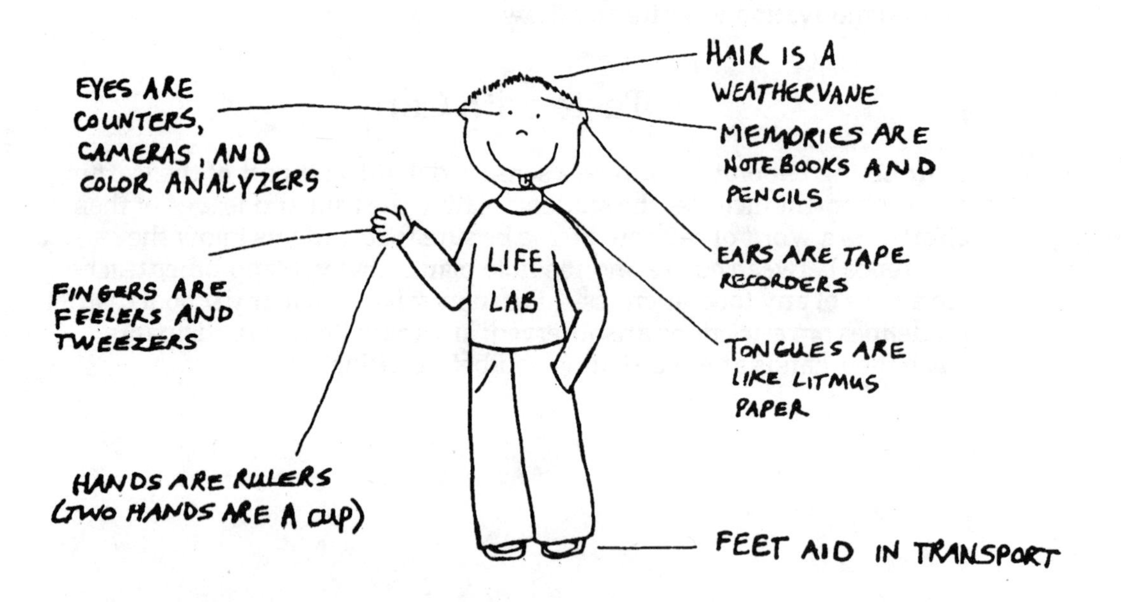

Kid (*Homo sapiens*): Belongs to many different families. Found in all parts of the world and in many different environments. Grows right before your eyes. Recently, kids have been found more often in gardens. Kids can be both beneficial and detrimental to garden areas, but the benefits far outweigh the detriments. Males and females of the species (sometimes called boys and girls) differ slightly physiologically and occasionally otherwise, but have much the same appearance and benefits. Kids are particularly nice to have around Life Lab for they add to the richness of the garden and are the most beautiful crop around.

Mary Cunningham

Breaking Ground

Green Acres School
Santa Cruz, CA 1979
(Before)

Green Acres School
Santa Cruz, CA 1983
(After)

Green Acres Elementary School
Live Oak School District
Santa Cruz, California

Life Lab inspires you to learn more about science!
Naima Leal, 4th grader

The discovery that a garden is a wonderful laboratory to experience science for both students and teachers was the motivation for starting the original Life Lab program. What is now a two-acre Living Laboratory, including a one-acre vegetable and flower garden, fresh water pond, science museum, solar greenhouse, and natural field study, started out as a small garden used by only a few classes in 1978.

Now twenty-five K-5 classes use their Life Lab to study a wide variety of life, earth, and physical sciences. Their Living Laboratory gives them the opportunity to practice the skills of observation, description, comparison, measurement, problem solving, and critical thinking.

Green Acres Life Lab is an example of the community and the local university actively forming a partnership with the school to enhance the education program. Parents raised funds for a barn and managed a successful 4-H program. University students developed many special projects with students, including a recycling center, science museum, and solar oven.

INTRODUCTION

Breaking Ground

SCIENCE + GARDENING = LIFE LAB

The Life Lab Science Program engages students in the exploration of science in an exciting and meaningful way, helping students create a sense of wonder about the world around them. The instructional approach is a combination of indoor and outdoor experiential activities taught within the context of the Living Laboratory. The Living Laboratory—an outdoor garden or planter box, an indoor growing center, or both—provides students with the opportunity to plan, create, and care for their own environment. In the process, they ask questions about their environment, and make observations, research topics, and set up experiments in search of answers. Drop by a Life Lab school any weekday, and here's what you may find: fourth graders carefully graphing the growth of their seedlings; third graders enthusiastically collecting and weighing their potato harvest; fifth graders attentively observing and recording the temperature of their compost pile; young scientists carefully testing their garden soil for nitrogen, phosphorus, and potassium; students using the Life Lab Guess-Test-Tell scientific method to prepare experiments in their classroom; boys and girls preparing nutritious snacks from their garden beds; and much, much more.

A large part of Life Lab learning takes place in small groups, where the students work together to solve problems and apply what they have learned. Opportunities to stimulate curiosity and share new knowledge are maximized as students explore science both in the classroom and the garden laboratory. The Life Lab Science Program has two basic goals:

- To assist schools in developing living laboratories in which young students can discover their world through scientific exploration
- To provide curriculum and inservice so that every teacher feels comfortable teaching science in a living laboratory

As educators, we have the opportunity to nurture a child's curiosity and desire to explore the world. We also have the responsibility to prepare the child for the world of tomorrow. Although we can't see that world, we can help students ask relevant questions; develop processes for thinking and searching for answers; and communicate, work, and live cooperatively. This is an exciting and important task, and as we teach students basic academic skills, we can incorporate learning processes that will help them be informed, knowledgeable, responsible citizens. That's what Life Lab is all about.

GETTING STARTED

Life Lab looks different at each school because the program can be tailored to fit specific needs and goals of an individual school and its surrounding community. However, all Life Labs share some common ingredients:

- *The Growing Classroom*, Life Lab's curriculum, provides you with year-round science activities. The activities are organized in units related to specific concepts and topics from "Cycles and Changes" to "The Living Earth/Soil" to "Climate" and "Food Choices." Each activity is keyed to let you know the appropriate grade level, whether it can be done indoors or outdoors, and its length. In addition, the introduction offers strategies for organizing and managing your Life Lab program from planning your garden laboratory to implementing the curriculum to involving administrators and your community.

The Life Lab Science Program offers awareness presentations, teacher inservice, and ongoing technical assistance in using the curriculum. The inservice is designed to provide educators with opportunities to experience activities from the students' point of view, learn basic science processes and gardening techniques to use with elementary students, and develop management strategies for implementing Life Lab at your school. Life Lab provides inservice nationwide through the U.S. Department of Education's National Diffusion Network. For more information, detach and return the appropriate postcard in this book.

■ *The Living Laboratory,* your Life Lab garden, provides the context for the Life Lab approach to science. Each time your students plant a new crop or set up a garden-based science experiment, they will incorporate basic science processes including observing, comparing, organizing, relating, inferring, and applying as they learn more about the growth habits of plants, their pollinators, pest problems, and so on. In Life Lab, students actively practice these processes in all of the topics they study. Students are encouraged to monitor the garden regularly, make observations, and record them in their Life Lab journals.

Living Laboratories come in all shapes and sizes. They thrive in both cold and warm climates, as well as in urban, suburban, and rural communities. Some are simple planter boxes located outside the classroom. Some are abandoned areas of the schoolgrounds that students transform into flourishing cornucopias. Some use greenhouses or coldframes (miniature greenhouses) to extend the growing season. All Living Laboratories provide an exciting environment for students to discover that *Science Is Alive.*

Ladson Elementary, Lodson, SC

■ *The Life Lab Center,* a learning center in your classroom, is an important component of your Life Lab Program. It may contain reference materials for students to research answers to their questions. It may have ongoing experiments and displays, as well as bulletin board space where students can keep charts and graphs of different data. It provides an opportunity for discoveries in the Living Laboratory to be pursued in greater depth. Also, 80 percent of the activities in this book can be done in the classroom Life Lab Center.

Could this be your school? Based upon the experience of educators around the country, we're sure that it can. With the enthusiasm and efforts of teachers, administrators, students, and your community, anything is possible. We encourage you to start small, dream big, and watch your efforts blossom! We are glad to assist you in anyway that we can.

Life Lab Science Program
1156 High Street
Santa Cruz, CA 95064
(408) 459-2001

PROGRAM IMPLEMENTATION AND MANAGEMENT

Since Life Lab integrates various teaching strategies (hands-on, experiential learning; utilization of indoor and outdoor classroom laboratories; small group and cooperative learning), it is important to take time to plan your Life Lab program. The key components to a successful program are

- Supportive administration
- Minimum of three teachers using the program at the site
- Regular meetings of a Life Lab steering committee
- Regular classroom instruction using *The Growing Classroom*
- Establishment of a community support committee

Following are some suggestions from successful Life Lab schools.

- Your principal is a key advocate for your science program. He or she should be an integral part of the initial planning and then be kept regularly informed of the program's progress. Support might include providing planning and inservice time, making presentations at faculty meetings, allocating funds for site development, involving community resources. We have seen principals invite school board members to observe the program and integrate it with the district's cafeteria by having student-grown vegetables incorporated into weekly menus and cafeteria waste saved for the garden compost! Your principal is an important ambassador.

- Set up a Life Lab steering committee. We strongly recommend a team teaching approach to Life Lab. If more than one teacher is involved, it is easier to manage the Living Laboratory, gather necessary supplies, brainstorm ideas, and gain school and community support. The planning committee can consist of several participating teachers, school volunteers, and an administrator. Monthly meetings can be used to plan the site development, organize grade-level use of the curriculum, and plan special Life Lab events. One school using Life Lab has a school-wide creative writing day based on students' Life Lab experiences, another has an annual school-wide plant sale, and many have organized annual Science Fairs and Olympiads.

LIFE LAB IN APRIL				
1 • PLANT CARROTS	2 • COMPOST • WATER	3 • PLANT BEETS	4 • WATER	5 • TRANS-PLANT LETTUCE
8 • SOW BEANS	9 • COMPOST BROCCOLI • WATER	10 • WEED SHARED AREAS	11 • WATER • SOW LETTUCE	12 • SOW RADISHES
15 • TRANS-PLANT PEAS	16 • WATER	17 • SOW CORN	18 • WEED • WATER	19 • DOUBLE DIG BEDS
22 • WATER	23 • TRANS-PLANT SQUASH	24 • WATER	25 • PLANT BASIL	26 • PLANT MINT
28 • WATER • WEED	30 • PLANT NASTURTIUM			

- Plan the management of the Living Laboratory. Detailed steps in establishing your Life Lab garden begin on p. 9. Developing the program is a longterm project. We recommend you start with the question: "What do we want Life Lab to look like in three years?" With that vision, you can plan annual goals with reasonable timelines and give your program the needed time to develop. For daily operation, it is important to develop a simple maintenance plan that includes a class-use schedule; a volunteer schedule; a watering schedule; a site development plan; a site maintenance plan for weeding, fertilizing, and so on; and a supply ordering system. An organized system with delegated responsibilities and a calendar schedule make management much easier. Some Life Labs have a volunteer or an aide coordinate the outdoor site. This greatly assists the teachers in their already busy days.

TEACHER Ms. Buckley WEEK OF April 5th
GRADE 5th

	1:00 - 2:00	2:00 - 2:45
MON.	SPELLING SOCIAL STUDIES	P.E. HEALTH
TUES.	LIFE LAB →	FLOWER POWER PART 1
WED.	SPELLING SOCIAL STUDIES	P.E. HEALTH
THURS.	LIFE LAB →	FLOWER POWER PART 2 MAGIC SPOTS (OUTDOORS)
FRI.	SPELLING P.E.	ART

■ Integrate Life Lab into your regular class program. You may choose to follow the Scope and Sequence in the Appendix or develop one for your school's needs. If your class is scheduled to use the Living Laboratory once a week, you may plan an outdoor lesson for that time and then two follow-up lessons for the classroom. The sample chart on this page shows one way to integrate these activities. From the beginning of your Life Lab implementation, reinforce cooperative, small group work with your students. The first unit of this curriculum, "Let's Work Together/Problem Solving and Communication," will prepare your students to learn from each other. This type of classroom management will allow you to be a resource facilitator while students work together on hands-on activities. For some activities, you may find it easier to divide the class in half. While you coordinate the hands-on activity with part of the class, the remainder of the class can work on a quiet lesson.

■ Life Lab needs will go through seasonal changes. For schools in cold climates, winter can be used to do many of the classroom lessons. Fall can be extended and spring can be given an early start by using a greenhouse or coldframe (see plans in Appendix). Summer will bring an abundant garden, which can be a problem if there are no students around. Schools have found many creative ways to care for summer gardens. The most popular way is to have a summer program. Some schools actually coordinate a science program as part of their summer school; others let city recreation programs use the garden for a children's environmental education camp. It is also possible to have students and their families care for the garden in exchange for the produce. Students can learn a great deal from returning to a garden that is wildly overgrown, too. We recommend planting short crops such as lettuce, radishes, carrots, and broccoli in early spring so that students can harvest them before school is out, and long crops such as pumpkins, corn, and tomatoes in late spring so that students can harvest them in the fall.

■ When your program is ready to involve the community through soliciting donations, asking for volunteer aides, or developing greater community awareness, we recommend you establish a community support committee to assist in coordinating these activities. Committee members are often people not associated with the elementary school and offer their skills and community outreach; for example, a landscaper built a pond at one site, an entomologist taught the world of insects to a group of children, and another committee raised donations to build a school barn. The community is often an untapped resource that is happy to share its skills with students.

Life Lab exists in big cities, in rural areas, in schools with no bare ground, and in schools with acres of land. All of these schools have adapted management strategies that work for them. Remember, you can tailor Life Lab to fit your school's needs.

Teaching Strategies

Life Lab encourages children to do science, so most of the activities in this curriculum guide provide learning situations in which students actively participate in science processes. Yes, your students will be getting their hands dirty in the garden laboratory! We have discovered that class management of experiential learning can be improved by incorporating the following teaching strategies, which can be applied both indoors and out.

SMALL GROUP COOPERATIVE LEARNING Hands-on learning can be difficult to manage while you keep an eye on 30 students who are trying to experience an activity. However, when you divide your class into small groups of three to five students, each group can share their knowledge and manage the activity together, and your role becomes a resource for five or six groups rather than 30 individuals. When structuring cooperative learning, it is important that each group share in the same project rather than each group member maintaining his or her own. The goal is to create interdependence within each group. A group may be double-digging one garden bed, collecting data for one experiment that requires shared results, or making one list of seeds to be ordered. The lessons in the first unit, "Let's Work Together/Problem Solving and Communication," will help orient your students to learning and solving problems together.

GUESS-TEST-TELL Science discovery and observations are supported by factual evidence. Scientists spend time gathering information by observing, reading, and asking questions. When they think they understand how something works, they conduct an experiment to test it. This process is called *the scientific method.* We have designed a simplified scientific method for our young scientists. Have students apply the following procedure whenever possible after they have explored a topic and formed a question:

- *Guess*—Suggest a reasonable answer to a question. (If I add compost to the garden, will my plants grow faster than if I don't add compost? Yes.)
- *Test*—Design a simple experiment for testing the guess. Experiments should have the test (such as a garden bed with compost and plants) and a control (part of the garden bed with no compost and the same type and number of plants). The test is a measured comparison and data collection of the plant growth for the composted and noncomposted areas.
- *Tell*—Give students an opportunity to analyze the results and draw conclusions in order to apply what they have learned and, of course, create more questions. (Students graph the average growth of plants in the two beds, analyze and report the results, and determine whether or not it is advantageous to use compost.)

For upper grade students, you may want to use a more sophisticated version of Guess-Test-Tell:

- *Information*—What do we want to find out? What do we already know?
- *Hypothesis*—What do we predict will happen during the experiment?
- *Experiment*—What will our test design be? What is the control? What materials will we use?
- *Results*—What happened in the experiment? What did our data tell us?
- *Conclusions*—What did we find out from the experiment? Are there more questions to ask now?
- *Application*—How can our information be used?

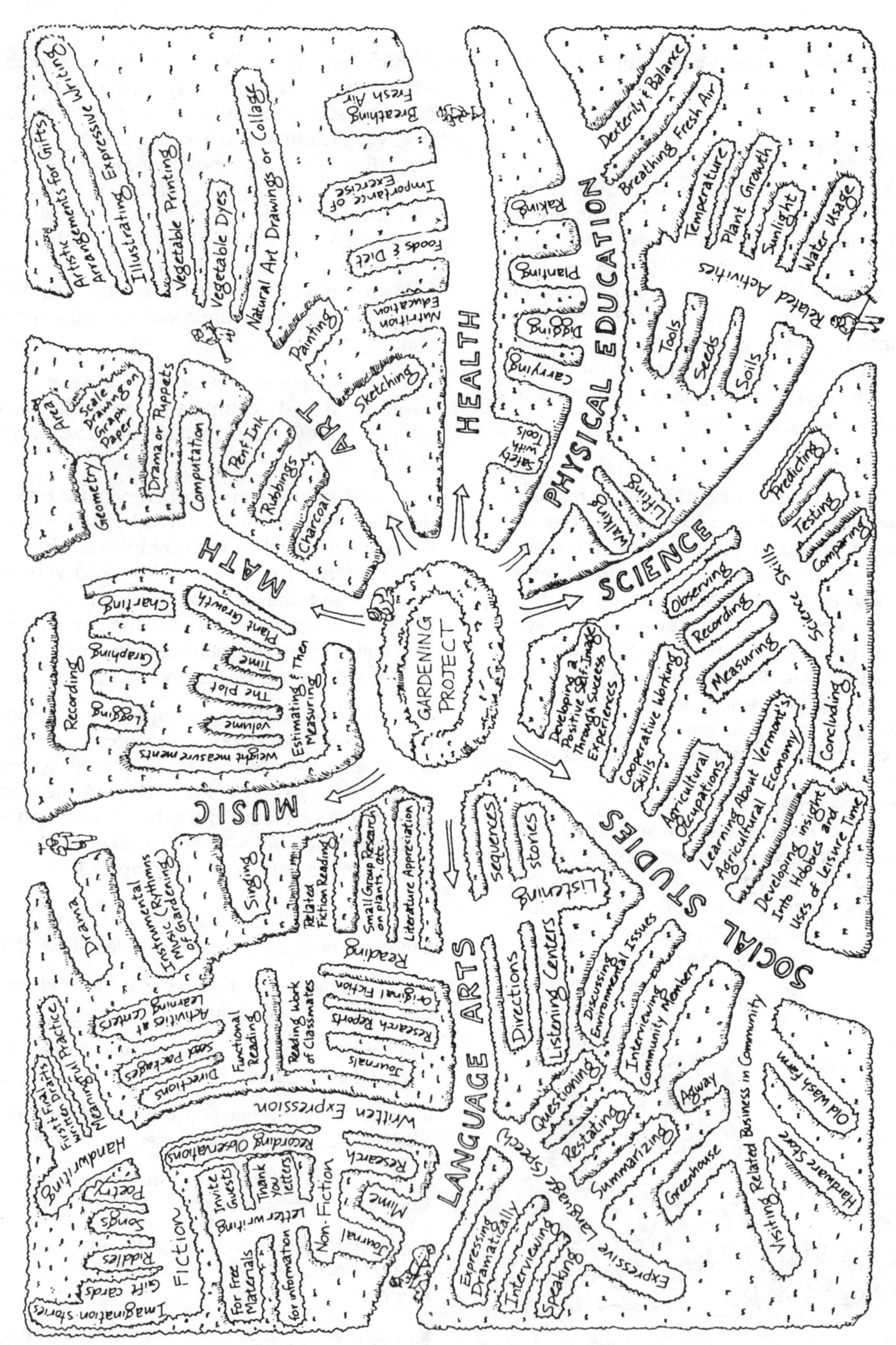

Garden study plan reprinted from *The Youth Gardening Book*, published by the National Gardening Association. For more information on youth gardening contact National Gardening Association, 180 Flynn Avenue, Burlington, Vermont 05401.

A fun way for students to remember the six important steps to this version of the scientific method is the mnemonic device "I Have Einstein's Rules Clear Always!"

From questions such as "How much did it rain yesterday?" to "What will be your favorite food at lunch today?" to extensive projects reinforcing and using the Guess-Test-Tell method, you will find that students develop a systematic method for learning critical thinking and analytical skills. In addition, they will be using a common format in many of the Life Lab activities and recording information in journals. (See Guess-Test-Tell blackline master, p. 391.)

RECORDING INFORMATION We strongly recommend that each student maintain a Life Lab journal to record observations, collect data, make analyses of his or her experiments, and keep records and drawings of the garden. The journal is a wonderful place to record Guess-Test-Tell experiments. The students, of course, will be reinforcing their writing skills, too. In addition, a classroom science bulletin board can be used for class data collection, drawings, and displays. Keep a list on the bulletin board throughout the unit of questions the students have about the topic being studied. By integrating small group cooperative learning with the Guess-Test-Tell method and information recording, you will develop a consistent format for teaching exciting hands-on activities. Your structure will give your students an understanding of your expectations of them as well as offer a means for you to evaluate their work.

INTEGRATING CURRICULUM Much of your Life Lab curriculum both in the Living Laboratory and the classroom will extend to subjects beyond science. Students who want to use references to gather more information will read and write; to analyze data, they will use math skills; to study the effects of erosion on soil, they will integrate social sciences; and to sit in the garden and draw observations, they will have special time for art. The potential for teaching your whole curriculum around selected common themes is one of the possibilities for your Life Lab program. The illustration on p. 8 shows many pathways of learning.

GROWING YOUR LIVING LABORATORY

After the first year it will be difficult to recall the original condition of your Living Laboratory. Suddenly it will be blossoming with beautiful flowers, flourishing with delicious, fresh produce, and active with exploring, young scientists.

There are three essential elements for the development of a successful garden lab:

- Start small and grow along with the program.
- Have teacher participation be on a voluntary basis.
- Develop a broad base of administrative and community support.

The gradual development of the Life Lab allows active participation of students, teachers, and community members at a rate that is comfortable for them. Thus, all participants help to create the learning environment and integrate hands-on science into their curriculum.

The original Life Lab site was 30 feet by 30 feet and had the interest and involvement of only a third of the faculty. A parent volunteered to rototill the land, and tools were purchased at the weekend flea market. The garden was originally used as an extracurricular project and was used by a handful of teachers as often as they wanted. Some teachers used it three times a week, others once a week, some with small groups, and others with the whole class.

Student Involvement

Together teachers and students using the original Life Lab experienced the joy of gardening. With the contagious enthusiasm of the students and the steady support and leadership of the principal, soon all of the classes were participating in the program. Many discoveries took place that year: the discovery of watching a spider spin its web; the discovery that spinach could taste good; the discovery that soil is alive. But one of the greatest discoveries was that the school garden was much more than a garden. It was a powerful learning tool. The garden lab provided a much needed context in which to apply academic science concepts.

Redding Elementary, Redding, CT

Be sure to involve the students from the beginning. The Living Laboratory should be owned and operated by the students. This helps to generate enthusiasm and make the students feel pride in their Life Lab garden. Math lessons can be derived from designing a garden plan and staking out garden beds and art lessons from painting garden signs. The more students are involved in the planning, the greater the opportunity for them to understand their potential for positive impact.

How can one classroom teacher manage 27 students planting seeds or preparing a garden bed? We have found that it is helpful to structure the Life Lab time so that there are a few activities taking place at once—for example, three groups of three students planting seeds; three groups of three students digging their classroom garden bed; and three groups of three students measuring and recording data in their Life Lab journals. Often volunteers (parents, college students, garden club members) can be solicited to assist with the garden activities. With well structured cooperative groups in which students understand their role in helping each other, the teacher can rotate from one activity to another while the students coordinate their groups.

Life Lab activities are much more effective when the teacher-student ratio is decreased. The hiring of a part-time Life Lab coordinator provides a support person for implementing the Life Lab program. The coordinator acts both as a teacher's aide and as a facilitator, preparing activities in advance and helping to maintain the garden. If the budget doesn't allow for the hiring of a coordinator, try forming a Life Lab volunteer group. Volunteers can be a tremendous asset and often bring extensive gardening expertise and creativity to your program. (See Volunteers, p. 18.)

Site Selection

Selecting the best site for your Life Lab garden is an important step. Following are four considerations for site selection:

- Minimum of six hours of full sun daily
- Easy access to water
- Easy access to classrooms
- Protection from vandalism

The other site determinants will depend upon the school's grounds and the participants' vision of the Living Laboratory. One school's plan may be for planter boxes outside of the classrooms while another's may be an expansive garden site. Remember, any site can be transformed into a Living Laboratory, even a dirt parking lot. So don't be discouraged if the soil seems more like dirt than like a friendly growing environment for plants; even terrible soil can be made fertile over time.

Site Planning

Once the Life Lab garden site is selected, it's time to design the layout. These beginning stages are great for engaging students in the development of their Life Lab. Many schools have each class design its own vision of what the Living Laboratory should look like and then pick and choose the best elements of each for the final plan. Dream big, but remember to start with a plan that is manageable for your school. Consider developing a three-year plan and implement a few components each year.

There are a variety of ways to organize the use of your garden lab. We try to foster cooperation in Life Lab and recommend that there be individual beds for each class to plan, plant, care for, and harvest together, as well as communal areas for the entire school to develop.

Following are some basic components of a Life Lab site:

- Classroom Beds—Beds that are 3-4 feet wide are easy for children to work from either side without stepping on the plants.
- Community Growing Area—Cut flower areas or herb and market gardens and specialty areas such as corn (popcorn too!) are great Life Lab extensions.
- Special Project Area—Dedicate a portion of the garden for individual, group, and class experiments and special projects. Signs can describe just what the scientists are testing.
- Compost Area—Composting is a fundamental garden and science activity with related lessons throughout the curriculum. Set aside an area for collecting compost materials and building compost piles.
- Greenhouse or Coldframe—Greenhouses or coldframes are used to start seedlings in a controlled environment and help to extend the growing season.
- Tool Shed—Tool sheds are important for protecting and storing tools and equipment.
- Outdoor Classroom/Meeting Areas—Design a shady area of the garden for gathering groups of students for discussions and to work on projects.

When planning your site, be sure to use the expertise of community resources. A local garden supply or garden club can assist you in laying out the garden plan and advise you on soil quality; an irrigation expert can make suggestions as to your water needs; a carpenter can help you build a simple tool shed.

Be sure to photograph the development of your Life Lab. Before and after pictures are a great way to document your growth and remember what those first few months were like. You'll be amazed at how fast it will grow!

Look at the garden map on p.12. It was originally developed as a working drawing for the development of a school site. Today it is used for weekly notations of specific work that needs to be done in the garden and is posted as a written form of communication for all Life Lab participants. This system works well for facilitating communication during a busy school day and helps to organize garden projects.

Include a Life Lab Garden sign in your planning. Signs can be simple and are best when they represent the uniqueness of your school's Life Lab. Sign planning can be a great class project. Get students to design a logo or drawing for your Life Lab Garden sign. Then form a construction committee to make the sign and place it in a prominent location.

An outdoor classroom laboratory is unique to most schools. It will generate enthusiasm from the teachers, parents, community, and most of all, the students. By involving the students from the beginning, you will keep the focus on active learning of concepts and application of skills and establish science in its rightful place as an integral part of their lives.

Tools

Before you start digging, you will need tools. The quantity will depend on your budget and the scale of your program. Think Quality when purchasing tools, even if they cost a little more. Well-made tools have a much longer life expectancy, especially with the wear and tear youngsters can give them. If your budget is limited, check your local flea markets and garage sales for bargains or local businesses for donations.

Please note that there are tools that are scaled to children's small sizes!

Following is a list of basic garden tools:

- Spades—to dig garden beds
- Spading Forks—to cultivate deeply, to break up dirt clods, and to aerate compost
- Iron Rake—to break up dirt clods, remove rocks, and shape garden beds
- Leaf Rake—to gather leaves and other organic matter
- Hoes—to weed and cultivate
- Shovels—to handle compost and other soil amendments

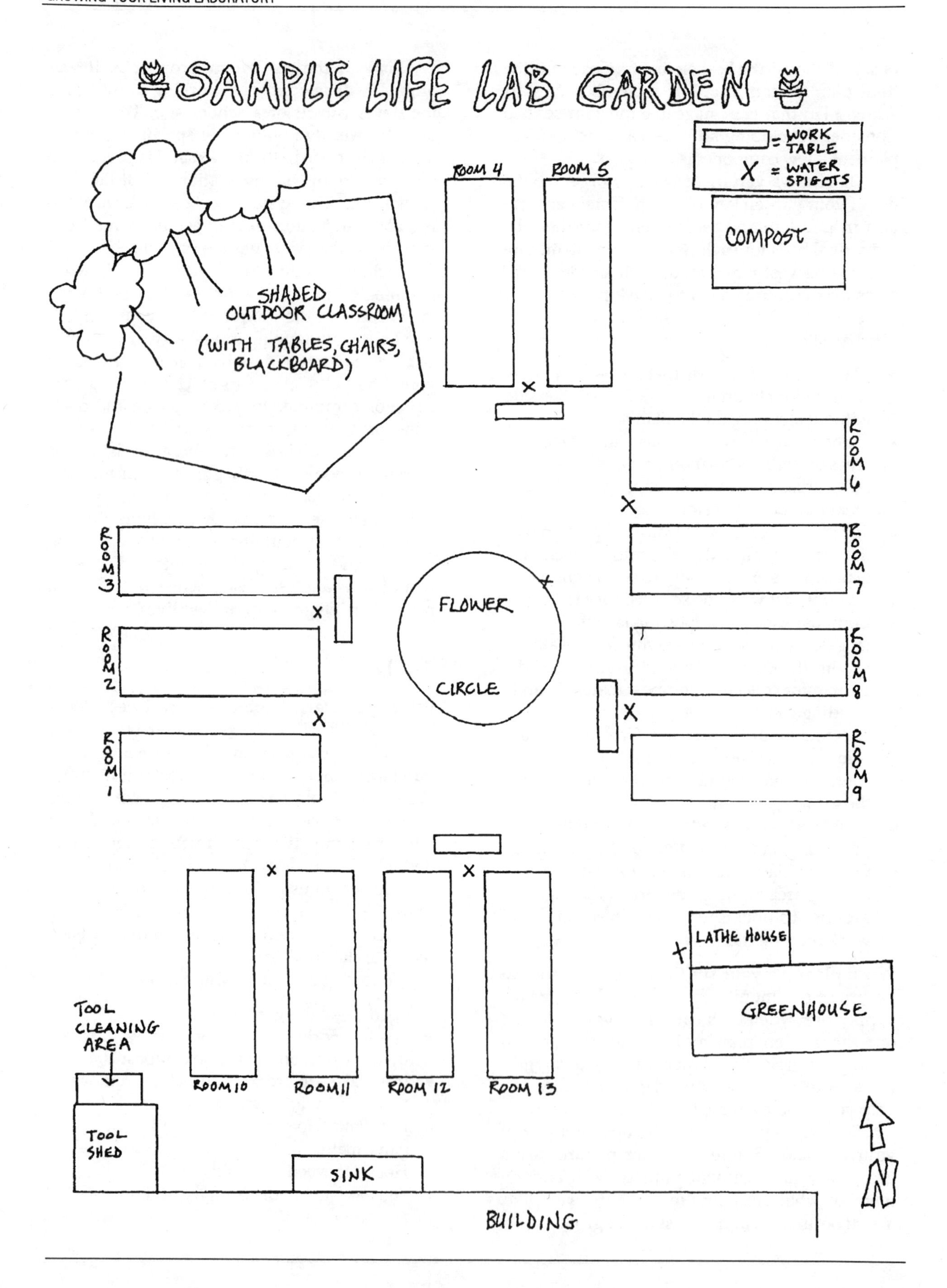
SAMPLE LIFE LAB GARDEN
= WORK TABLE
X = WATER SPIGOTS
COMPOST
ROOM 4
ROOM 5
SHADED OUTDOOR CLASSROOM (WITH TABLES, CHAIRS, BLACKBOARD)
ROOM 6
ROOM 7
ROOM 8
ROOM 9
ROOM 3
ROOM 2
ROOM 1
FLOWER CIRCLE
LATHE HOUSE
GREENHOUSE
ROOM 10
ROOM 11
ROOM 12
ROOM 13
TOOL CLEANING AREA
TOOL SHED
SINK
BUILDING
N

Robert F. Kennedy School, San Jose, CA

- Wheelbarrows—to transport materials
- Watering Cans/Hoses/Nozzles—to irrigate the site. Get watering equipment that will make watering as quick and easy as possible. Fine-spray nozzles that mimic a gentle spring rain are needed for seedlings, and fan-spray nozzles work well for established gardens. Soaker hoses are great for slow, deep water penetration. If your budget allows and your climate is dry, consider installing a drip-irrigation system.
- Trowels—to plant seeds and seedlings
- Stakes—to support climbing plants and trees
- Twine/String—to lay out straight lines for beds
- Sharp Knife—to harvest thick-stemmed crops
- Pruning Shears/Scissors—to cut flowers and trim trees and shrubs
- Harvest Baskets—to collect fruits, flowers, and vegetables
- Tool Cleaning Box—to keep equipment in top condition. Set this up near your tool storage area. Fill a box with sand and mix in just enough old motor oil (usually less than one quart) to lightly coat the sand grains. Once the tools have been cleaned off with a wire brush, dig them into this mixture before returning them to the tool shed. Coating your tools with this mixture will protect them from rust.

Life Lab's official supplier of tools and garden and science equipment is Let's Get Growing/ General Feed & Seed Co. To receive their catalogue, detach the appropriate postcard from this book. Their catalogue will advise you about start-up kits for your Living Laboratory and provide you with information on child-sized tools as well as most of the materials needed for this curriculum.

Note: A tool safety lesson can be found on p. 79; a complete materials list for this curriculum is in the Appendix, p. 463.

Preparing and Using the Living Laboratory

The original Life Lab site was so compacted that tractors were initially used to work the soil. It is hoped that the needs of your site will be less severe, and you can start digging right away.
The ground should be moist but not sopping. A rule of thumb is that if soil sticks to your boots or tools, it is too wet to dig. Digging soil when it is this wet will ruin the soil's structure, and you will be left with huge clods of dirt. Be patient and wait for the soil to dry out a bit. If your soil is very dry and dusty, water it for several hours and let the area sit for a day or two while the water percolates. To test for proper soil moisture, squeeze a ball of soil together in the palm of your hand. It should keep its shape when you open your hand, and crumble when touched.

Before you dig in, go over the checklist on p.14 of basic preparation steps as well as gardening procedures you may use. We have integrated the basic garden activities with the appropriate science units (for example: activities on soil preparation can be found in the The Living Earth/Soil unit). These activities will provide teacher background on the specific task as well as create a motivating format to involve the students. To easily find these activities, we have

listed them in the *Checklist for Your Living Laboratory* below. The tasks, whether preliminary or ongoing, are shown by season, and the related activity page numbers are provided.

Keeping Your Garden Healthy

There will be much to observe and experiment with while maintaining your garden. Most of the activities in this curriculum will provide opportunities to explore interactions in the Living Laboratory. The Garden Ecology unit will help students explore many plant interactions, and

Checklist for Your Living Laboratory

Task	Season*	Related Activities
Before you start		
Form Planning Committee	Any	
Select Site	Any	
Six hours sunlight		
Design Site	Any	
Designate garden areas		
Water system		
Tool shed		
Greenhouse/coldframe		
Outdoor classroom area		
Fencing		
Purchase Tools	Any	The Tools and Us , p. 79
Getting Started		
Preliminary Soil Test	Any	Soil Doctors, p. 86
Add General Soil Amendments	Sp/Su/Fall	A Soil Prescription, p. 88
Rototill Area	Sp/Su/Fall	
Order Seeds	Winter	ZIP Code Seeds, p. 116
Stake Garden Beds	Fall/Spring	
Gardening		
Plant Seeds in Flats	Fall/Spring	So What? Sow Seeds!, p. 156
Dig Garden Beds	Fall/Spring	Dig Me and Dig Me Again, p. 83
Test Garden Bed Soil	Fall/Spring	Soil Doctors, p. 86
Add Soil Amendments	Fall/Spring	A Soil Prescription, p. 88
Plant and Transplant	Fall/Spring	Transplanting...Or Let's Move 'Em Out!, p. 162
Water	As needed	Water We Doing?, p. 164
Weed	As needed	Weeding, Writing, Arithmetic, p. 169
Maintain and Observe	Always	
Harvest	As ready	
Composting	Fall	Let's Make A Compost Cake, p. 199
Cover Crops	Fall/Winter	The Matchmaker, p. 93
Mulch	Fall/Spring	What's to Worry?, p. 96

*** This may vary according to climate.**

the Garden Creatures unit will introduce students to many of the more mobile members of the garden family. When cultivating plants you also cultivate weeds, insects, and plant diseases.

A weed is any plant that is growing where you don't want it to be. (See related activity, p. 169.) The main problem weeds present is competition with your crop plants for nutrients, water, and sunlight. Cultivation (loosening or breaking up the soil around the plants) with a hoe can help keep weeds from becoming established. Mulching around your plants can also help to discourage weeds by blocking sunlight. Be selective in what weeds you remove. The presence of some weeds can actually be beneficial to your garden: deep-rooted weeds bring up nutrients from the subsoil; weeds with powerful roots can break up soil hardpans; weeds provide good composting material; weeds are often the preferred food of insect pests; and some weeds are edible! (Always accurately identify any plant before eating it.) So don't be overzealous in your weeding. Keep weeds from interfering with your crop plants, but experiment by letting some grow. You may find they're not so weedy after all.

There are also some plants that are found to have some broad-spectrum insect repellant properties. These include marigolds, asters, chrysanthemums, and many herbs, especially anise, coriander, basil, tansy, yarrow, and rue. Try interplanting these in your Living Laboratory. These plants will also add color to the garden and the herbs can be used in cooking and potpourri.

Not all garden insects are pests. Sometimes insect friends will take care of insect problems before you know there is a problem. These insect friends are called beneficial insects. Beneficial insects act as predators and parasites feeding on other detrimental insects. (See related activities in the unit Garden Creatures.)

Following are some preventative measures you can take to reduce pest problems:

- Remove plants that for one reason or another don't do well in your garden. Weak plants provide a focal point for a garden pest invasion that will eventually spread to healthy plants.

- Grow plants that are best suited to your climate conditions.

- Keep your garden area clean. Refuse from last season's crops provide the perfect haven for garden pests, such as snails, slugs, and earwigs.

- Rotate your crops. Avoid sequential planting of crops of the same family (for example, potato and tomato) in the same soil.

- Try to water early in the day. This helps to prevent the growth of water-borne organisms. Don't overwater.

If insect infestations do occur, there are some simple means of controlling pests on a small scale:

- Squash them (concussion control!).

- Spray the plant with water at a high pressure. This can effectively wash off insect populations such as aphids and drown them in the process.

- Grind up a mixture of pungent herbs such as garlic and cayenne, mix them with water and a non-phosphate soap, and evenly spray the plants. Be sure to get the underside of leaves.

Most of all, remember that your Living Laboratory experience is bound for the ultimate success of stimulating young minds to ask questions about the world around them—their enthusiasm

PEST CONTROL RECIPE

MATERIALS

6 CLOVES OF CRUSHED GARLIC
1 MINCED ONION
1 TABLESPOON DRIED HOT PEPPER
1 TEASPOON PURE SOAP (NOT DETERGENT)
4 TEASPOONS RUBBING ALCOHOL
1 GALLON HOT WATER
CHEESECLOTH OR NYLON STOCKING
STRING

ACTION

1. PUT ONION AND SPICES INTO CHEESECLOTH OR STOCKING, TIE STRING AROUND SACK TO CLOSE TIGHTLY.
2. SOAP IN THE WATER WILL MAKE THE SOLUTION STICK TO THE LEAVES. NAPATHA SOAP IS RECOMMENDED. ADD THE SOAP TO THE WATER. THE ALCOHOL WILL HELP TO DISSOLVE THE SOAP.
3. PUT THE CHEESECLOTH SACK IN THE HOT WATER AND LET IN "BREW" FOR A FEW DAYS. THEN USE THE SPRAY ON THE PLANTS.

is guaranteed—and a few snails and gophers just add to the opportunity to learn about the garden environment they are creating.

THE CLASSROOM LIFE LAB CENTER

You can start your classroom lab very simply and add to it as your program evolves. Most of the activities in this curriculum can be done in the classroom. They serve as ways of investigating processes that take place in the outdoor lab garden. Each activity is labeled with a key that designates it as an *indoor* or *outdoor* activity. We recommend you designate a specific area of the classroom as the "Life Lab Center," and start it with basic items:

- A Table—to keep projects/experiments in progress; display special projects and collections
- Bulletin Board Space—for class graphs and charts, murals, and so on; a current events section of the bulletin board can encourage students to bring in science-related articles in the news
- Bookshelves—for keeping science-related reference books and class science journals
- Basic Equipment—hand lenses, a simple scale, bug boxes, thermometers

As interest in the Life Lab Center evolves you may want to add the following items:

- Aquarium/Terrarium—for keeping garden creatures on short visits
- Stereoscope—a microscope that has two eye lenses; it is easy for young students to use and offers tremendous opportunities for discovery from observing pond water to watching a lacewing larva eat an aphid
- Grow Lab—Grow Lab is an indoor plant-growing center that allows for year-round plant projects that provide many opportunities for students to prepare controlled garden experiments. Grow Lab plans or the actual plant center can be ordered through Let's Get Growing. (See p. 480.) We highly recommend Grow Lab or some type of indoor growing center for schools in locations where there is a limited growing season.
- Computer and Modem—If your class is learning how to use computers, they can be simply applied to the Life Lab Center. Students

can keep data on the computer, providing an easy reference to garden records from year to year. A modem attachment to the computer and a regular phone line will allow you to share information via computer with other Life Lab schools. This can include weather information and experiment results, as well as different plants that are grown and enjoyed. Life Lab can give students an opportunity to apply relevant information to computer learning.

CULTIVATING SUPPORT AND GUARANTEEING SUCCESS

Experienced gardeners will tell you that in order to have a successful garden you need to build the soil, fertilize it well, and tend your plants with loving care. Your school's Life Lab will require that and more. Whether you are growing in planter boxes in the middle of a city or on an acre of land surrounded by farms, the Life Lab program is unique within the educational system.

This uniqueness requires that in addition to cultivating your plants you must cultivate support from teachers, students, parents, community volunteers, and the school administration. Together, you must work toward developing a shared sense of responsibility for this special program. Before long you'll find word-of-mouth (and the scent of roses and narcissus) attracting individuals who have not been involved with the schools since they were students.

Administration Support and Involvement

The introduction of a new program to a school is not easy. It is essential that the administration is involved with every level of the Life Lab adoption process. The principal plays a key role in insuring the integration of Life Lab into the school structure. The principal can play a key role by

- providing release time for Life Lab planning
- providing budget support
- integrating the program within the school, the district, and the community

> *"This garden looks beautiful and it's great that the students get to learn how to plant and grow their own food, but how does this relate to academic learning?"*
>
> *"This is the highlight academic program of our school district."*

These are both comments from school board members. Their difference in opinion is just one reason why it is so important to keep the district superintendent and school board members informed of the learning that takes place with this unique program. It is easy to look at a garden or plants growing in a classroom and appreciate their beauty, but the connection to academic learning is often less obvious. Invite your superintendent and school board members to visit the site during the school day and observe an activity. They will observe first-hand that concept building takes place at a rapid rate when students are active participants in their own education.

As a principal, teacher, community member, or science coordinator it is worth the effort to involve the district administration early on. We also recommend that you update these individuals on the program's progress by inviting them to Life Lab events and follow up with reports and presentations at school board meetings.

Volunteers

Volunteers are instrumental to the work of Life Lab. Their assistance is invaluable for the following reasons:

- Volunteers increase the adult-student ratio.
- Volunteers provide diversity for student interaction.
- Volunteers serve as spokespeople for the program.
- Volunteers offer a great deal of expertise.

Tap a variety of resources in search of volunteers:

- Your local educational institutions (universities, community colleges, high schools) are valuable resources. The teacher education, environmental studies, horticulture, and science departments often have work study, student intern, and community involvement programs.
- Parents and other community volunteers seeking work with children in an educational/gardening environment have been very helpful.
- Ask members of a neighborhood Senior Center to become regular volunteers.
- Encourage the local garden clubs, botanical gardens, and the County Cooperative Extension Office to offer their expertise.

Once you have lined up a volunteer group, you need to provide a Life Lab orientation for them. Each day you need to have a formal work schedule and clearly define their responsibilities. This will help them to be more comfortable in the new surroundings and more useful to the program.

Volunteers will come to you with varying degrees of experience. Some may have experience with children, some with gardening, and others with both. Provide a short workshop on the Life Lab approach to science and gardening. Include double-digging, seed sowing, flat preparation and transplanting, and watering techniques. A session on class management and discipline will also prove useful. Set a few basic Life Lab rules for everyone to follow:

1. Always walk in the garden.
2. Ask before using any tool or harvesting any crop.
3. Respect each other in the garden.

These basic rules will help to provide a consistent discipline policy. Review upcoming activities with the volunteers at regular group meetings. Let them observe teachers instructing different activities until they feel comfortable teaching a small group on their own. The number of students per volunteer should be as low as possible.

Evaluate volunteers' activities with them and develop a structure for them to use for disciplining students. (We have found that time-out works well, because most students want to actively participate in Life Lab; it's not a spectator sport.) Acknowledge their work and let them know how valuable they are to the program. Give them feedback about their work. The more they feel a part of the program, the more you will be able to depend on them. Managing volunteers can take much effort, but once the groundwork is laid, you will be rewarded with a steady flow of energetic help.

Donations

Donations can range from a solar greenhouse to a wheelbarrow to architectural drawings for building a barn. Donations can allow for expansion of your program well beyond your own financial means.

When pursuing needed donations, keep in mind the following basic steps.

1. Develop a specific project request.
2. Make a list of needed materials.
3. Make a list of businesses that could potentially supply the materials.
4. Make an appointment to go in person to meet with owners or managers.

In the meetings describe the Life Lab program and the proposed new project. Show the business persons the materials list and ask them whether they can donate a few things on the list. Once you have commitments for the project, show prospective contributors a list of donors. This will make them feel that they are joining together with other businesses to support the project. Each donation, no matter the size, should be followed with a thank-you letter. Thank-you letters handwritten by students are always nice. Keep a file of all donations.

Parents are another possible resource for needed materials. Keep them informed of what is going on at Life Lab. You'll be surprised how much they already know from their children. Send home letters with lists of materials needed. You will receive many things, from tools to refrigerators. Many parents own or work for businesses that would like to help Life Lab.

Establish on-going contact with garden-related businesses. For example, a local nursery could give you a call when they are going to discard plants, flats, or seedlings; landscape suppliers could provide help with plants and gardening expertise; hardware stores may be able to provide tools or irrigation supplies; stables can be a source for manure and straw.

Asking for donations will produce a ripple of benefits. Aside from directly enriching your program, it informs the community about what is happening at your school and demonstrates the important role they can play in helping to improve education. Once you develop a donation request format, involve parents and volunteers in the solicitations. Take advantage of the diversity in your community; it will greatly enhance your Life Lab program.

Green Acres School, Santa Cruz, CA

In addition to soliciting local donations, you can apply for grants from the state and federal government, foundations, and other organizations. Some possibilities include the following:

1. Your school district office will be able to inform you about possible federal and state funds. Examples of programs Life Labs have been funded through are
   - Title 2 or PL 98-377: For math and science program and/or staff development; amount per district is determined by ADA.
   - Chapter I: Available to schools with low socio-economic status.
   - Chapter II: Federal money administered by each state for materials/staff development.
   - State Funds may include environmental education funds, lottery money, mentor teacher programs, and so on.
2. National Diffusion Network of the U.S. Department of Education. Life Lab Science Program is funded by NDN to assist schools in establishing Life Lab programs. Each state has an NDN State Facilitator to help schools adopt NDN programs. Contact your State Facilitator as a resource for assistance in funding your Life Lab program. For more information contact Life Lab Science Program or the National Diffusion Network:

   U.S. Dept. of Education
   OERI/PIP/Recognition Division
   555 New Jersey Avenue, N.W.
   Washington, D.C. 20208
   (202) 357-6134
3. Community Foundation Grants. Contact your local United Way office to find out about community foundations in your area.
4. America the Beautiful Seeds provides free seeds for the cost of postage.
   Contact: ABS, 219 Shoreham Bldg., Washington, D.C. 20005. Telephone:(202) 638-1649.

Make New Friends

After generating a broad base of community interest for your Life Lab program, consider mobilizing your resources to form a community advisory committee. The involvement of educators, business people, parents, agriculturalists, and community residents can greatly enhance the success of your program. The group can work to generate resources to help support the continuation of the program as well as provide technical expertise.

It's helpful to focus the efforts of the committee. Try selecting a major project for the year as a focal point. Some examples of these projects are raising funds to construct a greenhouse, establishing an outdoor classroom area (tables and benches), or approaching local businesses and others for financial donations. The committee is an important liaison between the school and the community. Look within your existing school organizations for your support group. Groups such as the PTA and the School Site Councils can easily be advocates for your program.

Events

Events that take place at Life Lab serve four purposes:

1. To give the community an opportunity to find out about the program
2. To serve as a fundraiser
3. To give the Life Lab program an opportunity to thank the community for their participation and support
4. To provide a community service

Redding Elementary, Redding, CT

Happy Valley School, Santa Cruz, CA

The first important event is often a Life Lab Dedication/Ground Breaking Ceremony. This event helps to establish the program early on as a school- and community-based program. Other events include planting ceremonies; balloon launches; student Life Lab plays and skits; and Life Lab assemblies followed by pot-luck lunches, with food from the garden. One school has three annual events. They have a Harvest Festival in the fall, a Life Lab Science Fair in the spring, and a Saturday gardening workshop sometime during the year. Whatever format you choose, be sure to invite your local government representatives and the media to join in the fun.

Make your Life Lab program available in a variety of ways to people who don't normally associate with schools. Their participation will enable them to see the significant role schools play in the community. Life Lab can become a vehicle for community involvement with our most important resource—our children.

Let's Work Together

Problem Solving and Communication

Robert F. Kennedy School
San Jose, CA

Robert F. Kennedy School
San Jose, CA

Robert F. Kennedy School
Franklin-McKinley School District
San Jose, California

"It brings so many things together. The kids get a lot of science, math, reading and language combined in this one project. It brings them together in little groups, and it brings the teachers together because they feel like a Life Lab team."

—RFK principal Lynne Hopkins

Started in 1984 after two teachers attended a Life Lab in-service session, this inner-city school's Life Lab was selected as one of the nation's top ten youth gardens in 1988 by the National Gardening Association. The community was involved from the beginning. The district maintenance staff cleared the area and the school board provided funds for a chain link fence. Students from a nearby middle school constructed raised garden beds and a picnic table. Students from San Jose City College built a tool shed and local businesses donated needed supplies. Now thirteen classes (338 students) each plan, maintain, and experiment with their own garden bed—the garden beds have been given names such as "The Rainbow Riders" and "Gopher Haven."

During the day you may find a construction team of sixth-graders measuring, cutting, and hammering redwood flats. In another group fifth-graders read a plant experiment to fourth-graders, who are investigating which soil drains quickest: compost, sand, clay, or garden soil. At the worm box, other students examine jars full of worms chewing on greens. In the classroom each student keeps a garden journal to record all the learning that's going on, and teachers conduct follow-up activities related to the Living Laboratory.

Robert F. Kennedy School's unique character is enhanced by its multinational/multilingual student population. Most of the students are not native speakers of English. Nearly one-third are from Vietnamese and Cambodian refugee families; others are Filipino and Hispanic. Gardening helps all children feel at home. Life Lab is a safe place to take risks, and to learn.

UNIT INTRODUCTION

Let's Work Together

The Life Lab program will offer opportunities for your students to work together in new situations such as being outdoors in their Living Laboratory. It will also foster collaboration on short- and long-term projects, working independently, and sharing information and discoveries. This active learning approach enhances the students' problem-solving, communication, cooperation, concentration, and listening skills. Practice will develop important life skills that students will be able to use in other areas of study as well as in their everyday lives.

This first unit focuses on activities that will help your students develop these skills. Indoor, outdoor, active, and calm activities are included in this unit. Each activity strives to involve students in practicing one or more of the above-mentioned skills in a fun and novel manner. We strongly recommend that you emphasize these activities in the beginning of the year as a way of introducing the work habits you want students to develop. Many of the activities — such as "10-4 Good Buddy," "I Just Love Applause," "Tell It Like It Is," "Knots," and "Skis" — can be repeated regularly throughout the year to reinforce these skills. Many are short activities that can be used in a few spare moments. "Skis" requires advanced preparation to tranform two 2x4s into skis as described in the Appendix.

Participating in Life Lab means working with our environment. Students will be interacting with soil, plants, gophers, snails, birds, and insects. Most challenging will be their interactions with other people. Working with others gives everyone the opportunity to grow more than vegetables and learn more than science. Trust, respect, cooperation, and communication can be nurtured and harvested.

Activities	**Recommended Grade Level**
10-4 Good Buddy *(concentration and listening skills)*	2,3,4,5,6
I Just Love Applause *(communication skills)*	2,3,4
Lighthouse *(communication and cooperation skills)*	3,4,5
Tell It like It Is *(introduction of one-way and two-way communication)*	4,5,6
Count Off *(problem solving and communication skills)*	2,3
The Connected Circle *(cooperation and group problem solving)*	3,4,5,6
Knots *(cooperation and group problem solving)*	3,4,5,6
Skis *(cooperation and group problem solving)*	3,4,5,6
Sinking Ship *(communication and problem solving)*	4,5,6
Who Am I? *(communication and cooperation through problem solving)*	5,6

With People I Like

Capo four frets

```
           G   D            C  D
It's a new day we're living in

      G          D         C
Filled with bounty and praise.

        G              D             Am7  C
The cold and the darkness is driven away

           G               Am7       G
While morning light blesses our days.
```

Chorus

```
    G          Am             G         C
It's so nice to be here with people I like.

       G         Am          G
It's so nice to hear them all say,

                     Am          G          C
That it's so nice to be here with people I like

          G          Am          G
And with them to share a new day.
```

The people I like are the people I find
Who live with a smile of care.
Their looks and their lives are different than mine
But together this new day we share.

Chorus

Time spent apart seems to fade away
Smiling faces warm us inside.
We can talk without saying a word
Our love is deep and does not die.

Chorus

10-4 Good Buddy

Description

This activity uses a discussion format to focus on listening.

Objective

To practice the skills of concentration and listening.

Materials

None

Let's discuss our ideas of what science is. What are some things you would like to learn when we study science? (List responses on the chalkboard.) Does anyone have any pets they would like to study? What about favorite flowers? What are some of your favorite activities that you may be able to learn more about in science?

1. Choose one of the topics listed and start a class discussion about it. Example: What are some things we could learn in science about fish?
2. After the discussion is underway, interrupt the class and tell them that before anyone speaks she or he must first repeat what the previous speaker has said to that person's satisfaction. Other members of the class should notice whether or not the account is accurate.

What did you learn about science? How did the echoing rule affect the class? (Students will recognize that listening is often an active task, not a passive one. They may also realize that they are poor listeners because they become so absorbed in what they themselves are going to say. The exercise also helps students realize the amount they read into others' remarks.)

I Just Love Applause

Description

This activity is a game in which students must communicate a task without speaking.

Objective

To develop communication skills.

Materials

None

There are many ways to communicate our needs and desires to others. What are some ways that you use? How do you communicate with someone who is in the same or different part of the building? How do you communicate with someone in a different town? Why is it important to have different ways of communicating? Do you think it is possible to communicate without speaking? We are going to play a game to find out whether we can communicate without speaking.

1. Ask for a volunteer.
2. Have that person leave the room.
3. While the volunteer is out of the room, decide with the class what simple action you would like the volunteer to perform when he or she returns (such as sit down and take off shoes).
4. The class is to communicate the action nonverbally by clapping faster or slower as the person comes closer to or moves farther from the desired action. If you want the volunteer to walk to his or her desk and sit down, the class would clap if the volunteer began walking in the direction of the desk. The class would stop clapping if the volunteer turned away from the direction of the desk. This is similar to the common children's game called Hot and Cold.

Was this an easy or difficult way to communicate? What other methods do we use to communicate our needs and wants? How did it feel when the task was completed? What is feedback? How was feedback used in this game? Was listening an important part of this communication?

Lighthouse

Description

This activity asks one student (Lighthouse) to verbally lead another blindfolded student (Boat) through a maze of people.

Objective

To develop communication skills.

Materials

Blindfold

What are some important parts of good communication that we learned? (listening, different ways to communicate) If we are having a class discussion and one of us wants to share some ideas, what will help us to communicate? Let's find out.

1. Ask for a student who feels that he or she can communicate clearly to be the lighthouse.

2. Ask for another student who considers himself or herself to be a good listener to be the boat. Blindfold this student.
3. The remaining students will become obstacles in a bay. They can be bridges, logs, and so on. It is important that they are quiet and do not move during the activity.
4. Situate the lighthouse at one end of the playing area.
5. Place the boat at the other end.
6. The remaining students can take their places between the boat and the lighthouse as obstacles in the bay.
7. The task of the lighthouse is to verbally lead the blindfolded boat through the obstacles. The lighthouse should remain stationary. The lighthouse should give the boat explicit directions so that it will avoid the obstacles. It may help to have the boat keep a hand raised to assist in determining left from right.
8. The task is completed when the boat safely arrives at the lighthouse. Should the boat hit an obstacle and sink, choose another student to be blindfolded and begin again. The obstacles should remain still and silent.

How did it feel to be the boat? What did the boat have to do in order to stay afloat and reach the lighthouse? How did it feel to be the lighthouse? What did the lighthouse have to do in order to bring the boat in safely? What does "concentration" mean? What does "communication" mean? How did the obstacles help the lighthouse and boat communicate?

Tell It Like It Is

Description

In this activity, students practice one-way and two-way communication through drawings.

Objective

To introduce students to the concepts of one-way and two-way communication.

Materials

Pattern sheet, p. 392
A few sheets of blank paper per student

If you wanted to tell a story to the class, what would you do? What would help you communicate your story correctly? Do people always understand what you mean when you talk to them? How could you tell if the class heard the story the way you wanted them to? If you told your story and did not allow the class to ask questions, you would be using one-way communication. Two-way communication occurs when the listeners ask questions, give feedback, or take an active role in the discussion.

1. Divide the class into groups of six students. Explain that the first series of trials will involve only one-way communication.
2. Select a volunteer from each group. These students will use one-way communication to describe a drawing on the pattern sheet to their group. (Show only one pattern from the pattern sheet, so you can use the other patterns with other students. Do not duplicate the pattern sheet for the students.) The group tries to draw the pattern from the description. Group members may not ask questions or give the speaker clues concerning his or her success or lack of it. The speakers should not see the drawings while they are being made.
3. When the students in a group are done, have them share their drawings and compare them to the design that the speaker was describing.
4. Have students discuss what makes the task difficult.
5. Repeat the procedure several times.
6. Now have students use two-way communication in their groups by having the speakers watch the drawings as they are being made and instruct the group members to make changes in their drawings at any time. Students may ask questions of the speaker.

7. Have each group discuss how two-way communication differed from one-way communication.
8. Try both one-way and two-way communication with the whole class. Have one person describe a pattern to the whole class without seeing the students' drawings. For two-way communication, have one student describe a pattern as another student draws it on the chalkboard. The student giving the instructions can instruct the student at the board to make changes at any time.

What is the difference between one-way and two-way communication? Which did you prefer using? Why?

Count Off

Description

In this activity, students physically order themselves without speaking according to numbers on their backs.

Objective

To develop the skills of problem solving and cooperation.

Squares of wide masking tape for each student, numbered sequentially

Preparation

1. Prepare the number tape before the activity.
2. Clear a space for the students to line up (or circle) side by side.

Have you ever solved a problem by yourself? What did you do? Have you ever solved a problem with a friend or someone in your family? Do you think our whole class could work together to solve a problem? Do you think we could do it without talking? Let's try!

1. While students are sitting at their desks, attach a number to their back so that the number is visible. Be sure the order is mixed up.
2. Students should neither speak nor look at their own number; this is a nonverbal activity requiring the group to line up in numerical order.
3. Students may lead others to the proper place in line.
4. Use the perimeter of the classroom for the playing area and designate a beginning and an end point.
5. To inject a greater sense of challenge, inform the class that they are being timed (or ask them if they would like to be timed). Should anarchy prevail after five minutes, have everyone freeze. Remind them that the solution requires cooperation. Give them additional time to complete the activity.

How many people helped someone else get to the right place? How did you find your place? How did it feel to have to solve a problem without speaking?

The Connected Circle

Description

In this activity, each participant plays an integral part in forming a circle (of people) that is physically connected.

Objective

To develop skills of cooperation and group problem solving.

Materials

None

Name some ways that we have had fun together. Can you think of something we've done as a class that everyone helped to make work? Do you think we could all work together to form a circle? What about a circle in which we are all connected?

1. Have everyone in the group stand in a circle shoulder to shoulder.
2. Have everyone turn to the right.
3. Simultaneously have everyone very gently sit down on the lap of the person behind them. Everyone must do this at precisely the same moment.
4. Repeat until successful.

How were each of you important to the completion of the circle? What does it mean to cooperate? Can you give examples of people working together in order to accomplish something? What is the opposite of working together?

Knots

Description

Students form a knot with their connected hands. They are then challenged to untangle the knot without speaking or dropping hands.

Objective

To develop skills of cooperation and group problem solving.

Materials

None

What have we learned so far about working together? When you want to work with someone else, what are some things you can do to help cooperate? Do you think it is important to learn how to solve problems together? Can you think of examples of solving problems in your daily life? Do you want to be a good problem solver?

1. Divide the class into groups of six-to-eight students. Give each group instructions to form a knot:
2. Stand in a circle shoulder to shoulder.
3. Place your hands in the center.
4. Close your eyes.
5. Reach out and grasp two other hands, as if you are shaking hands. (Make sure that no one holds both hands with the same person, or holds the hand of a person right next to them.)
6. Now without speaking, or disconnecting hands, open your eyes.
7. Untangle the knot without speaking or breaking hands. (Occasionally a knot is too difficult to untangle. In that case, try again. A variation would be to allow students to speak while untangling the knot.)

How did it feel when you were stuck? Did everyone have to contribute to the solution of this task? What does cooperation mean? What is a problem? Can you name some ways people work together to solve problems?

Skis

Description

This game requires students to work together silently in order to move two ski-like pieces of wood.

Objective

To foster skills in cooperation and problem solving.

Materials

84 feet of strong cord
Two 10-foot 2x4s
24 eye hooks
Hammer

Preparation

Prepare the skis according to the plans in the Appendix, p. 451.

We're going to try to ski on grass! To do this activity, we're going to work in small groups. These groups will really need to cooperate well to make the skis work. What do we need to do to be cooperative? Who would like to try?

1. Have six students stand on the parallel skis with one foot on each ski. Have each student standing on the skis pick up a rope handle from the ski. Then give the following directions to the students.
2. This is a nonverbal game. That means there is no talking allowed until the end. The object of the game is for the six of you to move the skis that you are standing on a distance of ten feet. You cannot remove your feet from the skis. You cannot let go of the rope that you are holding. You have to work together, without speaking, and move these skis. Sound easy? Remember, no talking. O.K., go ahead and begin.
3. If after three minutes no progress has been made, you may choose to allow the group to speak.

What was needed for this game to work? What could you do next time to work together better?

Sinking Ship

Description

In this activity, students solve a problem in a cooperative manner through a simulated disaster.

Objective

To develop communication and problem-solving skills.

Materials

None

(Read the following story to the class.) As a field trip, your group has taken a boat ride to see some islands that lie off the Santa Cruz coast. The boat is now by a large island. We know that this island is deserted; there are no people on it. We do not know if there are animals on the island or if there is water on the island. From your boat, you can see that there are some trees and greenery on the island. Suddenly the boat scrapes along a large rock, which tears a hole in the boat's bottom. The boat will sink in 30 minutes. Fortunately, there is a small lifeboat that you can use to get to the island, but it is not big enough to sail on the open ocean back to Santa Cruz. There is room for all the people in your group and for five things that you can take with you from the larger boat.

1. On the chalkboard write the following list of things found on the boat:

5 jugs of water	10 flare kits
axe	bow and quiver of arrows
knife	canvas sail from the boat
first aid kit	fishing rod and tackle
pair of rabbits	1 box of kitchen matches

2. Divide the class into groups of four or five students. Have each group discuss what five things their group would take. Be sure each person in the group has time to talk and give reasons for each choice. Have one person in each group be the recorder.
3. Have each recorder report to the whole class. Why did each group choose their specific items?
4. Can the class come up with one list of five items?

How did the group come to a decision? What would have happened if only one item from the list had been allowed? Did you feel like your point of view was heard? Did you listen and respond to others?

Who Am I?

Description

Students will work together in small groups to solve a riddle by putting together written fragments of information.

Objective

To develop communication and cooperation skills through problem solving.

Materials

"Who Am I" Riddles, p. 393

Preparation

Prepare riddles so that each group of six has one square (bit) of information.

Each of you has skills to offer a group. What are some skills you think you have to offer? Can you think of a time when you have made a contribution to a group? When you are working in groups, each person has something important to contribute.

1. Divide the class into groups of six and pass out one of the six bits of information to a student in each group. Tell the students that there is a riddle to solve. They can tell their group what is on their paper or read it aloud, but they must not show it to others. Their paper, or bit, represents what they have to offer to the group.
2. Each group needs to determine what their riddle is and what the answer is. They have all of the information they need.

How did your group solve the riddle? How did you feel after you contributed to the group? Was it easy to talk in a small group?

Have each group prepare their own riddles to share with another group.

Answers:	Riddle 1: a flower	Riddle 3: ant
	Riddle 2: grasshopper	Riddle 4: the sun

We Are All Scientists

Awareness and Discovery

Happy Valley School
Santa Cruz, CA

Happy Valley School, Santa Cruz, CA

Happy Valley School
Happy Valley School District
Santa Cruz, California

Happy Valley School is located in Santa Cruz, California. It is a school with five classrooms, serving students from Kindergarten to sixth grade. There is much community support for the Life Lab program, which was started at Happy Valley in 1984. The teachers, principal, and surrounding community contributed to help students design and erect a greenhouse, tool shed, lath house, and five 20′X10′ raised garden beds. The garden beds are purposely arranged in front of each classroom. This gives the students the opportunity to pass by and notice what's growing at least five or six times a day. In the beginning of the Life Lab program, students from the University of California, Santa Cruz acted as teachers and led children through the outdoor gardening experience.

Each student at Happy Valley spends an hour per week in Life Lab. Half of that hour is spent working in the garden with the Life Lab teacher. Projects range from maintaining a worm bin to growing their own garden lunches. The children are quite familiar with sowing seeds, plant germination, composting, integrated pest management, mulching, and harvesting. Many students plan and design their own gardens within the large raised garden beds. The students look forward to Life Lab time each week.

The children are exposed to food and nutrition in a way that is unique in these modern times. Although Happy Valley School is about two miles from town, there is easy access to the large grocery stores and supermarkets that are representative of where most people shop. The children in Happy Valley's Life Lab have come to realize that all food does not come pre-wrapped and washed. Their delight in eating fresh, raw cabbage and broccoli from the stalk is an experience not easily duplicated in most science programs.

UNIT INTRODUCTION

We Are All Scientists

Science opens the world around us for our discovery. The scientific tools we learn allow us to explore how physical components function, how living organisms grow, and how the living and nonliving interact. The key tool is observation. Creating an awareness in young students of the importance of their senses and encouraging their practice of observation will fine-tune a skill useable in every area of study and living. Learning to be expert users of our senses provides us with new knowledge and fosters questions for further exploration.

The activities in this unit focus on the idea that our senses are our tools for discovering the world. Activities begin with simple awareness of and practice using each of the five senses, and progress toward higher-level application with use of multiple senses and mapping skills. These observation skills will be practiced over and over again with the active learning approach of Life Lab as you study different subject areas. You can extend the students' observation skills by encouraging them to take quiet time in the Living Laboratory, sitting in one spot and drawing in detail things they see.

"I think, therefore I am," Descartes said. Our ability to taste, see, hear, touch, and smell tells us that the rest of the world exists. Being outdoors in the Living Laboratory appeals to all of the senses — the snap of a fresh carrot, the taste of a fresh tomato, the sight of morning dew or a spider spinning its web, the smell of flowers, and the feel of well-prepared soil. The more we sense, the more we learn, and the more we are.

Activities	**Recommended Grade Level**
Sharp Eyes *(observation skills)*	2,3,4
Candid Camera *(observation skills)*	2,3,4
The Unnature Trail *(observation skills)*	3,4,5,6
Ear-Ye, Ear-Ye *(sense of hearing)*	2,3,4,5
Big Ears *(sense of hearing)*	3,4,5,6
Only the Nose Knows *(a sense of smell)*	4,5,6
Six of One, Half Dozen of the Other *(sense of touch)*	3,4,5,6
Everyone Needs a Rock *(sensory awareness skills)*	2,3
See No Evil, Hear No Evil *(to develop all five senses)*	2,3,4
Mystery Powders *(to integrate senses with problem-solving skills)*	4,5,6
Little Munchkins *(to develop observation and recording skills through mapping)*	4,6
Burma Shave *(to integrate sensory awareness skills)*	4,6
On Location *(to apply sensory awareness skills through mapping)*	5,6

Take the Time to Wonder

G
What's the reason for the dance of the red-tailed hawk at noon?

C G
Or the reason for the robin's song or laughing of the loon?

C D
You've got to take the time to wonder, take the time to care...

G C G
If you listen very closely the answers you will hear...

Chorus

D G
There's nothing like the sunshine to tantalize my soul.

D G
There's nothing like the first evening star to set my heart aglow.

C D
There's nothing like a rainbow to put color in my mind.

G C G
There's nothing like a sunset to show me the reason why.

I look up in the evening sky, a thousand stars above.
A child asks me what it is on this planet that I love.
I love the sparkle of the dew drops, the giant redwood tree
The music of the blowing wind, the magic of the sea.

Chorus

Did you ever see a whale breach, the blood red sky at dawn?
Did you ever hear a coyote howl, did you ever hear his song?
You've got to take the time to wonder, stop, and you will see,
The beauty of the Mother Earth, she does it naturally.

Chorus

Sharp Eyes

Description

This activity helps demonstrate how eyes serve as important information-gathering tools. Students work in pairs to observe changes their partners make.

Objective

To develop the skill of observation.

Materials

None

When you entered the classroom this morning did you notice anything different? What part of your body did you use to notice the change? (eyes) Your eyes are your observers. Do you think your eyes can help you observe small changes someone makes?

1. Have students form two lines facing each other. Each student should be standing directly opposite another.
2. Give the pairs time to observe each other, noting color of clothing, rings and other jewelry, tied shoelaces, and so on.
3. After they have had sufficient time to observe, have students turn away from each other and change one thing about their appearance, such as move a ring to another finger, untie a shoe, unbutton a sleeve cuff. Emphasize subtlety.
4. Have students face each other again. Can each member of a pair tell what the other changed?

What sense was important in this activity? Why? How did you detect what change was made?

Candid Camera

Description

This activity helps demonstrate the importance of our sense of sight in connecting us with the world we inhabit by having us become cameras.

Objective

To develop the skill of observation.

Materials

None

We are going to go outside. When we go outside we will meet in a group and I will tell you about the activity. As soon as we are outside, I want you to be good observers and be able to tell the class what you've seen. (Gather the class outdoors.) What have you seen so far? What part of your body helps you to know what is in the world around you? We are going to play an observation game in which we will learn how important our eyes are in connecting us with the world.

1. Have students work in pairs. One is the photographer, the other is the camera. The photographer focuses the camera (whose eyes are closed) by pointing it at the subject. The camera must keep its eyes shut tightly until the photographer exposes the picture by lightly pressing on the camera's shoulder.
2. Use one pair to demonstrate the procedure to the class.
3. Have each photographer take three pictures and then switch roles with the cameras.
4. As the pairs complete their task, have them gather in groups of six and have each camera describe the pictures he or she took.

What would it be like not to have the sense of sight? What sense do you make the most use of? Can you name something that you have looked at that made you feel good? Can you name something that you have looked at that made you feel bad? Can you name something that you have looked at that made you feel angry? What would it be like to have all of our senses taken away?

The Unnature Trail

Description

Students walk through a natural setting and silently observe things that are out of place.

Objective

To develop visual awareness.

Materials

15 to 20 objects (synthetic and natural; some big; some small; some easily observed; some easily camouflaged, such as shoelaces, pine cones, plastic toys, rubber bands)

This activity is adapted from *Sharing Nature With Children*, by Joseph Bharat Cornell, 1979, Ananda Publications.

Preparation

Choose a 40 to 50 foot section of isolated playground or garden that preferably has one or more trees and place along it 15 - 20 objects. Some of these should stand out brightly; others should blend with the surroundings and therefore be more difficult to pick out. Keep the number of objects you have planted secret. Place objects from ground level up to ten feet high.

Do you think you would notice something if it was out of place? What types of objects would be easy to notice? (those whose color was different than the background; large objects) Which would be difficult? (objects that blend in) Let's find out how well you can spot things that don't belong in a certain setting.

1. Have each student explore the section of playground or garden, trying to spot (but not pick up) as many out-of-place objects as he or she can. It is important for students not to give away their findings by pointing, jumping up and down, shouting, and so on to others in the group to keep the interest level high. Put a time limit on the walk.
2. When they reach the end of the trail, they can whisper in your ear how many objects they saw. If no one saw all of the objects, tell everyone that there are still more objects to find. Then let them start over. Repeat as often as necessary or until interest fades.

What was the most difficult object to see? Why? What was the easiest? Why? What are some things our eyes can tell us about the world around us? Are you more aware of your surroundings when you are in a new place?

Ear-Ye, Ear-Ye

Description

Students try to identify objects dropped behind a sheet. This hearing activity demonstrates the way students use their senses to make judgments about the world around them.

Objective

To develop the sense of hearing.

Materials

Various objects from around the classroom (light, heavy, metal, plastic)
One old sheet

Can anyone tell me something that is happening outside of the classroom right now? How do you know? (sense of hearing) Name some sounds that are easy to identify. Can you identify sounds of objects that are very familiar to you?

1. Ask two students to hold the sheet so that it is impossible for any of the other students to see behind it.
2. Tell the class that you are going to drop different objects behind the sheet. Their job is to use their sense of hearing to identify the object being dropped.
3. Pick one object and drop it. Repeat.
4. Ask that hands be raised when students have a possible answer. Allow adequate time for all students to consider the problem before soliciting answers.
5. List guesses on chalkboard. Ask students to give reasons for their guesses. After students have an opportunity to discuss the possibilities, identify the object.
6. Repeat with another object.

What helped you decide what each object was? How can hearing protect you (or alert you to danger)? How would it be to live without the sense of hearing?

Big Ears

Description

This activity introduces the concept of animal sound communication through a game of hide and seek. Secret partners wear blindfolds or bag masks and try to find each other using their sound signals and their sense of hearing.

Objective

To have students concentrate on using their sense of hearing.

Materials

11 paper bags with ear holes cut out, or 11 blindfolds
Five different pairs of noisemakers
One noisemaker unlike any other

Preparation

This activity is best done in groups of 11. The nonparticipating students can form a boundary by making a large circle around the group.

Imagine that you are an animal with poor vision or are active only at night when vision is restricted. What other senses might you rely on to survive? What animals do you need to stay away from? (animals that may eat you—your predators) What animals may help you survive? (animals that are the same species as you) In this activity you will use your hearing to find your partners and avoid being eaten by the predator.

1. Choose 11 students to participate. Have the rest of the class form a large circle around them.
2. Have all the participants put on a bag mask or blindfold.
3. Hand out the paired noisemakers. No one is to know what noisemaker anyone else has.
4. Give one student the unique noisemaker and let this student know that he or she is the predator with a tap on the shoulder.
5. Spread the players over the playing area.
6. Give the signal to sound off. The object of the game is for each student to find the animal that makes the same noise as he or she does (the secret partner) and then to stay together and avoid being tagged by the predator (eaten) until the game is over.
7. Students can sound off only when standing still, not while moving.
8. If tagged by the predator, students take off their masks and move to the edge of the circle. Predators must be sure the prey know they have been tagged. If partners find each other before the predator does, they must stay together and continue to avoid the predator until the game is over.
9. The game is over when the leader calls time.
10. Allow about five minutes for the game and then choose new participants.

What helped you to identify your partner? to identify the predator? How can hearing be important to animals? Are there any animals you can identify by hearing? What animals can you name that depend on their hearing for survival?

Only the Nose Knows

Description

Students use their sense of smell to identify mystery fragrances.

Objective

To develop an awareness of the sense of smell.

Materials

Cloth
Ammonia
Various fragrances (peppermint, vanilla, cumin, ginger, cloves, bay leaves, coffee, cocoa)
One opaque container (film container) for each fragrance

Preparation

Place one fragrance in each container. Number the containers and make up a code sheet so that you can easily identify the fragrances.

What senses serve as important information-gathering tools? Tell me which sense you will use to identify this substance. (With students sitting very still, release in one corner of the room ammonia poured on a cloth. Perfume may also be used. Ask students to raise their hands as soon as the odor is detected. Have them note the progress of the diffusion of the odor through the air across the room.)

Of what use is our sense of smell? What kind of messages does our sense of smell give us about the environment? Can you think of any times when your sense of smell told you something before your eyes did (such as knowing what you were having for dinner or discovering a fire)? What would it be like to be unable to smell anything? Can you think of any animals that would have trouble surviving without a sense of smell? Can you name some good odors and some bad odors?

1. Have students number from 1 to 8 in their journals or on a piece of paper.
2. Pass out the mystery fragrances. Tell students that they are smell detectives.

3. Have students use their sense of smell to identify the mystery fragrances. This should be done individually.
4. As the students do their detective work, put graphs on the chalkboard for each container. (See illustration for different options.)

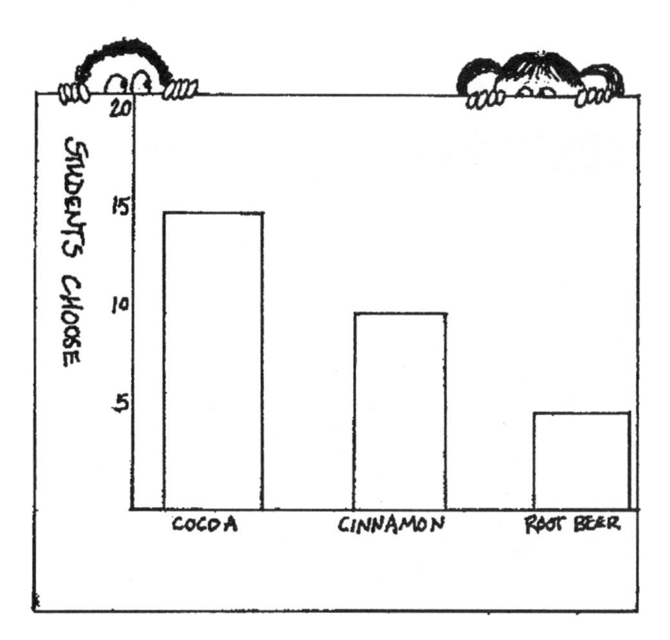

5. When the students have completed the exercise, discuss the mental connection they made with past experiences that helped them to identify each fragrance. Graph different choices for each fragrance on the board. Have students offer a number of suggestions for each fragrance and then graph the number of students that chose those identifications. After you complete the graphs, go over the answers.

Name instances in which your sense of smell told you something before your other senses did. What would it be like not to have the sense of smell? What happens to your sense of smell when you have a cold? What information did the graphs give you? Did they reveal the correct answers?

Six of One, Half Dozen of the Other

Description

Groups of students find and classify contrasting textures in the natural environment.

Objective

To use the sense of touch to identify and classify objects in the garden laboratory.

Materials

One egg carton per group of three

Preparation

Across the top of each egg carton write "Textures." Label half of the bottom of each carton with one texture, such as "rough," and the other half with an opposite texture, "smooth." Label each carton with different opposites. Examples: wet-dry, scratchy-soft, squishy-solid, fuzzy-smooth, hard-powdery.

What parts of your body can you use to identify different textures? (fingers, hands) What do many animals use to feel things in front of them? (whiskers, noses) How about using your noses to feel things? Today we're going to explore the garden for different textures. What are some textures you expect to feel in the garden? (List responses.) Which of the textures listed are the opposite of each other? What are some other examples of opposites?

1. Divide the class into groups of three. Tell the class that each group will get a special collecting container in which to collect 12 items. They shouldn't let any other group see the secret information on the bottom of the carton.
2. Distribute the cartons and demonstrate to each group how the opposites should be placed. On the bottom of the carton are secret words that tell what textures to collect. Every group will be collecting different opposites.
3. Remind students to handle everything gently and to take only small specimens. Allow enough time for students to explore the site and gather the items.
4. When groups are finished, have them exchange cartons and try to determine the opposite textures that the other group collected without looking on the bottom of the carton.
5. Discuss strategies that groups used for identifying the other group's classification.

What things that you collected felt the scratchiest, the softest, the wettest? To identify the textures you first used your sense of sight. How did you know from looking what objects would feel soft? Hard? Squishy? What textures did you find on young plants? What season do you think of when you feel dry, scratchy things? Soft, wet things? Where did you find warm things? Cold things?

Everyone Needs a Rock

Description

Students use all of their senses to explore and identify pet rocks.

Objective

To develop sensory-awareness skills.

Materials

One rock per student
One blindfold per student

If you lost your favorite toy, what would you do to try to find it? Would you use any of your senses? What if you were looking in a big toy box and couldn't see to the bottom. What senses would you use to find out if your toy was in there? Do you think you could identify a rock by using your senses? (Take a tally of the class guesses.)

1. Have students find a rock that is about the size of a golf ball. Have students bring their rocks and sit cross-legged in circles of six to eight students.
2. Ask each student to make a sensory exploration of his or her rock. For example:
   - Look for the number of colors in your rock.
   - Feel for sharp points, smooth places. Is your rock cold against your cheek? Is it a heavy or light rock?
   - Smell three parts of your rock. Any surprises?
   - Without tasting your rock, imagine what it would be like to eat it.
   - Listen to your rock and see if it can tell you something about its life. Was it always where you found it? Tap your rock with your fingernail. What sound does it make?

3. Have students imagine getting smaller and smaller until they are so tiny that they can explore their rock as if it were a small planet. Where would they plant a garden? Is there a bit of dirt on the rock? Where would they locate a lookout tower? Where would they locate a cave for hiding? Where would they locate a valley to collect rainwater? Where would they build a house?
4. Collect the rocks. Blindfold the students or have them close their eyes. Redistribute the rocks at random. Ask students to feel the rock. When you say "Pass" have students pass the rock to their right if it is not theirs. Then have them feel the rocks again and pass them. Repeat until everyone has their rock. Then have students remove the blindfolds and discuss what they could tell about the rocks without looking.
5. End by having the students hide their rocks somewhere nearby so that they can find them again at a later time.

List adjectives used in describing the rocks. How many colors were in your rock? How much do you think your rock weighed?

Have students categorize rocks by characteristics such as size, texture, shape, and color.

See No Evil, Hear No Evil

Description

Groups of four students identify objects in the outdoor environment, each student representing a different sense.

Objective

To develop the senses.

Materials

Three blindfolds for each group of four students

What would it be like to have just one sense? Which sense would you choose? Do you think you could identify objects using just that one sense? In this activity, we will each use only one sense to try to identify different objects outdoors.

1. Divide students into groups of four.
2. Choose one person in each group to be the eyes. The other three people are then blindfolded, and each person chooses a role: one is the nose, one is the ears, and one is the hands, so that four senses are represented.
3. Have each group form a line with the eyes in the front.
4. The eyes lead the senses to three different objects (plant, tree, compost pile).
5. Each object must be investigated by each person independently in terms of his or her sense.
6. Do not allow discussion yet. Once everyone has formed an idea of what the objects are, the groups return to the classroom and sit down together.
7. Have each group discuss the identification of each object. Have the senses share what they think the objects are and why. After discussion, the eyes can identify the objects.

How did you feel using only one of your senses? How important were the other members of the group to you? Which sense did we not use?

1. Have each group think of a way to share what they have discovered through collage, pantomime, verse, and so on.
2. Some animals have poor sight (nocturnal animals, subterranean animals). How do they use their other senses?

Mystery Powders

Description

Students use their senses to identify six different white powders.

Objective

To develop sensory and problem-solving skills.

Materials

Six containers such as pie tins
Flour
Sugar
Salt
Powdered milk
Baking soda
Cement (plaster of Paris)

Preparation

Put each substance in a different container and number the containers 1-6. Make a code chart for yourself so that you can easily identify each substance. Place the containers in different locations of the room.

We have a mystery in our classroom. As a matter of fact, we have six mysteries. There are six powders in our classroom. They all look somewhat alike. I would like all of you to be detectives and help me identify these powders. What senses can you use to identify them? Why is it bad to taste unknown substances?

1. Have students number a sheet of paper from 1 to 6 on the left-hand side.
2. Explain to students that placed around the room are six containers with numbers. Each contains a different white powder. Have students use their senses to discover what is in the containers. Five students may be at each station. No talking!

3. When students have identified what is in a container, have them write the name of the powder next to the proper number. Have them spend no more than three minutes at each station. Have them return to their seats when they have finished.
4. When most students have finished, end the testing part of the activity.
5. Ask students to discuss what it was that helped them identify the powders, but do not have them reveal their answers. Explore the process students used to arrive at a conclusion. List adjectives that describe the contents of each container. Then list all of the different answers given.
6. Make a class graph for each container. (See illustration below.) Graph the number of students who selected various choices.
7. Identify each powder, leaving the cement for last.

CONTENTS	CONTAINERS					
	1	2	3	4	5	6
flour	12					
sugar	0					
salt	0					
powdered milk	3					
baking soda	2					
Plaster of Paris	2					
cornstarch	3					
cakemix	4					

Discuss the five senses and what they are. Explain that our senses can profit from exercise and concentration. It is important that we develop our senses so that we may make judgments about our environment. Our senses connect our bodies to the outside world.

Little Munchkins

Description

In this outdoor activity, students work in pairs using careful observation skills as they pretend they are miniature people on a 100-inch hike.

Objective

To develop observation and recording skills.

Materials

String
Life Lab journals
Hand lenses

Preparation

Cut one piece of string 100 inches long for each pair of students.

Have you ever imagined that you were very small? What would it be like to take a hike through the garden if you were only as tall as your thumb? What would the plants look like? What would rocks and sticks look like? Would you notice things that you don't notice now?

1. Divide the class into pairs and go to the garden and find a place that might seem like a forest to little munchkins.
2. Give each pair of students a piece of string and tell them they are now little munchkins and that they are going to take a hike that is as long as their piece of string.
3. Have each pair place the string in a straight line or in a winding curve. Then have them follow the string, carefully recording everything that lies along the path. Have them look closely at the plants on the route to find differences in leaf shapes, edges, textures, and colors.
4. Have each pair of students decide how they want to record what they find on their hike. Then have them take turns being recorder. Have them record the different kinds of plants they find and how many of each one. Suggest that they draw pictures of the different things they see. Have them describe any other creatures they encounter and describe the texture of the soil. Remind students of their point of view: they are less than two inches tall. Even tiny bugs are going to seem huge to a munchkin.
5. When the hikes are completed, have three teams join together and share their findings, identifying similarities and differences.

Give an example of a texture found on your hike. Tell what it felt like to look closely at a small area. Classify the many things that you observed and recorded.

Burma Shave

Description

In this self-guided, cooperative group activity, students carry out instructions and answer questions that are written on 3X5 cards at different locations in the garden.

Objective

To develop awareness and observation skills.

Materials

3 X 5 cards
Life Lab journals

Preparation

1. Prepare 3X5 cards with challenging instructions and questions such as the following: Listen for three human-made sounds. Listen for three sounds not made by humans. Look for three different animal homes. Who lives in them? Hot and thirsty? Find a drop of water. Should we build a restaurant here so that you can buy lunch? How would a restaurant change this place? Smell five things before going to the next card. Run to the next card. Who do you think lives in this hole? Can you see any waste of water near here? Find a seed. What do you think was here 50 years ago? Find something that feels rough. Feel this.
2. Scout a particular area of the schoolyard or garden for your Burma Shave trail. Place the cards along the trail in advance or as the first group progresses through the activity.

What do you use to make observations? (senses) How do you help your senses tune in to specific observations? Do you have ways that help you focus? We're going to go on a walk that you've been on lots of times. This time, though, you will find cards along the way. When your group gets to a card, I want one person to read

what it says out loud, and each person in the group to follow the instructions or answer the question on the card. Then one person will record the group response. Do you think the cards will help you be better observers?

1. Divide the class into groups of four. Assign one person in each group to be the reader, another the recorder.
2. Lead the groups to the beginning of the Burma Shave trail. Students should remain at least 20 steps from the group in front of them.
3. Have the last group pick up the cards.
4. When everyone has finished, review the cards with the students. Once the strategy is learned, use it again in different places. Each time, students will get more out of the activity.

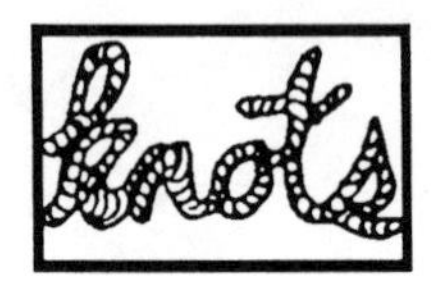

What are some new things you found out about this area? What senses did you use? What would it have been like to do this activity without your sense of hearing? Sight? Touch?

On Location

Description

In this activity, students map different characteristics of the school or garden site to help create awareness of the area.

Objective

To reinforce observation skills and to develop mapping skills.

Materials

One map outline per student

Preparation

Prepare a ditto of an outline of the boundaries of your Life Lab garden.

If you were going on a trip and weren't sure what roads to take, what could you use to find out? How else are maps useful? (to help find where places are located, to give information about the geography of a location, and so on) What if you were going to visit a special friend who lived in another town and you wanted to show her or him what your Life Lab garden looked like but you didn't have a camera? Could a map of the garden help? We could also use a map of the garden to record information. Let's make a list of the kind of information we can record on a map. (location of class beds, traffic flow, direct-sun and shade spots, noisy and quiet spots, favorite spots)

1. Give each student or pair of students a map of the perimeter of the school or garden site. Have students investigate the site, adding details. These maps can be records of their own observations and discoveries rather than a collection of information they have merely checked.
2. Have students map locations of objects, buildings, fence gates, garden beds, and so on. Help them develop ways to measure distances and scale them on the map without using measuring instruments.
3. Now have students work together in groups of three or four. Assign one of the map uses listed on the board during discussion to each group. Have each group use their maps to describe that specific use. One group might use arrows to mark the typical traffic flow in the garden, another might map insect homes, and so on.
4. Have each group share their discoveries with the whole class.

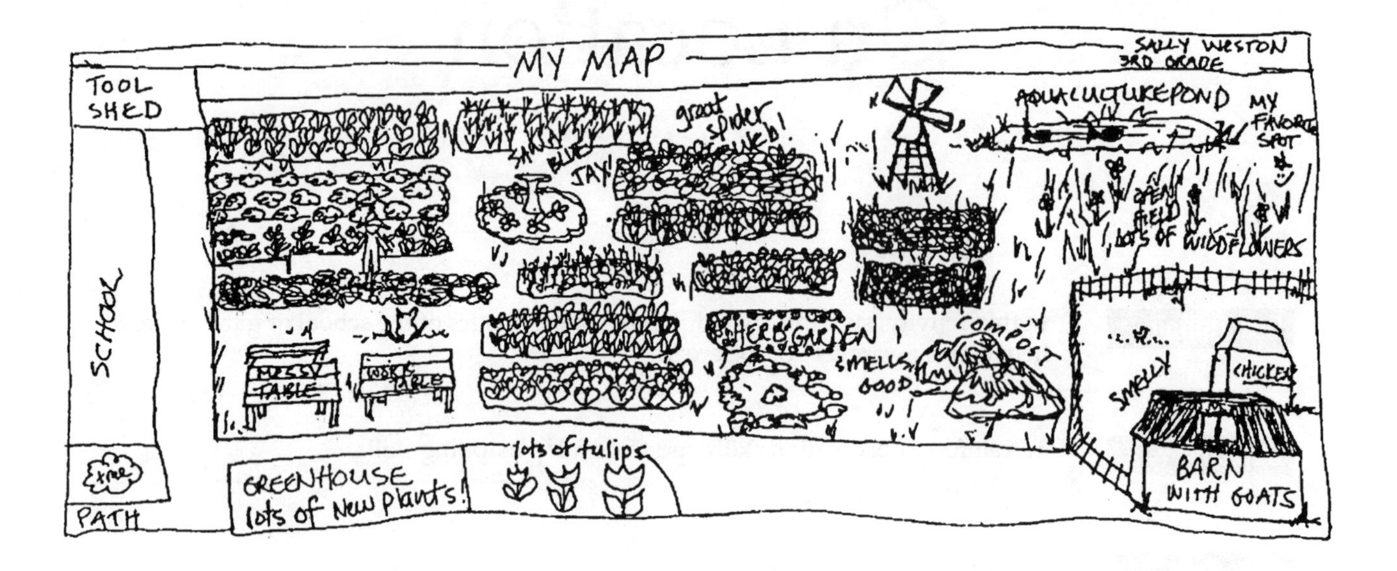

Do people use the land around our school in different ways?

1. Before groups share their discoveries, have each group exchange maps with another. Have the other group describe the information recorded on the map.
2. Choose something that changes throughout the year in the garden. Have students keep a map record of the changes.

The Living Earth

Soil

E. Ruth Sheldon School
Fairfield, CA

Robert F. Kennedy School
San Jose, CA

Hall Elementary, Watsonville, CA

E. Ruth Sheldon Elementary
Fairfield-Suisun Unified School District
Fairfield, California

Opening day of the E. Ruth Sheldon Life Lab was in March, 1987. We followed the Life Lab staff's recommendation to "start small." An initial grant allowed 8 classrooms to use the curriculum, fence off an area, purchase materials, and begin planting in wine barrels.

Our program has grown to include all 19 classrooms, 23 raised beds, a fully-equipped tool shed, and a greenhouse. Each classroom has Life Lab instruction during the week in the garden and/or in the classroom. The curriculum is taught by the classroom teacher and an assistant. Our four Kindergarten teachers spend part of each day working as these assistants. Students keep a record of their Life Lab activities and discoveries in special journals.

Raising funds and building our site has taken the energy of our whole school community. For example, our 16' X 24' greenhouse was constructed on seven Saturdays by students and 35 volunteer parents, staff, and community members. We take pride in our Life Lab and believe it has increased our enthusiasm for our school by giving us something to build together, to beautify our school and, most importantly, to make science come alive.

UNIT INTRODUCTION
The Living Earth

Soil is our starting point for applying observation skills to science. Soil is the top layer of the earth's surface—the layer we depend on for food, fiber for clothing, wood for warming, materials for shelter. Soil is integral to our survival, yet we tend to take it for granted. Our study integrates physical, life, and earth sciences, and analyzes the physical components of soil and how they change, tests its chemicals, and determines how the physical and chemical components affect life and growth in the soil.

The Living Laboratory depends on healthy soil. Students start with simple observations and progress to data collection and experimentation as they continually apply what they learn to the preparation, caretaking, and improvement of their own garden soil. Activities start with exploring how soil is made and lead to tests for determining soil quality. Students analyze their soil nutrients, learn how to prepare their soil for planting, and improve their soil nutrients through composting and cover cropping. Current societal issues concerning soil, such as compaction and erosion, are investigated.

Most of the activities in this unit are best done outdoors at a time of preparation of the soil. "The Matchmaker" requires planting cover crop legume seeds (bell beans, vetch) 6-10 weeks in advance. "Soil Doctors" and "A Soil Prescription" need a simple soil test kit. "A Soil Prescription" and "What Good Is Compost?" require ready-to-use compost. "To Dig or Not to Dig...," "What Good Is Compost?," and "What's To Worry?" are garden-based experiments and require data collection over an extended time.

"The civilized nations—Greece, Rome, England—have been sustained by the primitive forests which anciently rotted where they stood. They survive as long as the soil is not exhausted," wrote Henry David Thoreau. Perhaps soil provides roots for more than just plants.

Activities	Recommended Grade Level
Space Travelers *(exploration of soil ingredients)*	2,3,4,5
Sensual Soil *(soil exploration and creative writing)*	3,4,5,6
The Nitty-Gritty *(exploration of quality and composition of garden soil)*	4,5,6
Water, Water Everywhere *(water-holding and draining capacity of soil)*	4,5,6
Living in the Soil *(creating a healthy soil)*	3,4,5,6
Tools and Us *(proper tool use and care)*	2,3,4,5,6
To Dig or Not to Dig: That Is the Question *(compaction and plant growth)*	3,4
Dig Me and Dig Me Again *(soil preparation technique)*	2,3,4,5,6
Soil Doctors *(soil testing for major plant nutrients)*	3,4,5,6
A Soil Prescription *(soil amendments)*	5,6
What Good Is Compost? *(compost garden experiment)*	3,4,5,6
The Matchmaker *(nitrogen fixing cover crops)*	3,4,5,6
What's to Worry? *(uses of mulch in the garden)*	5,6
Splash *(rain and soil erosion)*	4,5,6
Day at the Races *(erosion of different areas)*	5,6

Dirt Made My Lunch

Am **C**
Dirt is a word we often use

Am **C**
When we talk about the earth beneath our shoes.

Am **C**
It's a place where plants can sink their toes

F **G**
And in a little while a garden grows.

Chorus

C **F** **C**
Dirt made my lunch, dirt made my lunch.

F **C**
Thank you dirt, thanks a bunch

F **C**
For my salad, my sandwich, my milk, my munch.

G **C**
Dirt made my lunch.

A farmer's plow will tickle the ground.
You know the earth has laughed when wheat is found.
The grain is taken and flour is ground.
For making a sandwich to munch on down.

Chorus

A stubby green beard grows upon the land.
Out of the soil the grass will stand.
But under hoof it must bow
For making milk by way of a cow.

Chorus

Space Travelers

Description

Students work in small groups as space travelers trying to decipher the composition of soil.

Objective

To explore the composition of various soils.

Teacher Background

Soil is something all of us take for granted. However, it is one of the necessary life-sustaining ingredients of our planet. And soil is exciting! It varies dramatically within a small area. When students explore the surface soil (topsoil) they will discover many living things—roots, earthworms, insects. In addition, the topsoil contains humus (the high nutrient component of the soil that is formed by decayed organic matter) and rock particles. As students dig deeper, the soil composition changes. Soil is formed by natural processes that wear away rock and break it into tiny particles. This wear can be caused by rain, wind, glaciers, and plants. Soil formation is a very slow process. It takes over 100 years to produce 2.5 cm (1 inch) of topsoil!

Materials

Two trowels per team of three
One hand lens per team
Life Lab journals

Imagine that you are scientists journeying to the planet Earth aboard the Star Ship Life Lab. You receive the following message from President Gorgo Buerhing, the Head of State of your planet: You have been chosen to make a most important journey. The future of our beloved planet is in danger. As scientists, you well know how the planet we love has become so polluted that we are no longer able to produce our own food. Our astronomers have detected a very faraway planet called Earth. It appears to be lush, green, fertile, and productive. Our computers have been analyzing the reasons for this and have concluded that the secret appears to be a dead, brown-grey substance called soil. It is difficult for us to believe that all of their food comes from this substance. Your mission as scientists is to find this substance, dissect it, and record for our computer each and every ingredient. This will enable our planet to manufacture soil and save us from the tragedy that is about to befall us. Upon landing, divide into groups of three with two dissecters and a recorder in each team. Use the special tool [trowel] our engineers have designed especially for this purpose. Remember: It is crucial to the success of this mission that each and every substance found in the soil be recorded. Good luck to all of you.

1. Divide students into groups of three and have them explore soil in different areas of the garden and school yard. Have each team investigate just one spot.
2. Upon completion of the task, ask teams to compare and contrast the soils they investigated. Ask them to list the qualities of the soil. Have the groups discuss the ingredients they found: crushed rocks, crumpled leaves, twigs, clay, sand, and so on.
3. Assign ingredients to each team and ask them to return with a small quantity of each ingredient.
4. Upon their return, challenge teams to use the raw ingredients to manufacture soil by scraping rocks together, breaking twigs apart, and so on. When the frustration level of the students is reached, ask them whether soil can be made by hand. Why not? Explain that each inch of topsoil requires over 100 years to form. Bacteria, fungi, and other living things slowly decompose nutrients, such as leaves and twigs, recycling them into soil. Soil is alive. Over 100 billion microorganisms live in a pound of soil. Our hands and tools cannot equal the power of the bacteria and fungi.

Will the super computer on our home planet be able to manufacture soil? How is soil important to Earthlings' lives? Is soil alive? How? Do all materials in soil decompose at the same rate? What do earthworms do for the soil? What would be the result of covering, washing away, and stripping all of our soil?

Sensual Soil

Description

People often mask many of their sensory experiences, focusing only on the visual. Through exploring soil in this activity, students use most of their senses to discover qualities of different soils.

Objective

To explore with our senses different kinds of soil.

Materials

Four containers with different types of soil: clay, compost, sand, garden soil
Four lunch bags
Scrap paper
Four large pieces of construction paper

Preparation

Set up four stations in the classroom or outdoors. At each station place a container of soil, one lunch bag, scrap paper, pencils, and one sheet of construction paper. Make sure it is easy for 1/4 of the class to gather around one station.

Look at this container I am holding. What tools do you have to explore what is in this container? (senses) What are some words you would use to describe something that you saw? Heard? Smelled? Felt? In this lesson you will spend a few minutes at each station. Each of you will look very closely at the soil at each station; smell a clump; rub it with your fingers near your ears to hear what it sounds like. Then each of you will choose a word to describe the soil at the station, write it on a slip of paper, and put the paper in the lunch bag.

1. Divide the class into four groups. Each group will spend a few minutes at each soil station exploring each sample.
2. At each station, ask each student to write on a scrap of paper a descriptive word about the sample and place it in the bag.
3. After the groups have been to all stations, give each one a bag of words. Have them use the words in the bag to compose a Soil Poem, and have them copy it onto a large sheet of paper. The poem will be a random ordering of the words in the bag. It could even have a title if the students are so inspired.
4. Have one person from each group read their poem to the class.

5. Post the poems and attach the proper classification (sand, compost, garden soil, or clay) to each one. Explain that students will learn more about these soils as part of the Life Lab program.

Which soil had the strongest smell? Which felt the weirdest? Which felt smooth and slippery? Which felt gritty and coarse? Which made the loudest sound? Compare the two soils that seemed very different from each other. Compare the two soils that seemed the most similar.

The Nitty-Gritty

Description

Through a simple process, students separate soil into its three major components: sand, silt, clay.

Objective

To explore the composition of garden soil and determine its quality.

Teacher Background

Soil is composed of a blend of various-sized particles. The proportion of sand to silt to clay determines the quality of the soil. Sand, silt, and clay may seem to be uniformly categorized as small particles, but there is a great difference in the size of each of them, and this difference affects soil quality. If a particle of sand were the size of a basketball, then silt would be the size of a softball, and clay would be the size of a golfball.

Gardeners describe soil types in many ways: heavy, light, sandy, clay, loam, rich loam, poor soil, and so on. Scientists and horticulturists classify soil types by the proportion of sand, silt, and clay particles they contain, based on the sizes of mineral particles. The texture of the soil is determined by the blend of these various-sized particles. Classifying the soils in our garden will give us some indication of the problems we are likely to encounter in working with them: "Soil that has too much clay is hard to work" and "Soil that has too much sand dries out fast." Through the years it is possible to change the texture of soil by adding amendments such as sand and compost to balance the proportions.

Materials

One glass jar with a lid per group of five
One piece of masking tape per group
One trowel per group
Markers
Soil samples gathered by student groups during activity
Water
One clay/silt/sand chart (Blackline Master) per group, p. 394

What have we learned about how soils are made? (They are made when materials break down.) Do you think all soils are the same? (no) Why would some be different from others? (They are made from different types of materials, different weathering processes, different climates.) Do you think all soils are good for growing food? (no) What might make some soils better than others? (good drainage, ability to hold nutrients, easy to dig, lots of living things) In this activity, we are going to do a simple demonstration to determine the parts of our soil and find out whether it will be hard or easy to dig and if it holds water. Can anyone predict what these soil parts might be? (Write all predictions on the chalkboard.)

1. Divide the class into groups of five. Give each group their materials.
2. Fill each quart jar about 2/3 full of water.
3. Demonstrate how to take a soil sample. First dig a few inches below the surface. Then carefully scoop up soil for the sample.
4. Help each group select a different location in the garden and school yard to take soil samples.
5. Have each group add soil to their jar until it is almost full. Then have them put the lid on the jar.
6. Have groups label the jar lids with the group name and soil location.
7. Have students shake each jar vigorously. Let the soil settle. Have each group observe their jar. What do they see happening? (In a short time the heaviest sand particles sink to the bottom and the sand layer becomes visible, but the silt and clay particles will take hours to settle.)
8. Place the jars in a location where they may be easily observed. Be sure no one lifts the jars to observe them.
9. In 24 hours, the soil will be completely layered. Have each group describe the layers. Which layer is on the bottom? (one with the heaviest, biggest particles) Is that the same for each group? Which layer is the thickest? (Answers may vary.) How do you think the thickest layer will affect your soil for gardening?
10. Each group can use the clay/silt/sand chart, p. 394, to determine their soil name. Then have them mark off the layers on a piece of paper held up to the jar, as shown below, and compare each one to the chart. If the particles divide into about 40% sand, 40% silt, and 20% clay, the soil is called "loam"—a very good kind of soil to have. If the soil falls into other classifications, you may want to add sand or organic matter to change its classification.

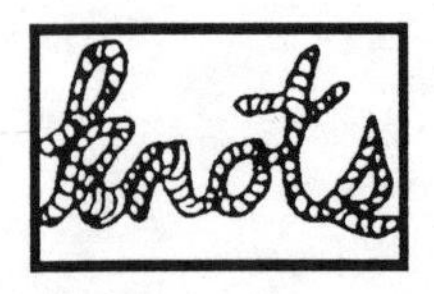

Were all of the soils the same? How did they differ? What are the three different particles in soil? Which is the biggest? Which is the smallest? What do you predict will make your soil better for gardening? Why? Which soil sample will be easiest to dig? Which will not let water drain?

Keep a soil history from year to year for comparison.

Water, Water Everywhere

Description

Students time the passage of water through different soil samples to illustrate the water-holding and drainage capacities of different soils.

Objective

To compare and explain water-holding capacities of different soils.

Teacher Background

Soil requires both water and oxygen to support strong plant growth. The ideal soil is a combination of sand and clay. Sand provides fast drainage and good aeration but fails in the water-holding department. Clay is tops in water-holding ability but dangerously low in supplying air to the soil. As water fills the spaces between soil particles, it drives out air. In soils with a high proportion of clay, water remains a long time in the pore spaces, and plant roots are deprived of oxygen for many hours. This temporary lack of oxygen can be very damaging to some plants; it is easy to drown plants in a clay soil.

Materials

Four lamp chimneys
Four dry soil samples of very different soils: sand, garden soil, compost, clay
Screen or cheesecloth
Strong tape
Four quart jars
Measuring cup
Clock or watch

Preparation

Set up chimneys as in diagram. Label each of the soils.

How can soils be different from each other? How can some of these differences affect how fast water drains from the soil? (Bigger particles will allow faster drainage.) Why is water drainage important to plants? (If water drains too fast, the plants will not get enough; if it drains too slow, they may drown.) Let's make some predictions. Of the four soils we have, which do you think will drain the fastest? Which will drain the slowest? Which will drain all of the water? Which will hold some water and stay moist? (Write predictions on the chalkboard.) How can we test these predictions?

1. Have the students pour a pint of water through each of the four inverted lamp chimneys, timing the length of time it takes the water to drip into each quart jar and how much water comes out from each sample. Record the results.
2. Have the students compare the water-holding capacity of the four samples by figuring the amount of water left in the soil after it finishes dripping.
3. Ask the students to test the drainage ability of the soils by saturating the soil sample with water until it is holding as much as it can, adding a pint of water to each sample. Record the length of time it takes the water to move completely through the saturated soils, stop dripping, and collect in the jars. Compare the results with the predictions.

Which soils would you plant a seed in? Why? Which sample could possibly drown your plant? Why? Which sample would you not plant a seed in? Why? Which soil would you not want your plant to be in during a drought?

Living in the Soil

Description

Students make their own healthy soil mixture.

Objective

To discover and describe the components of a healthy soil.

Materials

A hard surface to work on, preferably cleared ground or pavement. It should be large enough for each group to have room to mix soil and situated near a flat mixing or compost area that has piles of compost, topsoil, and sand.
Two hand trowels per group
One bucket or gallon pot per group
Seed flats
Seeds to sow

When you look at a plant, what do you see? You only see half the story! Each plant travels at least as far underground as it grows above ground. Plant roots are always growing, pushing through the soil, absorbing all the nutrients and water the plant needs in order to live. Without this secret underground life, many plants couldn't live. What kind of soil do plants like? It should be porous (water can get through it), but still be able to hold some water and nutrients; it should be moist but not sopping; it should have organic matter (such as compost) in it, and it should be something plants can really sink their roots into! Do you think you can make a soil that plants would like?

1. Ask students to work in groups of four.
2. Have each group go to different spots in the garden to collect a potful of soil with their trowels and buckets. Have them dump the soil on the hard surface in their own separate piles.
3. When students are reassembled and seated next to their soil piles, ask them to imagine they are a little plant. What kind of soil would they like to grow in? Hard and compact? Light and fluffy? Lots of nutritious food, such as compost, or just plain sand?
4. Each group should decide how to modify their soil to make it more suitable for a little plant. What would they like their soil to be like? What can they add to make it that way? Sand? Compost? Soil from another location?

5. Have them test their soil by adding water. A test for good soil consistency is to moisten the soil, then take a handful and squeeze it in your fist. When you open your fist, the soil should hold together but then crumble apart when lightly touched.
6. Give each group the opportunity to improve their soil mixture.
7. Ask them to compare each other's piles of soil. Then have the class discuss the things each group had to do to improve their soil and why.
8. Have students fill a few flats with their improved soils and sow some seeds!

How could you improve your garden soil? List the reasons why beach sand alone is not a good environment for most plants. How would you describe the ingredients of a good soil mix? How can we test whether a soil mix is good?

Keep records of soil mixtures and germination and growth rates.

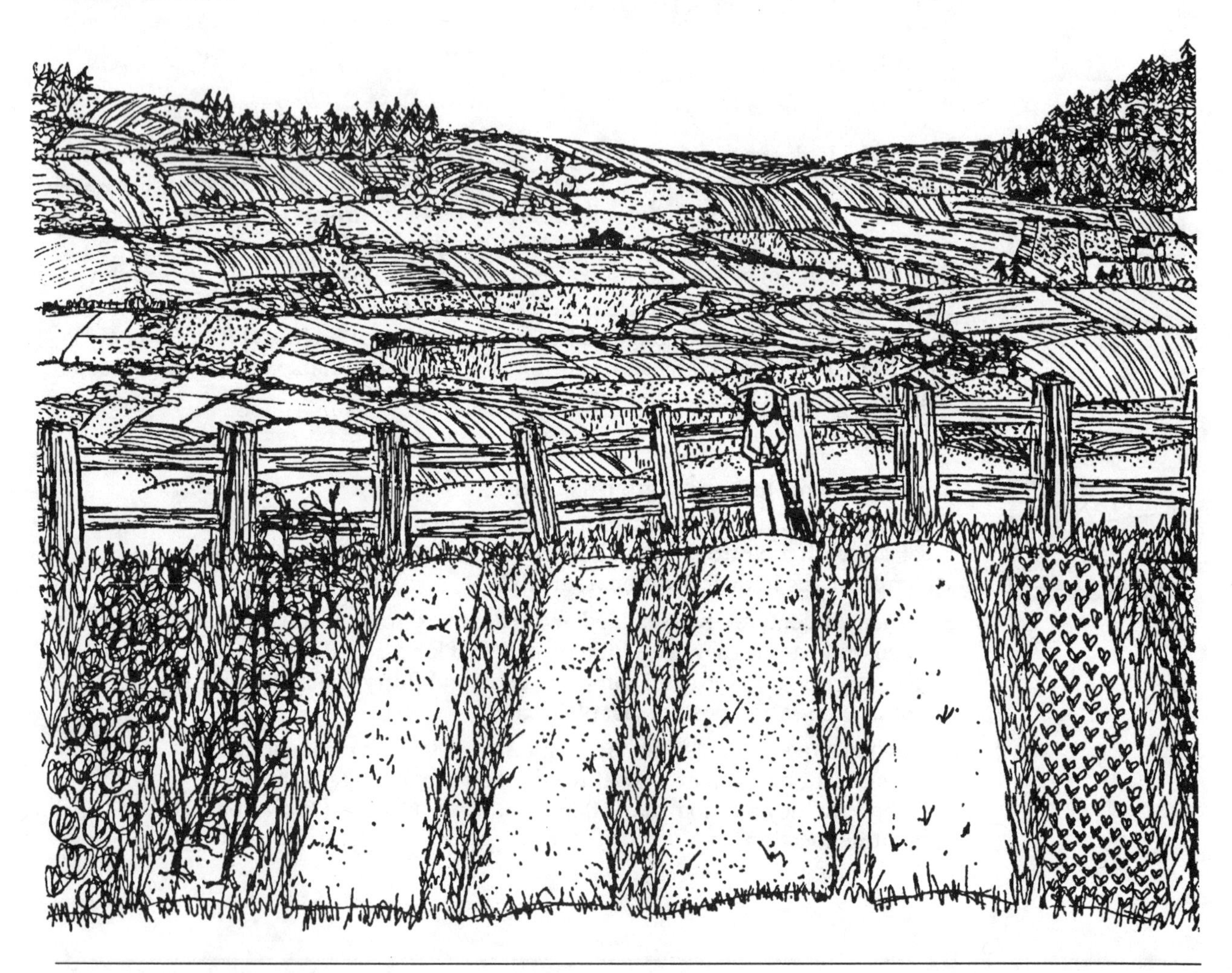

Tools and Us

Description

This activity uses role playing to demonstrate safe tool use in the garden.

Objective

To learn proper tool use and care.

Materials

One of each kind of tool that will be used in the garden per student if possible
Tool cleaning materials: Wire brushes and a box filled with sand and motor oil

What are some examples of tools? Why do we have so many tools in our society? If we didn't have a shovel, how would we dig a hole? Throughout history tools have changed a lot. Prehistoric humans developed simple tools to aid in hunting. Some animals use tools. One species of monkey dips a piece of grass in termite and ant nests and then uses this handy tool to eat those ants and termites that attach themselves to it. Sea otters use a rock or other hard object to crack shellfish for their tasty meals. What are some examples of tools used by people or animals? (List responses on the chalkboard.) Would we be able to garden easily without tools?

1. Show students the tool storage area. Go over the names of the tools. Demonstrate proper storage. Have each student take a tool so that at least one of every tool is represented.
2. Use role playing to teach the following safety rules for tool use and care:
   - Spading forks and shovels should be used only by those with shoes on. No sandals.
   - Walk when carrying tools.
   - Keep tools below shoulder level.
   - Take your time. Wait until other people are out of the way.
   - Do not use tools in crowded areas.

- Always clean a tool before you put it away.
- Never leave tools on the ground. Always stand them up or return them to the storage area.
- Always walk with the wheelbarrow. Do not use it for rides.

3. Choose a proper tool use. Ask for a volunteer, and whisper the action to the student. Have the student role play the action. Ask the group to identify the action. Repeat with different examples. Then have the teacher role play improper use for students to identify.

4. After role playing, demonstrate tool cleaning. Scrape the tool with a rough brush. Make sure all dirt is removed. Then dip it in a container of sand that has been soaked with about a quart of used motor oil.

Why did people develop tools? What are some examples of tools no longer used? How will you use tools in the garden?

As the group works in the garden, any student may call "freeze" if they observe an unsafe or inappropriate tool use. It may then be discussed.

To Dig or Not to Dig: That Is the Question

Description

Students conduct a garden experiment on the effects of compaction on plant growth by monitoring seedling growth in compacted and loosened soil.

Objective

To discover the effect of compaction on plant growth.

Teacher Background

The next two activities demonstrate the importance of planting in loosened soil. In compacted soil or dense, closely packed soil, there is less room for air; it is difficult for water to drain, so roots rot; and seedlings have a hard time pushing through the soil.

Materials

Approximately 20 of the same variety of seedling, such as lettuce or broccoli
A 4 ft X 8 ft garden bed or area that has not been prepared
String
Markers
Spading forks

What are some ways we've learned soil is important for plant growth? Close your eyes for a minute. Imagine you are a small seedling just transplanted into soil. Picture what it is like to be in the soil. What are your roots doing? When you are ready, open your eyes. Describe what it is like to be a seedling in the soil. As gardeners, what can we do to prepare the soil for seedlings? (List responses on the chalkboard.)

1. Have the class or group gather around the selected garden area.
2. Ask each student to stick out one finger and pretend it has become a seed. Have them try to push their seed-fingers into a compacted area. They should really push around and feel how hard it is. Discuss what it would be like to grow in this soil, and how they would like to change it.
3. Divide the garden bed in half and place a string between the two sections.
4. Have students use spading forks to loosen the soil to one foot in depth in one section, labeled Bed A.
5. Now ask students to sink their seed-fingers into the loosened section. Ask them to plant their finger-seeds where they think they would grow the best. Record the predictions.

6. Have students plant Bed A and the compacted section (labeled Bed B) with the same amounts of the same kind of seedlings. Try to keep all other factors the same: The soil should be similar, one side should not have more nutrients than the other, watering should be the same, and so on.
7. Have students make weekly notes on the progress of the plants, recording in a chart which plants grow faster, get bigger, look healthier, have less insect or disease damage. Did more plants survive in one section than in the other?
8. When the crops mature, have students compare their charts.

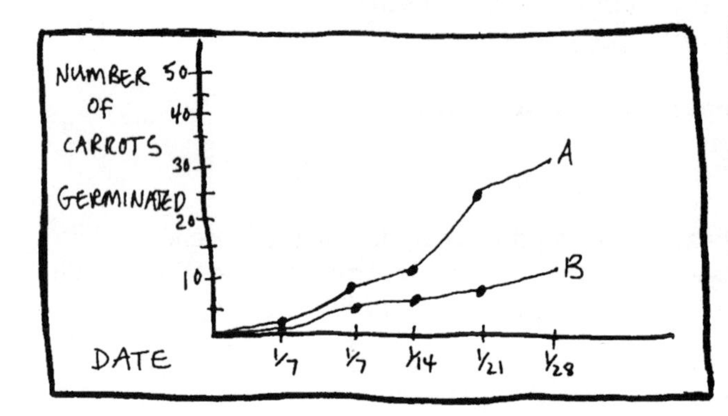

LENGTH OF GROWN CARROT
— IN INCHES —

BED A	BED B
6	3
8	2
4	4
7	4
7	3
6	5
AVERAGE $6\frac{1}{3}$"	$3\frac{1}{2}$"

Which bed did better? Why? How do soils become compacted? (people walking, cows grazing, machinery, and so on) How can we prevent compaction in our garden beds? (Don't walk on them or jump through them.)

Dig Me and Dig Me Again

Description

This activity demonstrates the double-digging method of soil preparation.

Objective

To demonstrate and practice soil preparation.

Teacher Background

The purpose of the double-digging method is to loosen the soil to a depth of 24 inches to allow the roots to grow easily and to improve aeration and water drainage. If your soil is compacted or heavy (high clay content), we recommend using this labor intensive method to loosen the soil. If you have a light, sandy soil it may be sufficient to loosen the soil in a simpler manner. The double-digging method forms beds that are approximately three feet wide. This width allows students to easily reach the center without ever having to walk on the planted area.

Materials

Shovels
Spades
Spading forks
Steel rakes
One flat filled with soil
Two spoons
One fork

If our plants are growing in packed soil, what will happen to them? Can you think of ways we can loosen hard, packed soil? We are going to learn a method called double-digging. When we use this method we will plant our crops in beds. The beds are for the plants. The people-paths around the beds are for us to use while working with tools and wheelbarrows, planting, and harvesting. There are many different methods used to prepare a garden plot. Our two steps will be loosening the soil and adding nutrients to the soil.

1. Demonstrate double-digging using a flat filled with soil (the bed), two spoons (the shovels), and one fork (the spading fork) before working out in the garden. Refer to the illustration for step-by-step directions. After students have practiced in the flat, and understand the method, divide them into groups of two or three. Have one pair start at an end of each bed and work toward each other. If the bed is long enough, two can also start in the middle. Dig on!
2. Help students lay out the proposed bed with string or chalk.

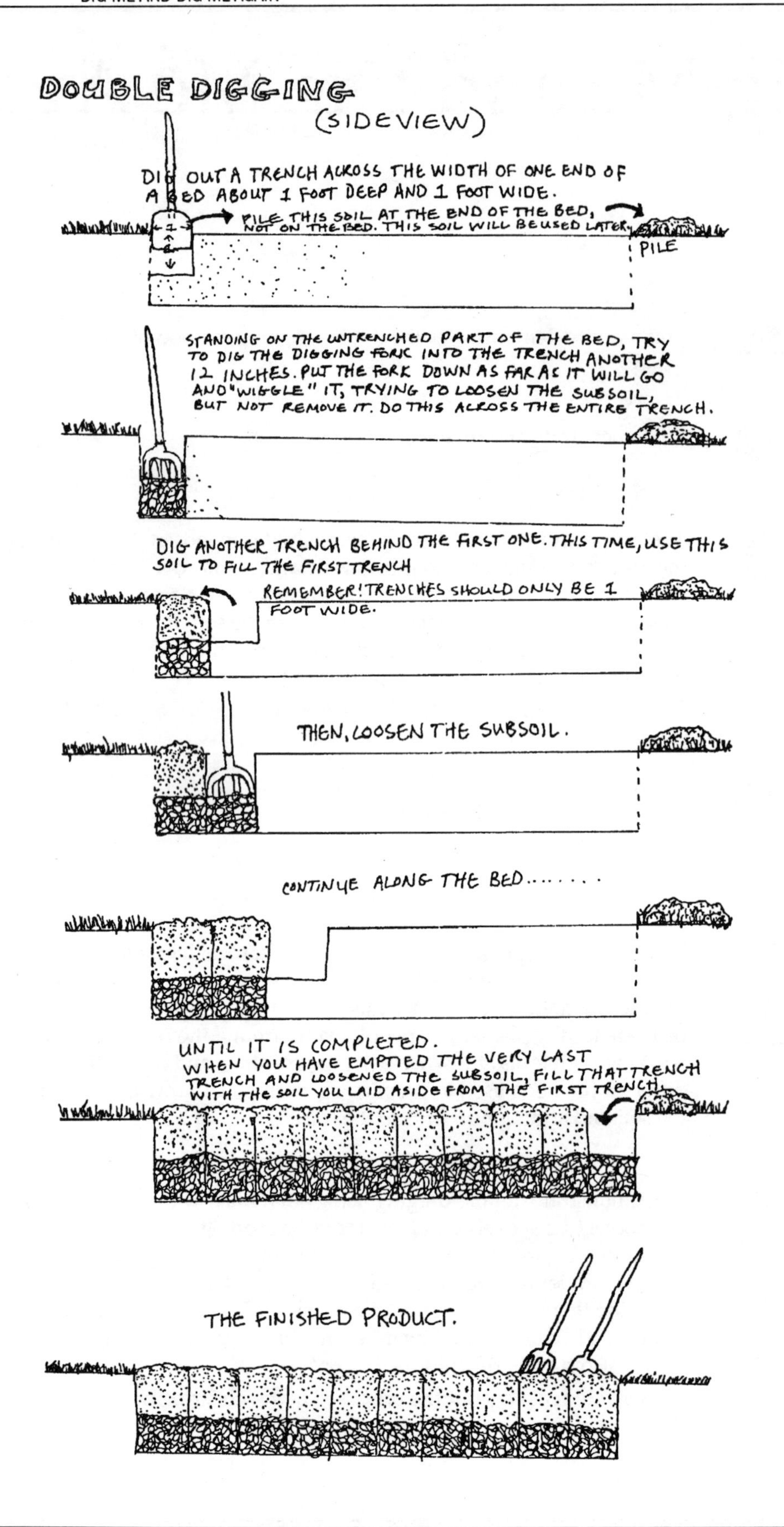
DOUBLE DIGGING
(SIDEVIEW)
DIG OUT A TRENCH ACROSS THE WIDTH OF ONE END OF A BED ABOUT 1 FOOT DEEP AND 1 FOOT WIDE.
PILE THIS SOIL AT THE END OF THE BED, NOT ON THE BED. THIS SOIL WILL BE USED LATER.
PILE
STANDING ON THE UNTRENCHED PART OF THE BED, TRY TO DIG THE DIGGING FORK INTO THE TRENCH ANOTHER 12 INCHES. PUT THE FORK DOWN AS FAR AS IT WILL GO AND "WIGGLE" IT, TRYING TO LOOSEN THE SUBSOIL, BUT NOT REMOVE IT. DO THIS ACROSS THE ENTIRE TRENCH.
DIG ANOTHER TRENCH BEHIND THE FIRST ONE. THIS TIME, USE THIS SOIL TO FILL THE FIRST TRENCH
REMEMBER! TRENCHES SHOULD ONLY BE 1 FOOT WIDE.
THEN, LOOSEN THE SUBSOIL.
CONTINUE ALONG THE BED.......
UNTIL IT IS COMPLETED.
WHEN YOU HAVE EMPTIED THE VERY LAST TRENCH AND LOOSENED THE SUBSOIL, FILL THAT TRENCH WITH THE SOIL YOU LAID ASIDE FROM THE FIRST TRENCH.
THE FINISHED PRODUCT.

3. Have pairs of students dig out a trench across the width of one end of a bed about one foot deep and one foot wide. Have them pile this topsoil at the end of the bed. (Do not have them pile it on the bed; this soil will be used to fill in the last trench.) If the soil is too hard, have them dig as deep as possible.
4. Standing on the untrenched part of the bed, have students try to dig the spading fork into the trench another 12 inches. Tell them to put the fork down as far as it will go and wiggle it, trying to loosen the subsoil, but not to remove it. Have them do this across the entire trench.
5. Have pairs dig another trench next to the first one, using the topsoil to fill the first trench. Remember the trenches should be only one foot wide. Then have them loosen the subsoil of this trench. Continue along the bed until it is completed.
6. When students have emptied the very last trench and loosened the subsoil, have them fill that trench with the soil they set aside from the first trench.
7. Have students shape the bed using a steel rake, so that it is shaped with a gentle arch and the surface is smooth.

Imagine you are a little seedling trying to grow your roots in a hard, compacted path. Think of your fingers as roots. Would it be easy to poke them through the soil? Now imagine you are growing in a double-dug bed. Which place do your roots prefer? Why?

Soil Doctors

Description

Students use a simple soil test kit to determine mineral content of soil.

Objective

To understand that plants need certain minerals for healthy growth.

Teacher Background

Soil contains minerals and nutrients necessary for plant growth. Sometimes shortages of certain minerals exist and soil needs to be supplemented. The soil test kit is simple to use and allows students to determine whether garden soil is missing any major nutrients (nitrogen, phosphorous, and potassium). This activity also demonstrates how science tests often let us gather information that is not visible. In the soil test, chemicals are used to gather this information. You may want to establish a soil nutrient record for your garden bed and have students compare the nutrients before and after harvest and from year to year.

Materials

Soil test kit
Soil sample from garden
Life Lab journals

Plants are like people. They need vitamins, minerals, and nutrients to live. Where do we get our vitamins and minerals? (from food we eat) When people feel they are not getting enough of a certain mineral or vitamin, they change their diet. They may decide to see a doctor who may prescribe vitamins or minerals to supplement their diet.

Where do plants get their minerals? (from the soil) They manufacture their own vitamins. We can all be soil doctors and find out if our garden soil has enough of the major minerals needed for healthy plant growth. There are about 12 minerals needed by plants. With our soil testing kit, we can test for the three very basic and most important nutrients: nitrogen, phosphorous, and potassium. *Nitrogen* is what makes the plant green and what makes it grow. *Phosphorus* is for strong roots. *Potassium* helps overall strength and disease resistance. How can this information be helpful to us?

1. Follow the simple directions in the test book accompanying the kit. This activity is best done with a group of seven or eight students at a time. Involve every student by assigning a task to each: reader, test tube holder, soil mixer.
2. Test as many of the three nutrients as you wish. Each group can test a different one.
3. Use the color chart in the kit to assess your results. Have students record results in their Life Lab journals.
4. Discuss what your soil has and what it needs for healthy plants. Use your results for the following activity, A Soil Prescription.

Name the three important minerals that we tested for. Why do plants need adequate quantities of these minerals? Name two sources of vitamins, nutrients, and minerals for people.

Based on our soil analysis, how would you predict that plants would grow in our garden soil?

1. Have students test garden plots to determine what they need to add. Compost helps to add nitrogen; bonemeal adds phosphorus; and wood ash adds potassium. Have them test again after improving the soil and compare their results.
2. Keep records of the garden plot to compare each year.

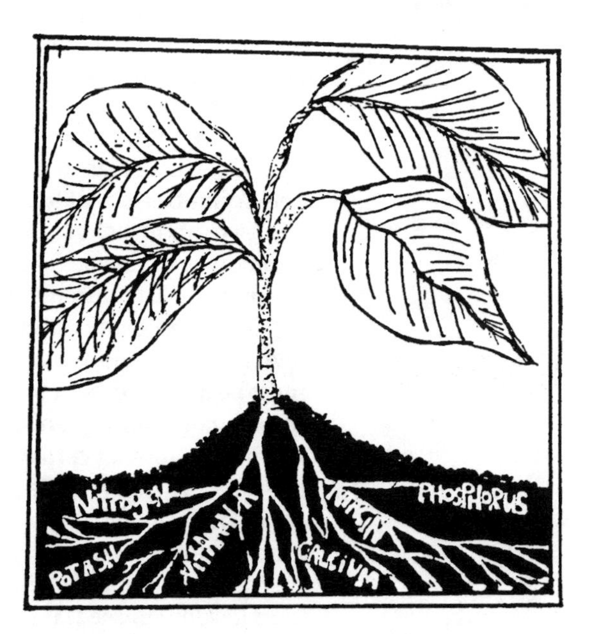

A Soil Prescription

Description

Students test prepared compost for nutrients and develop a compost recipe that provides the soil with needed nutrients without purchasing commercial fertilizer or soil amendments.

Objective

To develop skills in determining how nutrients can be added to soil.

Teacher Background

This project is divided into three parts. First, students will test prepared compost for nutrients prior to adding it to the soil. Second, students will write recipes for a compost pile that meets their soil's needs and can be created from free materials in the community. The final, optional activity consists of testing the soil for nutrients two weeks after the compost has been added.

There are many commercial petroleum-based and organic soil amendments that can be bought to improve the soil's nutrient balance. It is recommended that your school garden be grown without purchasing these nutrients, particularly the petroleum-based materials. They are at best an extra expense and at worst energy intensive and wasteful of natural resources. Recycling through composting, in addition to reducing gardening costs, provides a hands-on understanding of how nature recycles nutrients into the soil.

There are many different materials that can be used in compost, and each has its own strengths and weaknesses. Think of the compost as your soil's diet. The diet must be balanced between materials that are strong in nitrogen and those strong in carbon, between wet and dry ingredients, and between acidic and basic ingredients. It is important that certain elements, such as phosphorous and potassium, be provided. A ready-made compost is created based upon the compost-maker's assumptions regarding what nutrients the soil needs and also upon what materials are available to the compost maker. (See Let's Make A Compost Cake, p. 199.) In the first activity, students will test ready-to-use compost (purchased or made earlier in your garden) to see if it meets the soil's needs. In the second activity, students will create a custom compost recipe, or fill the soil doctor's prescription, by using locally available, free materials.

Materials

Ready-to-use compost
Soil test kit
Several copies of the local yellow pages
One Soil Prescription blackline master per group of three, pp. 395-396
Life Lab journals

Preparation

Have completed soil test from class garden (see Soil Doctors, p. 86).

What was the result of our soil test? How can we provide the minerals that our garden needs? (You may develop the discussion by asking how we get the minerals and vitamins that our bodies need.) Some ways to feed our garden are more expensive than others, and some ways include more waste, through packaging and transportation and advertising, than others do.

Could we get all the minerals that our soil needs without spending any money on fertilizers? Can you think of anything that is thrown away that might provide the nutrients that the garden needs?

What is compost? How does compost speed up the natural process of decay of organic materials? Finished compost is rich in nutrients. Compost reduces the wasteful disposal of soil nutrients at land disposal sites and returns them to the soil to provide nutrients for plants.

Part One:

1. Have students test ready-to-use compost with the soil test kit, using the color chart in the kit to assess their results. Have them record results in their Life Lab journals.
2. On the chalkboard or a large sheet of paper help students compare the test results for the garden's soil with the test results for the compost. Ask, How does the compost fill nutrient needs of the garden soil? Does the compost neglect any of the garden's nutrient needs? Does the garden need a lot or just a little bit of compost to fill its needs?
3. Have students mix compost into the top few inches of the garden soil.

How do you think adding compost has changed the nutrients in the soil? How could you test this? Where does compost get its nutrients? How do you think the insects and other animals in the garden will be affected by the change in soil composition? (The change in soil composition will change the habitat, making the garden more comfortable for some creatures and less comfortable for others.)

Test the garden soil again, about two weeks after you have added compost to the soil. Compare the original soil test with the new soil test.

Part Two:

1. Distribute the blackline master to groups of students. Ask, Which of the compost materials are available at school? Which of the compost materials are available nearby, and could be obtained at no charge?
2. Give each group time to write a recipe for compost on their worksheets. Place copies of the phone book in several locations around the room, so that students can look for possible local contributors of free compost material.
3. After each group has written a recipe for compost, write a class recipe by taking suggestions for ingredients from the class. You may want to establish approximate amounts of each material once a complete recipe has been written.

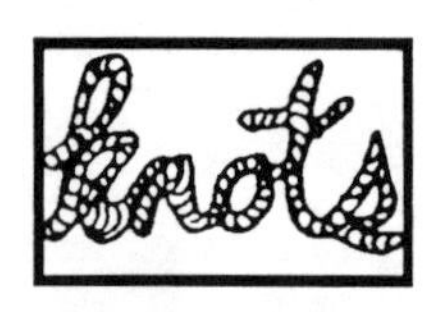

What happens to compost ingredients if they are not used to make compost? Why do you think people buy fertilizers instead of using compost?

Have students make a compost pile based on the class recipe. The recipe will probably have to be adjusted depending upon what materials are available in the community. You will find that once the compost has been started, you will be making constant changes to the recipe, depending upon available materials in need of recycling and upon the productivity of the compost. (See activity on composting, p. 91.)

What Good Is Compost?

Description

Students grow two identical crops, one in a bed with compost and one in a bed without compost.

Objective

To determine the effects of compost on plant growth.

Materials

Garden bed or planter box
Ready-to-use compost
Seedlings
String
Markers

Preparation

Prepare a garden bed by digging the soil. Do not add any soil amendments.

Plants have to eat too! They need a good balanced diet, just like people. Compost provides a healthy combination of important nutrients, including nitrogen, potassium, and phosphorus. How could we design an experiment to see if compost helps plants grow? (Record ideas and design plan.) How will we tell if compost makes a difference? (Design charts for measuring the difference.) Let's make some predictions about how growth will compare between plants in the bed without compost and plants in the bed with compost. (Record predictions.)

1. Have students divide the bed in half and mark it off with string.
2. Have them fertilize one half with ample amounts of compost. Tell them to dig the compost into the top few inches of soil and leave the other half alone.

3. Have them plant the whole bed with one crop, or plant several kinds of the same types of crops in each bed. (It doesn't matter which crops you choose, but try to pick at least one root crop, one leaf crop, and one fruit crop.)
4. Have students make charts comparing the success of each crop in the two beds. They could compare speed of growth, health as they grow, and final size when they are harvested.

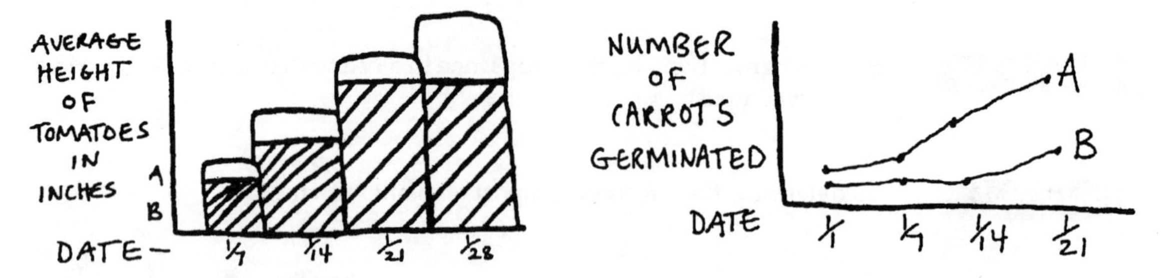

5. Have students set up a maintenance schedule so that both beds receive the same care.
6. Have students set up a data collection schedule so that measurements and information are collected regularly.

Summarize your information at the end of the experiment. Which bed did better? Why did it do better? How do your results compare with your predictions? Do people also grow better with better nutrition? Do they grow bigger, faster, and have fewer "pest" problems? Did this experiment give you ideas for other tests?

The Matchmaker

Description

In Part One of this activity students test soil for nitrogen and plant nitrogen-fixing cover crops. Ten weeks later in Part Two, students examine nitrogen-fixing nodules on the roots of legume cover crop and retest soil.

Objective

To examine bacteria nodules to learn one way that nitrogen from the air is fixed into the soil.

Teacher Background

Although nitrogen is the most abundant gas in air and is a very important nutrient for both plants and animals, it is very difficult to use. There is one very large family of plants, the Leguminacae, that are able to use nitrogen from the air. This group of plants includes beans, peas, alfalfa, vetch, and clover. When selecting seeds to plant for this activity, check with your County Agricultural Extension to determine which legume cover crops will grow well in your area.

Legumes have a special relationship with a bacteria called *rhizobia.* Rhizobia live on the roots of legumes. The rhizobia take nitrogen from the air and fix it to the soil. The roots of the plants can then take up this nitrogen, and the plant can use it. In exchange for the nitrogen, the plants give the rhizobia carbohydrates.

Farmers and gardeners very often rotate their crops with legumes to help replenish the soil. It is a very good idea to plant a cover crop of legumes after the fall harvest to help protect and add nitrogen to the soil. There is also inoculant available that may be used to coat seeds before planting. The inoculant is actually rhizobia. If you are not sure that rhizobia are in the soil, you may want to use the inoculant to guarantee success.

Rhizobia will form nodules on the roots of the legumes. They are very easy to see. The maximum number of nodules will be on the roots when the plant flowers. However, you can find them on the plant throughout its growth. Each nodule is a cluster of rhizobia.

When using the legume as a cover crop, cut the plants at time of flower and allow the roots to decompose in the soil.

Materials

Part One:
A selection of different seeds from the Leguminacae family: bell beans, fava beans, red clover, alfalfa, peas, purple vetch
Legume cover crop to plant, minimum two seeds per student
Soil test kit
Garden plant markers and grease pencil
Garden area for planting cover crop
Inoculant (optional)

Part Two:
Two or three varieties of mature legume plants—at least one for each student
Hand lenses
Microscope (optional)
Life Lab journals
Soil test kit
Trowels

Preparation

Select garden area for planting cover crop. There does not need to be much soil preparation for these seeds, and it is an ideal way to incorporate new land in the garden.

We've learned that plants need nutrients from the soil. What are some ways that we can add nutrients to the soil? (compost, fertilizers) All plants need some nutrients. How can plants that don't have people to help them get their nutrients? (natural decomposition) There is one family of plants that is especially good at getting nitrogen. This special family is very large, and has many plants in it that we know: beans, peas, clover. They made a deal with a tiny organism—a bacteria. The bacteria grows on the roots of plants in this family. This special bacteria can take nitrogen from the air (yes! there is a lot of nitrogen in the air) and bring it into the soil for the plants to use. In exchange for the nitrogen, the plants give the bacteria carbohydrate food that they need! How can these plants help us add nitrogen to the soil? (Discuss cover crop plants.)

Part One:

1. Show the class the sample of cover crop seeds. Have them identify those that are familiar.
2. Write the word *Legumes* on the chalkboard. Explain that all of the sample seeds belong to this special family, and that legumes let the very helpful bacteria, rhizobia, grow on their roots.
3. If you have the inoculant, display it also. The inoculant is rhizobia. It grows only on this family of plants.
4. Follow the directions to prepare the seeds to be planted with the inoculant.
5. Have students test the soil in the area to be cover cropped for nitrogen. Record the results.
6. Have students plant the legume cover crop seeds in the garden area.
7. Have them label the sections for the different types of seeds.
8. Have students water as needed.

Part Two (when plants are flowering):

1. In the garden, have students carefully dig out different samples of the legumes they are growing. Tell them to be sure to mark each plant to remember what kind it is.

2. Have students compare the nodules on the different types of legumes. Ask, Do some have more? Compare sizes and colors of nodules.

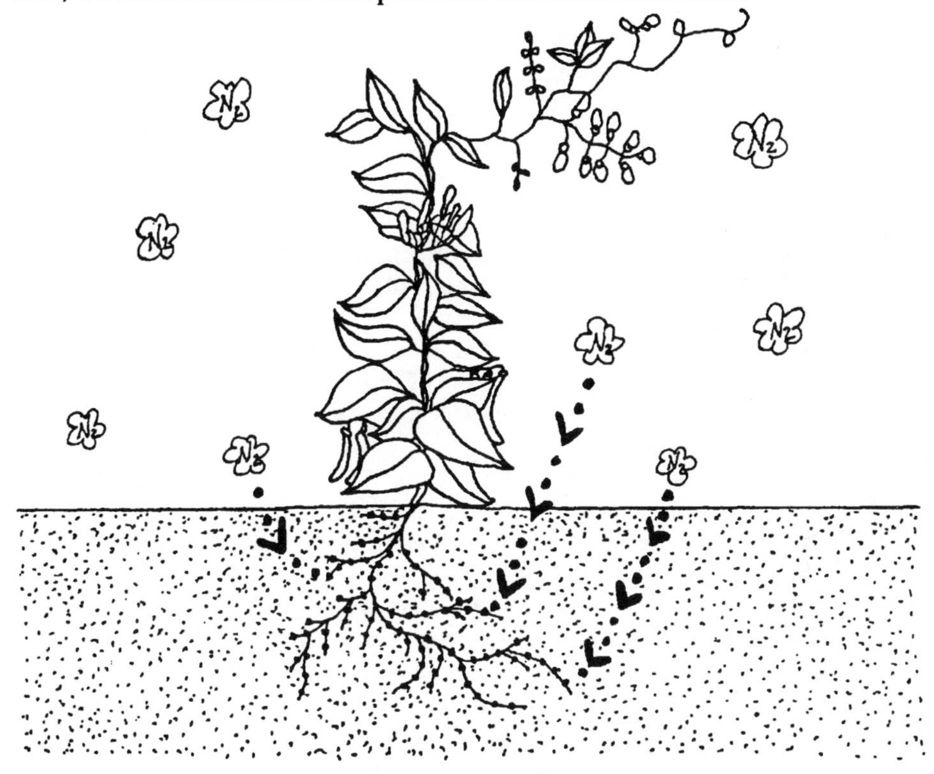

3. Have students remove some nodules and examine them with the hand lens.
4. (Optional) Set up a slide of one nodule opened and have students look at it under a microscope.
5. Discuss with students how legumes can be used by farmers and gardeners to help bring nitrogen into the soil. Discuss crop rotation and cover cropping. Ask, When would be a good time to plant a cover crop in the garden?
6. After plants are cut (great compost material!) and roots have decomposed in the soil, have students test the nitrogen in the soil and compare it to the earlier test.

What is one way to get nitrogen from the air into the soil? Why is it important to have nitrogen in the soil? What are some plants that help fix nitrogen from the air into the soil? Plan a cover crop rotation for your garden.

1. Have students compare the roots of a legume with the roots of a nonlegume.
2. Once the cover crop begins to grow, have students harvest one plant a week and count the number of nodules, graphing the change in numbers each week.
3. Have students plan an experiment comparing different cover crops. What criteria will be used for the comparison?

What's to Worry?

Description

Students design and test ways to protect plants from frost damage. This activity should be used during cool weather.

Objective

To describe the impact of mulch on soil by comparing it to the impact of frost on protected and nonprotected plants.

Teacher Background

The unpredictability of weather is what makes all farmers gamblers. The history of agriculture is a story of attempts to control weather through technology. Examples include irrigation systems, wind generators to circulate air to prevent frost damage, and greenhouses to extend growing seasons. An early fall frost or late spring frost can easily destroy a crop. There are several common ways to protect plants from frost. Mulch (organic materials applied loosely over the soil and around plants) insulates the soil and keeps it warm. Covering the plants with plastic heats up the soil during the day and traps some of the heat at night. Wind generators circulate the air and keep the cold air from falling onto the plants. It is important to point out to the class that farmers must work with the weather and often must simply accept what weather changes may do to a crop, although widespread frost damage to food crops may result in higher food prices that affect all of us.

Mulch can also be used to reduce evaporation of moisture from soil and reduce weed growth. Some of the more common mulches are: straw, leaves, aged animal manure, and ground tree bark. A mulch layer should be 3-6 inches deep.

Materials

Two experimental garden beds planted with the same crops
Two minimum-maximum thermometers mounted on short stakes, or one soil thermometer
Organic materials for mulch: straw, leaves
Life Lab journals

Preparation

Prepare two garden beds for the experiment by planting both with the same number of the same crop, such as kale or broccoli.

What impact does the weather in our area have on local farmers? Are there times of the year when they cannot grow crops? Is there always enough rain to water the crops? Can wind be a problem? How do farmers try to solve these problems? (greenhouses, not growing crops all year, irrigation, tree windbreaks)

A common problem for farmers is an early frost in the fall or a late frost in the spring. Frost can kill some crops.

Since this is the time of year when we have cool temperatures, let's design some ways to protect plants from the cold. (Discuss ideas.)

1. Have students design an experiment to test the effect of mulch on soil. Ask, What is our hypothesis? How will we test it? What data (information) will we want from the experiment? How will we get it? How will this data help us draw a conclusion?
2. Have the class use the mulch materials to set up the experimental bed.
3. Have students label the experimental and control beds and make a sign with the hypothesis.
4. Demonstrate the maximum-minimum thermometers to the class. Place one in each garden bed. Make sure the thermometers are low to the ground. The thermometers will record the coldest and hottest temperatures since the last time they were read. If these thermometers are not available, a simple soil thermometer can be used to read the soil temperature as early in the morning as possible.
5. Establish a time each day for the recording of temperatures and observations.
6. Record data on a class chart for two weeks (See illustration.)
7. Have students compare results from each bed. Based on their observations of plant health and growth, do they conclude that the protection actually helps the plant?

Did temperatures differ between the experimental and control beds? Why or why not? Was there a difference in plant growth between the two beds? Would a farmer be able to use this method of plant protection? What methods do farmers use in our area to protect their plants from frost? Did you observe any other ways mulch affected the plant growth and the soil?

Have students test mulch for its ability to hold water in the soil and to slow weed growth.

Splash

Description

Students build a simple device from milk cartons to observe the effects of raindrops on soil erosion.

Objective

To measure and graph the relationship of the force of moving water to the rate of soil erosion.

Teacher Background

Erosion is the natural process of soil being moved by water or wind. It is of serious concern today because of the enormous amounts of topsoil being washed off farmland. This activity demonstrates and compares the impact of hard rain and soft rain on the soil. The soil splashed onto the milk cartons represents "moved" or "eroded" soil. The harder the rain, the bigger the splash. You can extend the activity by testing different types of soil, including freshly dug soil, compost, sand, clay, and plant-covered soil.

Materials

One half-gallon milk carton per group of four
Enough sand or pebbles to weigh down the cartons
Two large sheets of white construction paper per group
Tape
One ruler per group
One watering can with a sprinkler head per group
Life Lab journals

When a raindrop hits hard ground, such as rock or concrete, what happens to the ground? What would happen if a raindrop struck a part of the earth's surface made out of soft soil? What would happen to the soil if the drop were bigger or coming out of the sky faster? (Record all predictions.) Imagine now the real situation of millions of raindrops striking the land.

Erosion is the name given to the movement of small rocks, sand, and soil from one place to another, either by wind or water. In this activity, you will build a special device called a splashboard and will use it to investigate the part that raindrops play in erosion.

1. Divide the class into groups of four.
2. Demonstrate splashboard construction:
   - Cut the top off of the milk carton.
   - Put sand or pebbles in the carton until it is 1/3 full.

- Wrap a piece of white paper around the outside of the carton and tape the ends together. Do not tape the paper to the carton.
- With a crayon, write the word *slow* along the top of one side, *medium* along the top of the next side, and *fast* along the top of the third side. Do not write on the taped side.

3. After all the groups have finished building their splashboards, show them (without water) how they will be used and have them guess what will happen. Record predictions.
4. Have each group place its splashboard outside on open soil in an area at least two feet in diameter. Be sure the splashboards are standing straight.
5. Have students fill watering cans and create a mini-rainstorm over the soil in front of the side marked slow. Gently pour water from the can from about knee height. Do not pour water directly on the paper, but rather on the soil as close to it as possible. Any soil splashed up by the water drops will stain the paper.
6. Leaving the splashboard in place, have students repeat the procedure in front of the medium side and then the fast side, pouring from waist and shoulder height respectively.
7. When students have finished, have them slip the paper off, open it up, and measure and compare the soil splashing. When dry, the papers can be used to illustrate how graphs can be pictures of how nature works.
8. Discuss the results. With the splashboard sheet opened up and dry, have students trace a line along the top edge of the splashing, showing that as the water drops moved faster, bits of soil were heaved higher into the air.

What did you learn from this experiment? Why hasn't all the soil on the earth washed away? What helps to keep it in place even in a heavy rain? (Plants provide a protective cover, with their roots holding on to the soil.) How could you use your splashboards to test your ideas?

1. Have students put splashboards on different surfaces—sand, grass, in a garden, on pavement—and compare splashes.
2. Have students tour the school grounds and look for evidence of splash erosion. Does soil splashing along the base of the school buildings give any clues about the direction of the storm?

A Day at the Races

Description

Students prepare soil flats using five different conservation techniques and compare water flow and soil loss.

Objective

To demonstrate soil erosion and ways to conserve soil.

Teacher Background

Throughout history, different means of soil conservation have been used. Terracing has been used to keep soil on hillsides by creating level platforms that step down the hill. These platforms are then used for farming. Terraces take a long time to build. Contour farming is a simpler practice, in which the planting takes place in rows across the hill rather than along the slope. However, the key to soil conservation is probably to mimic nature and to keep the soil rooted in with plants.

Materials

Five shoe boxes trimmed to five centimeters deep, V-notched on one end, and lined with plastic
Five watering cans
Five measuring cups
Sod
Soil
Water
Five blocks
Clock or watch

Why is soil important to plants? Have you ever seen soil washed away by rain or rivers? What do you think will happen to soil if it gets washed away year after year? Do you have ideas for saving soil and keeping it where it is? (List responses on the chalkboard.) Let's design a soil race and see who the real winners are.

1. Divide the class into five groups.
2. Distribute the materials.
3. Have each group fill a box with soil prepared in the following ways:

   Group 1: Fill the box with moist soil and pack down tightly.

   Group 2: Fill the box with sod.

   Group 3: Fill box with moist soil and, using fingers, make packed furrows across the slope. (Furrows run the width of the box.)

Group 4: Fill the box with moist soil and, using fingers, make furrows up and down the slope. (Furrows run the length of the box.)

Group 5: Fill the box with soil and, using a ruler, make steps (terraces) across the slope.

If other ideas were suggested during discussion, another group can be added to test the ideas.

4. Have each group use the blocks to line up their boxes on an incline and place measuring cups beneath the V-notches to catch the water that drains off.
5. Have one student from each group simultaneously sprinkle a measured amount of water from about thirty centimeters (twelve inches) above each box, pouring steadily for five seconds.
6. Have groups record how long water continues to flow out of the V-notch.
7. Let the water in the cups settle and measure the sediment in each.

Discuss which box lost the most soil. Which lost the most water? Which methods were most effective in controlling erosion in the experiment? What other methods might help conserve the soil?

Have students walk around school grounds and identify evidence of soil erosion. How can it be prevented?

Growing

Robert F. Kennedy School
San Jose, CA

Martin Luther King
Jr. School
Washington, D.C.

Martin Luther King
Jr. School
Washington, D.C.

Pacific Elementary School
Davenport, CA

The Open School
Los Angeles, CA

Martin Luther King, Jr. Elementary School
District of Columbia Public School District
Washington, D.C.

"We have math, reading, and spelling while we are having science everyday."

Martin Luther King, Jr. Elementary Student

Martin Luther King, Jr. Elementary School (MLK) is an urban school located in the southeast section of Washington, D.C. The school's Living Laboratory includes the two fifth grade classrooms, a section of the front campus, and a small segment of one of the city's playgrounds situated behind the school.

Life Lab is adopted through the school system's Mathematics/Science Initiative and partially funded through the District Facilitator Project. It is also an extension of a beautification project initiated by the school's science resource teacher.

We grow a variety of plants in our indoor garden such as green beans, white potatoes, tomatoes, collards, and herbs. Radish seeds were also planted on numerous occasions; however, the students' curiosity never permitted those plants to mature. Instead we now have a collection of dried, uprooted radish plants in stages from one to fifteen days after germination.

This down-to-earth program has captured the interest of MLK students as well as their teachers because it provides numerous opportunities for hands-on experiences, the practical application of skills, and can be adapted to incorporate a variety of subject areas. The science resource teacher overheard an excited Life Lab student say that she has science everyday. When schoolmates from other classes responded, "Well, we have math everyday," the Life Lab student replied, "We have math, reading, and spelling while we are having science everyday."

The Open School—Center for Individualization
Los Angeles Unified School District
Los Angeles, California

Spring in the Garden

Flowers bloom,
Worms wobble by your feet,
Fish jump up.

Bright flowers burst out
and we meet new friends.
The garden is a friend.
I rest there.

We water,
We dig,
We sit and write,
relaxed.

Tova Katz, grade three

In 1984, two teachers initiated a Life Lab program and integrated it with their science, language arts, fine arts, and social studies curriculum. Until 1987, the Life Lab garden flourished as students worked the soil and carried out their garden experiments. Then the school was relocated. Dr. Ruben Zacariás, Associate Superintendent, insured there would be a garden at the new site.

For a second time, The Open School in central Los Angeles created a 30′ X 60′ garden where none had existed. The 300 students (who arrive each morning by bus from racially diverse areas) broke the tarmac-covered schoolyard, and dug a new pond; fruits, vegetables, and flowers that could not be transplanted from the old garden were planted or sowed again. Students in the third and fourth grades formed design committees to draw plans for the layout of the garden beds, from which a single plan was chosen. Parents and friends donated topsoil, plants, a greenhouse, bricks to edge the beds, and fish for the pond. Soon the "growing classroom," named the Ruben Zacarias Life Lab Garden, was back on the schoolyard for all students of The Open School to use and enjoy.

UNIT INTRODUCTION
Growing

The mysteries of growing, beginning with the basic ingredients of life, are the focus of this unit. Plant anatomy from seeds, roots, stems, and leaves to flowers and fruits will be investigated. Students will explore plants as living systems and observe germination, water and nutrient transportation, transpiration, and photosynthesis. The reaction of plants to various environmental cues such as light and gravity are demonstrated.

Most of the activities in this unit can be done inside the classroom. Students will practice observation, classifying, data collection, and experimentation. The concepts learned in this unit will directly apply to planning and preparing the Living Laboratory. Starting with root view boxes in the classroom, students will be able to see the variation of root growth. They will learn more about different crops as they study seed catalogues and plan their seed order.

To manage this unit, we suggest you divide your class into four groups. Many of the activities, such as "Let's Get to the Root of This," can be repeated by each of the four groups. The four activities that demonstrate plant responses to environmental cues ("Glass Seed Sandwich," "Run, Root, Run," "Let's Get A Handle On This," "Which Way Did It Grow?") can each be done by one group and the results shared with the whole class.

Two activities in this unit, "Let's Get to the Root of This" (root view boxes) and "ZIP Code Seeds" (seed ordering) require advance preparation. You will want to plant the root view boxes as early as possible to have good root observation during the unit. Seed catalogues need to be ordered six weeks in advance.

Three activities require data collection for a few weeks: "Star Food," "Seed to Earth, Seed to Earth, Do You Read Me?," and "Room to Live."

Understanding that the growth of living organisms depends on nonliving physical components of our planet develops a sense of interdependence in our studies. To understand the similarities and diversity in plant growth teaches us to observe variations of basic principles. Investigating the plant as a whole system lets us appreciate the interaction of the parts to support the whole and, thus, we grow along with the plants.

Activities	**Recommended Grade Level**
Bioburgers *(illustrates the importance of air, water, soil, and sun to life)*	2,3,4,5,6
Seedy Character *(classification skills and seed anatomy)*	2,3,4,5,6
Let's Get to the Root of This *(demonstrates root growth and function)*	3,4,5,6
ZIP Code Seeds *(seed ordering)*	4,5,6
Adapt-A-Seed *(study and invent seed dispersal mechanisms)*	4,5,6
Seed Power *(water absorption experiment with seeds)*	2,3,4
It's Getting Stuffy in Here *(depth of seed planting experiment)*	2,3,4
Lotus Seeds *(a story about seed longevity)*	2,3,4,5,6
Growing, Growing, Gone *(seed/plant development experiment)*	4,5,6
Glass Seed Sandwich *(root growth demonstration)*	4,5,6

Activities	Recommended Grade Level
Run, Root, Run *(root growth behavior demonstration)*	4,5,6
Let's Get a Handle on This *(climbing plants experiment)*	4,5,6
Which Way Did It Grow? *(effects of gravity on plants)*	4,5,6
Sugar Factories *(photosynthesis; plant production of food)*	4,5,6
Sipping Through a Straw *(observation of water absorption by plants)*	3,4,5,6
Plant Sweat *(transpiration experiment)*	4,5,6
Magical Mystery Tour *(demonstration of CO_2 utilization by plants)*	5,6
Plants Need Light Too *(demonstrates the need for light in plant food production)*	4,5,6
Plant Food Magic *(how plants make food)*	4,5,6
Star Food *(garden experiment on the effects of light on plant growth)*	5,6
Room to Live *(indoor experiment on the effects of crowding on plant growth)*	2,3,4
Seed to Earth, Seed to Earth, Do You Read Me? *(effects of different soil types on seed/plant growth)*	2,3,4
Stem, Root, Leaf, or Fruit? *(classification of produce by the parts we eat)*	2,3,4

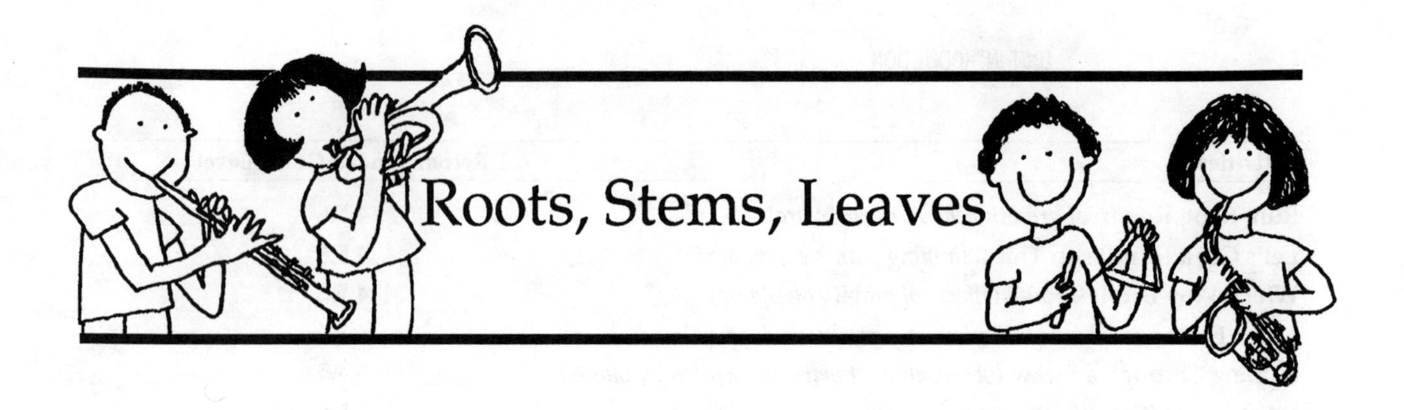

Roots, Stems, Leaves

Chorus I (2 times):

C
Roots, stems, leaves, flowers, fruits, and seeds

C G C
Roots, stems, leaves, flowers, fruits, and seeds

Chorus II:

C
That's six parts, six parts, six plant parts

G C
That plants and people need.

Verse I

C
Roots hold the plant in the ground

G C
They gather up the water that falls around

C
And there's a root inside of me

G C
Because a carrot is a root that I eat.

Chorus II

Verse 2

A stem is an elevator growing up from the ground.
The water goes up and the sugar back down.
And there's a stem inside of me
Because celery is a stem that I eat.

Chorus II

Verse 3

The leaves are the kitchens where the food is done.
They breathe the air and catch rays from the sun.
And there's a leaf inside of me
Because lettuce is a leaf that I eat.

Chorus II

Chorus I (2 times)

Chorus II

Verse 4

The flowers are dressed so colorfully
They hold the pollen and attract the bees.
And there's a flower inside me
Because cauliflower is a flower that I eat.

Chorus II

Verse 5

The fruit gets ripe, then it falls down
It holds seeds and feeds the ground.
And there's a fruit inside of me
Because an apple is a fruit that I eat.

Chorus II

Verse 6

The seeds get buried in the earth
And the cycle starts again with a new plant's birth
And there are seeds inside of me
Because the sunflower is a seed that I eat.

Chorus II

Verse 7

Now you know what this whole world needs.
It's roots, stems, leaves, flowers, fruits, and seeds.
There are six plant parts inside of me
Because a garden salad is what I eat.

Chorus II

Chorus I (2 times)

Chorus II

Bioburgers

Description

This activity draws connections between a common fast food and the Big Four essentials of life: air, water, sun, and soil. Students will trace the ingredients of a hamburger to their origins of air, water, sun, and soil.

Objective

To illustrate the importance of air, water, sun, and soil for life.

Materials

Poster or sketch of a hamburger with bun, cheese, lettuce, and so on.

How many of you like hamburgers? What are the basic ingredients of a hamburger? Would you believe that if we are really good detectives we can trace each part of a hamburger, including the bun, the cheese, the meat, and so on, back to the same four ingredients? As a matter of fact, these same four ingredients make it possible for each of us to live! Want to try?

1. Design a Food Flow Chart by taping the hamburger poster to the center of the chalkboard.
2. Help the students trace the origin of the major ingredients—meat, bun, cheese, tomato—of the hamburger. Begin by asking the students, "Where did the meat come from?" Continue questioning until they reach the source of the ingredients: sun, soil, air, and water. Record responses on the chalkboard. Give guidance where needed. Your chart may be a simpler version of the one illustrated in this activity.

How long could you live without any one of the Big Four essentials of life? Can you identify a living thing that can survive without the Big Four? (anaerobic bacteria; fish live without soil; some seaweeds live without soil)

1. Ask the students to plan a meal that does not in some way depend on the Big Four. Is this difficult?
2. Have each class member use his or her favorite meal and make a food flow chart tracing the food back to its beginning. Where does all the food ultimately come from?
3. Suggest that students try the same thing, tracing the packaging required at each step. Where does all of the packaging come from? Where does it all ultimately go?

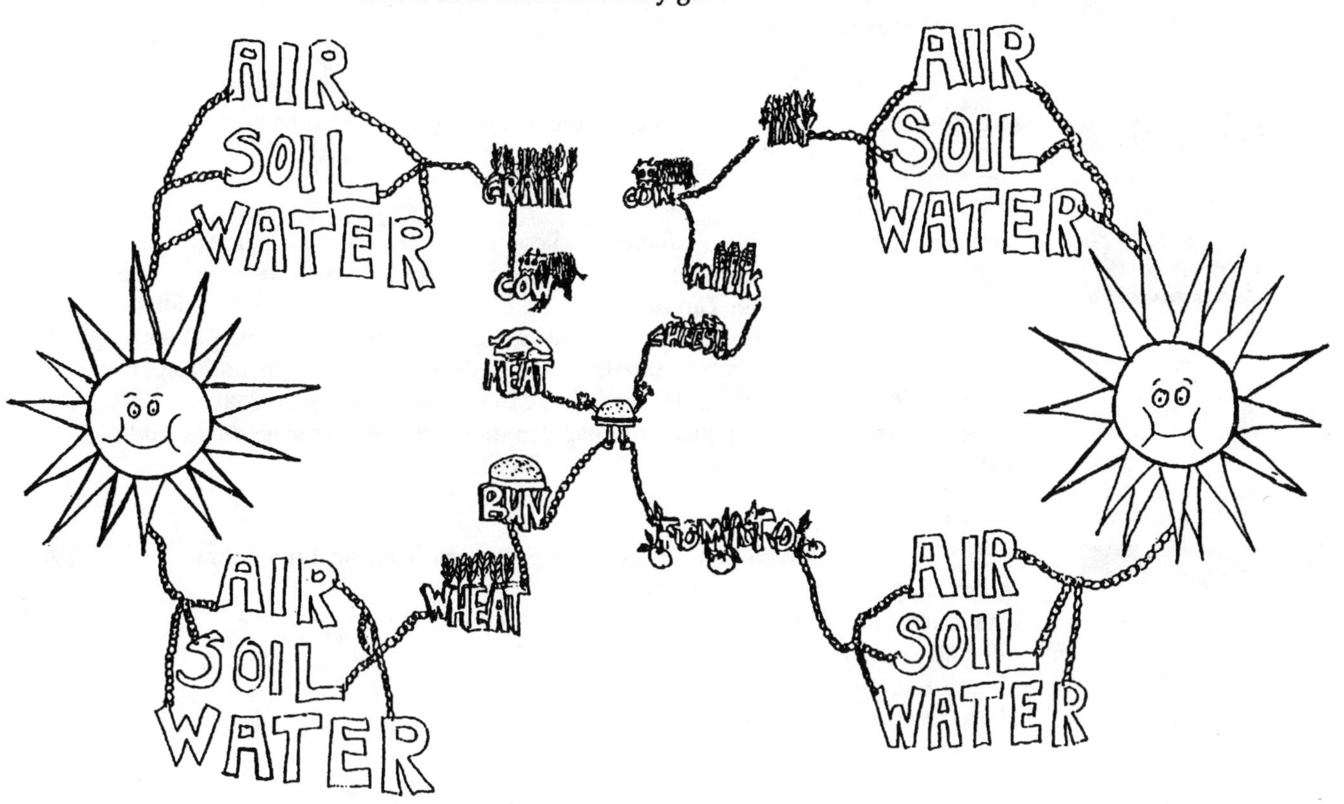

Seedy Character

Description

In Part One, students examine and classify different type seeds. In Part Two, soaked pinto beans are dissected.

Objective

To observe, classify, and identify different kinds of seeds and seed parts.

Teacher Background

Seeds come in different sizes, shapes, and colors. Some are edible, some are not. The reason for the diversity is related to the seeds' need to disperse and grow. Some seeds are light so that they can be carried by the wind; some float; some stick to animal fur; some are brightly colored to attract birds who carry them to other locations. Others are eaten by animals and then deposited in the ground as part of the animals' waste. However, all seeds have the same parts: a seed coat to protect it; an embryo that is the baby plant; endosperm that is the food that feeds the embryo until it is a seedling and can make its own food.

Materials

A tray containing a variety of seeds: coconut, avocado, apple, nuts, beans, pumpkin, popcorn
One soaked pinto bean per student
Life Lab journals
Magnifying glasses
Black construction paper

Preparation

Soak pinto beans in water one day prior to the activity.

Some seeds are very nutritious. They are very rich in protein, minerals, fats, and vitamins. Why are they so nutritious? Life comes from seeds. A whole plant grows from a seed. The core of every apple has seeds that could grow into apple-bearing trees.

Part One:

1. Divide students into groups of three and have one person in each group be the recorder.
2. Have each group examine the seeds on the tray and list three general observations about the seeds. Discuss the observations. Have students guess why the seeds have the characteristics they observe.

3. Have each group classify the seeds: size, color, texture, edibility.
4. Give each group a handful of seeds and have them sort the seeds on a piece of black construction paper according to one of the classifications.
5. Have each group display and describe their classification to the rest of the class. Can the class guess the classification scheme used?

Part Two:

1. Pass out a pinto bean to each student and have the class follow your step-by-step dissection.
2. • Peel off the outer skin or seed coat.
   • Split pinto bean in half lengthwise.
   • Identify the following parts: seed coat (outer protection of seed—usually paper-thin); embryo (part that will grow into plant); root system and shoot system that will grow from embryo; food (surrounds the embryo for use until it is big enough to produce its own food).
3. Have students make scientific illustrations of the seed and its parts in their Life Lab journals.

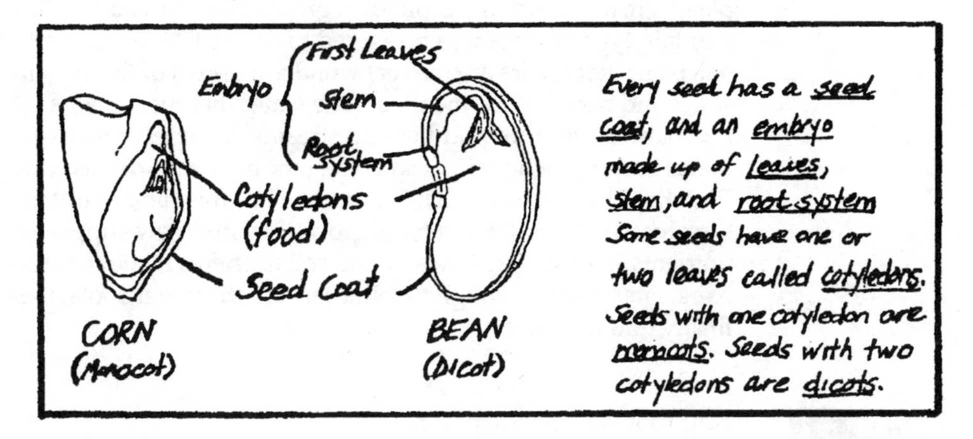

What is the function of the seed coat? The food? The root system? The shoot system? Can a seed sprout without soil? Why? Why are seeds different in size, shape, color, and so on? How is classification helpful?

1. Have students go into the garden and collect seeds from different plants. The seeds are ready when seed pods are brown and dry. Describe and categorize home-grown seeds.
2. Cook with seeds. Have students research how different cultures cook with seeds by reviewing ethnic cookbooks. Have an international seed meal, and taste your learning!
3. Have students bring in seeds from home and prepare a class display of different seeds.
4. Have students glue seeds onto black paper and write a description of each kind. Make a class display.
5. Use the seeds for art projects.

Let's Get to the Root of This

Description

Students will observe root growth in a root view box.

Objective

To illustrate root growth, function, and variation.

Teacher Background

Each group will start with this activity and then conduct different activities to learn about plant growth. This activity is designed to take place throughout the unit. The root view box is easy to construct and provides a unique opportunity to watch how plants grow beneath the ground. A grown plant can have miles of root hairs. In fact, a certain type of grass was measured to have 6,600 miles of root hairs coming from one plant. Root hairs are the real wonder workers of the system. They absorb all the water and minerals for the plant. The water that most plants take in through their root hairs is not from puddles and streams. It comes from a thin coating of water that is around each grain of soil. The root hairs absorb this water into the plant. Each tiny drop of water that is absorbed has the mineral nutrients from the soil dissolved in it. As the root hairs grow, they find new supplies of water and nutrients. This supply of water and soil nutrients is moved upward from the root hairs. First it goes through pipelines inside the larger roots, then into the stem, and finally into the leaves.

Materials

Four root view boxes, p. 448
Seeds with fibrous and tap root systems: carrots, lettuce, radishes, marigolds
Potting soil
Labels
Grease pencils

Preparation

Construct four root view boxes or have each group construct its own. You may choose to plant the seeds in the boxes a few weeks prior to the unit.

What do you imagine plants look like below ground? Do different types of plants look different below ground? Do we eat any parts of plants that grow below ground? (carrots, radishes) Why do you think plants have roots? (Record responses.) With our root view boxes, we are going to be able to observe plants as they grow beneath the soil.

1. To plant root view boxes, fill them with soil. Plant a variety of seeds in each box, but cluster and label them. Plant seeds close to the front of the box near the glass so that the roots may be observed. As the plants grow, you may need to thin them. Water the plants as needed.
2. Observe growth weekly. Have students record the growth by drawing pictures every week or by measuring the plant growth above and below ground.
3. Have students look very closely at the roots. Ask, Do you notice any very thin, hair-like threads on the large roots? Ask, What do you think happens there?
4. Explain that *geotropism* is the inherent capacity of plants to direct their roots downward toward the pull of gravity. Plants germinated in complete darkness and lacking any environmental cues such as light or wind will send their roots downward.
5. When plants mature, have students harvest them and identify the parts.

What would happen to the roots if we turned this box on its side for a couple of days? Why? Which of these roots are edible? Why do different plants have different root systems? What do roots do for the plant? How might these roots appear if they were growing in sand? Clay? Could we grow plants in other substances such as cotton or styrofoam? Why? What do nonsoil substances lack? What would happen if a plant grew its roots into the air or sideways along the top of the soil? How would the roots look on a planet without gravity? Are roots strong? Have you ever seen roots growing through cement? Can you name any places where you have seen roots growing through a strong material?

Have students set up experiments to test some of the above questions.

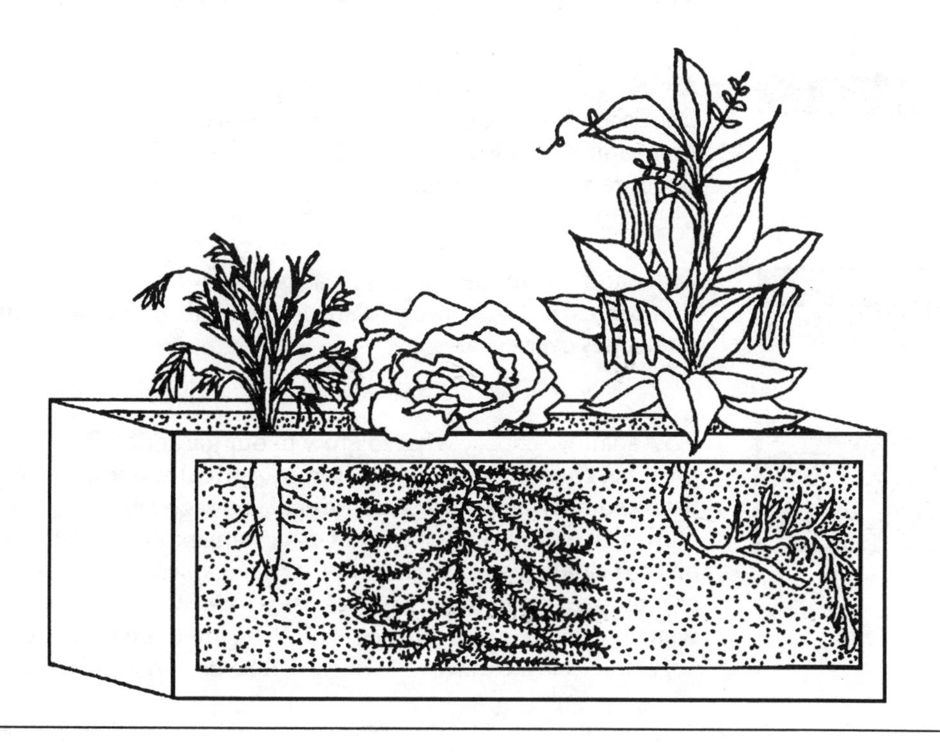

ZIP Code Seeds

Description

Students will choose a variety of seeds to order from catalogues based upon climate and food, and aesthetic preferences.

Objective

To learn to apply knowledge of climate, plant varieties, consumer preferences, and the ability to estimate quantities in compiling seed orders.

Teacher Background

It is less expensive to purchase seeds directly from seed companies, and it is fun for students to look through many beautiful seed catalogues. Order a variety of catalogues to include popular varieties as well as specialty seeds that grow well in your particular climate. Order catalogues from companies selling commercial hybrid seeds as well as open-pollinated seeds. Hybrid seeds have been bred for dominance of certain traits for one season only. The seeds from their mature plants will not reproduce well. Open-pollinated seeds will enable you to develop varieties that grow best in your particular region over successive seasons.

Materials

A variety of seed catalogues for groups of three
One Seed Ordering Chart per group of three, p. 397
Companion Planting Guide, p. 455
List of recommended vegetables and flowers (contact County Agricultural Ext.)

Preparation

Order and obtain catalogues at least six weeks in advance. (See Appendix, Seed Companies, p. 477.) Once you are on the mailing lists, you will automatically get catalogues every year.

How shall we choose what to grow in our garden? What flowers and vegetables grow best here? (Consult the list from Agricultural Extension or develop your own list by comparing your annual rainfall, average temperatures, type of soil, number of frost-free days, and amount of direct sun exposure with the needs recommended on seed catalogues or packets.) What vegetables do you like to eat? Which plants are companions to those we have listed? (See Companion Planting Guide.) Do we want to grow these vegetables also? Why are there so many varieties of one vegetable? How can we tell which will be best to grow in our conditions?

Should we grow seeds for next year's garden? If so, which seeds shall we grow? The ones we choose to grow for seeds must be open-pollinated seeds. Why? What do you think will be the easiest to grow? Fun? Challenging? Which season are we planting for, cold or warm? Do we want a variety of edible root/stem/leaf/flower/fruit/seed plants?

1. Divide the class into groups of three students.
2. Give each group a copy of the Seed Ordering Chart. Instruct them to fill in the specific characteristics with the answers from the class discussion. Using these characteristics, ask each group to go through their catalogue and compile their own list of seeds to grow. Encourage each group to use different criteria, such as variety, cost, open-pollination. (You may want to limit the number of vegetables, flowers, and herbs each group is to select.)
3. As a class, compile one master list of seeds to be ordered and indicate the particular catalogue or company to be used.
4. Ask the students in their small groups to fill out an order blank for their seed company. Obtain a check or money order and mail to the company. You may want to combine your order with another class to get a wholesale price. Allow sufficient time for the seeds to arrive.

Are there any seed companies located near here? Do you think their seeds may be better suited for our garden? Why? How could the class contribute to the development of varieties of vegetables that are especially suited for the region? (by letting the healthiest plant from a nonhybrid seed go to seed, collecting and storing the seeds, planting them next year, repeating this process over time. This is called artificial selection.)

1. Have students make a class collage by cutting pictures out of seed catalogues. Divide the collage by plant parts that you can eat.
2. See the activity on plant varieties, What's in a Name? (p. 167).
3. Have students keep a record of planting dates and days to maturity of plants in the garden and compare this information to that listed in the catalogues and seed packages.
4. Have a speaker who knows about seeds, such as someone from a seed company or a gardener who grows his or her own seeds, come and talk to the class.
5. Visit a nearby seed company's greenhouses, garden plots, or trial grounds.

Adapt-a-Seed

Description

Students will use human-made materials to adapt seeds for different means of dispersal, such as flying and floating.

Objective

To discover some plant adaptations for survival.

Materials

Seeds: bean, pumpkin, corn, pea
Construction paper
Tape
Paste
Collection of materials such as rubberbands, toothpicks, balloons, scissors, pencils, plastic bags, cork, cotton, feathers, tacks, metal springs, wire

What does a new plant grow from? (a seed) Where do the seeds come from? (Plants grow seeds to reproduce more of the same kind of plant.) How does a seed get to a certain place? Where did the weed seeds in the garden come from? Have you ever seen seeds flying in the air? (dandelion fluff) Floating on water? (coconuts) Being carried by a dog? (burrs, foxtails) If you examine these seeds, you will see features that help them travel in a special way. A seed has one purpose: to become a new plant. Can you design seeds to travel in different ways?

1. Divide the class into small groups.
2. Give each group seeds to be adapted.
3. Ask each group to adapt their seed to float on water at least five minutes; be thrown at least two feet away from the parent plant; attract a bird or animal; hitchhike on an animal or person for 20 feet; or fly at least three feet. When dispersal inventions are complete, have students demonstrate how they work.

Why do seeds have dispersal mechanisms? Predict what might happen if maple seeds fell straight to the ground and grew right under the mother maple?

Cut open a pepper and count the number of seeds inside. How many seeds are in one pepper plant? How many pepper plants could grow from the seeds in that one pepper? If one pepper produces 30 peppers, how many plants could be grown from all the seeds of those 30 peppers? Why don't peppers cover the earth?

Seed Power

Description

Students will fill a seed jar with water and observe what happens. Start this activity first thing in the morning.

Objective

To demonstrate the response of seeds when mixed with water.

Materials

Four small bottles with cork stoppers
Enough pea seeds to fill each bottle
Water

What do seeds need to start to grow? (water) Describe what you think might happen as a seed fills with water. (expands until it pops open) We are going to put these pea seeds inside these bottles. Then we'll fill the bottles with water, and put the stoppers on. Can you guess what will happen? (Record predictions.)

1. Give each group one bottle and stopper, pea seeds, and water.
2. Have students fill each small bottle with seeds. Then have them fill the bottle with water. Have them seal the bottle with a stopper or with a piece of plastic held tight by a rubberband.
3. Have each group observe their bottles, recording their observations each hour.
4. Discuss students' observations and compare them with their predictions. (After six hours the swelling seeds will pop off the cork or lift up the plastic cover. Water moves into the seed cells through the seed walls. This swells the seeds and puts pressure on the container. This principle was used to stretch tight leather shoes years ago.)

What did the water do to the seeds? If the seeds were in the ground, what do you think would happen? Why do you think a germinating seed is so strong?

It's Getting Stuffy in Here

Description

Students will conduct a simple indoor experiment to determine how the depth of a seed planting affects germination.

Objective

To demonstrate how the depth of planting affects germination.

Teacher Background

Most seeds will sprout in the dark. All they need is moisture, the right temperature, and air. The plant will make its own food once it has leaves and light. Until then it uses the food energy stored in the seed. If the food energy in the seed is used up before the sprout breaks through the soil, then the sprout will die.

Materials

Four clear plastic containers with holes for drainage
Soil
Beans or pea seeds

Seeds have the ability to grow into plants. There are two main ingredients that are essential for seeds to start growing. We have already learned one of them. What is it? (water) Let's do an experiment to determine the second ingredient. Before we start, let's make some guesses about what the second necessary ingredient is. (Record all predictions.)

1. Divide the materials among each group.
2. Have students complete the following steps to test some of the predictions:
   - Fill each container with about one inch of soil.
   - Press the soil down and then put a few seeds next to the glass on the inside.
   - Put one more inch of soil in the container and press it down.
   - Plant a few more seeds next to the glass.
   - Repeat with one more inch of soil and more seeds.
   - Moisten the soil with water but don't add too much. Pour off any extra water.

3. Discuss what this experiment could be testing. Let students share their ideas.
4. Place the containers in a warm place.

5. After a few days, have students observe which layer of seeds sprouted best. Tell them to be sure to keep the soil moist during the experiment but not too wet. (If the soil is too wet, the seeds may not germinate at all. Seeds usually germinate but grow poorly in wet soil mainly due to the lack of air.)
6. Help students interpret their results. Why did seeds in the top layer grow the best?

What do seeds need to start growing? How do they get each of these? When you plant seeds in the garden, what will be important to remember?

Have students find out how plants such as seaweed grow under water.

Lotus Seeds

Description

Students hear a true story about the tenacity of seeds.

Objective

To realize that seeds can be very hardy over long periods of time.

Materials

None

What comes from a seed? When a plant begins to grow from a seed, the process is called *germination*. What does a seed need to germinate? (water, air, proper temperature) Some seeds can wait a very long time until germinating. How long do you think a seed can last? (Record predictions.)

Read the following story to the class:

Once there was a beautiful lake in China. In its waters grew the sacred lotus plant. Each year, seeds from the lotus flower fell into the water and sank to the muddy bottom. Over many years, the lake dried up. The lotus seeds, which were very hard and covered with a tough outer skin, stayed buried in the dry lakebed.

Many years passed. The land that had once been a lake was used for farming. A scientist came and began to dig in the farmland, for he was interested in its history. He found the lotus seeds. As an experiment he decided to try to sprout the seeds in his laboratory in Washington, D.C. He put them in strong acid to dissolve the hard seed coats and then planted them. He saved a few seeds to test by radiocarbon dating, to find out how old the seeds were.

While the seeds were in the soil, he found out that they were over one thousand years old! After learning that news, he doubted that they would ever germinate. Then one morning in June, 1952, a tiny sprout poked through the soil. The thousand-year-old seeds had sprouted! Today they are still growing in the Kenilworth Aquatic Gardens of Washington, D.C.

Of course, not all seeds last that long. The hard coats and the fact that the seeds were buried so deeply helped these lotus seeds to survive.

Describe what you think the scientist did to sprout the seeds. Why did the seeds last so long? Can you design an experiment to test the germination of seeds? (Try it with one-year-old seeds, two-year-old seeds, and three-year-old seeds.)

(Adapted from Ladybugs and Lettuce Leaves, Project Outside/Inside, Sommerville Public Schools)

Growing, Growing, Gone

Description

This indoor activity demonstrates the different stages of development as groups stop and observe seed growth on progressive days.

Objective

To demonstrate that seeds have different stages of development.

Materials

12 each of four different types of seeds: lima beans, corn, pumpkin, radish
Five jars
Rubbing alcohol
Paper towels
Tweezers

Look at these four seed types. Use your Life Lab journal to draw how you think one of the seeds will look when it starts growing after two days, four days, one week, and two weeks. Label these drawings "Growing Predictions for (name of seed)." We can set up a test for our predictions by putting the seeds in water and removing them every other day. If we place them in rubbing alcohol when we remove the seeds from the water, they will stop growing. We can then mount them and observe the changes in growth.

1. Each group needs one type of seed, a jar, and water.
2. Have students put the seeds in a jar of water and soak them for 24 hours.
3. Have students pour off the water and replace it with wet towels.
4. Tell them to keep the seeds against the side of the jar, approximately two inches from the bottom.
5. Have students take one seed out of the jar every other day. Have them place the seed in rubbing alcohol to stop its growth. Be sure that they label the cup of rubbing alcohol.
6. After five days in alcohol, have students remove the seed, dry it, mount it on paper, and label it.
7. Have students describe the appearance of the seed, its root growth, its top growth as changes occur, and the appearance of the leaves.

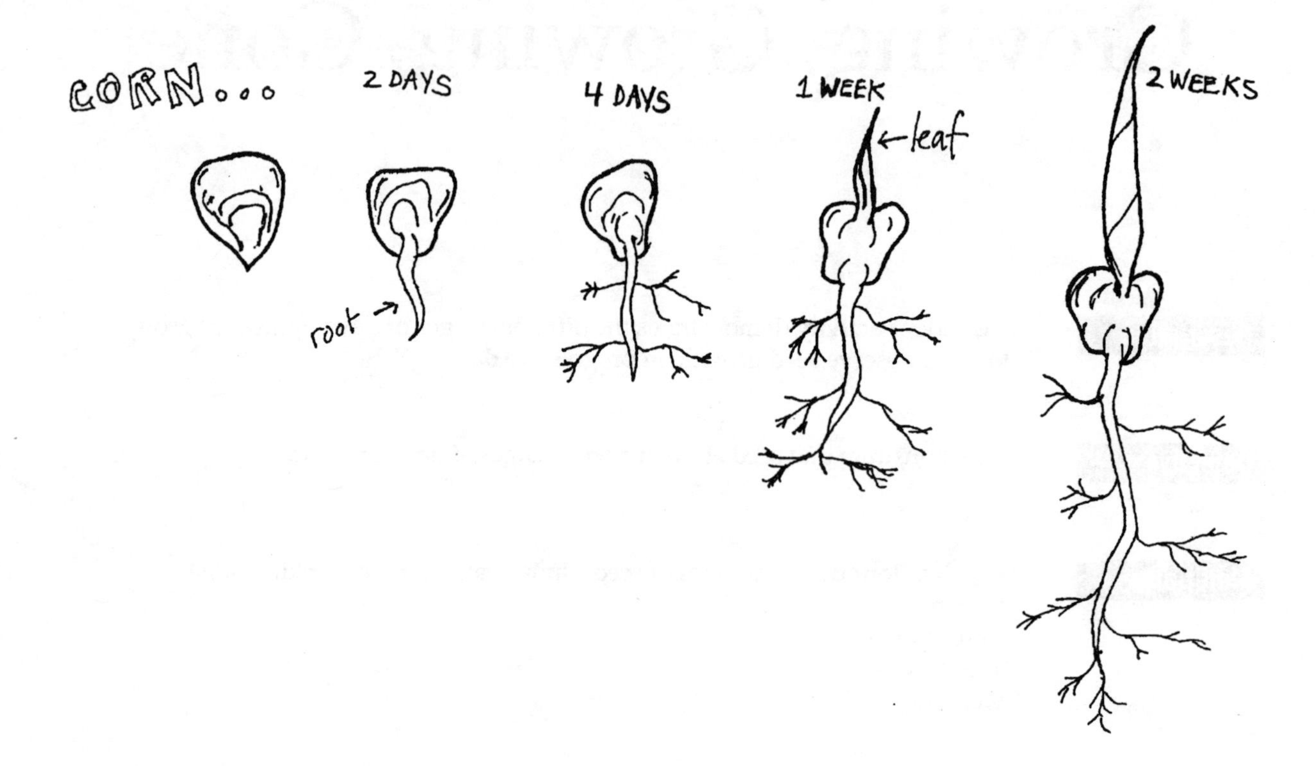

From where did the seed get its nutrients? Describe what would have happened to the seed if the water had not been drained prior to sprouting. Identify the visible parts of the sprouted seed. (root, leaf, stem) Compare and contrast different seed and germination characteristics. (timing, color, rate of growth, and so on)

1. Have students measure the seeds' growth and draw them to scale, comparing the measurements to their predictions.
2. Have students make relief displays of the seeds' development at different stages.

Glass Seed Sandwich

Description

Students set up a simple demonstration to observe how roots react to water.

Note: The next four activities demonstrate different traits (tropisms) of plants. We recommend that you have one of the four groups set up and monitor each demonstration and then share their observations and results with the rest of the class. This activity may be done by Group 1. The class discussion is the same for all four activities.

Objective

To observe plant root growth in the presence of water.

Materials

Two pieces of glass (approximately 8″ x 10″)
Paper towels
Radish seeds
Tape

Preparation

It helps to soak the radish seeds for a day prior to use.

Let's list some things we know about how seeds and plants grow. (Record all responses.) Even though plants may look different, there are certain growing characteristics that all plants share. We are going to set up four different demonstrations to determine some of these traits. Each group will monitor one of these demonstrations. When we are done we will see if we can add to our list of things we know about seed and plant growth.

1. Have students prepare the glass seed sandwich by doing the following steps:
   - Soak a folded piece of paper towel in water and squeeze out excess water.
   - Fold another dry piece the same size as the wet piece.
   - Place the towels on one of the pieces of glass, leaving a 1/2-inch space between them.
   - Place radish seeds in the space between the two towels, being careful to keep the seeds dry.
   - Place the other sheet of glass on top and tape the sides of both pieces of glass together.

2. After the seeds sprout, have students observe the direction of root growth. Have them draw conclusions about what they observe.
3. Have the group present its findings to the class.

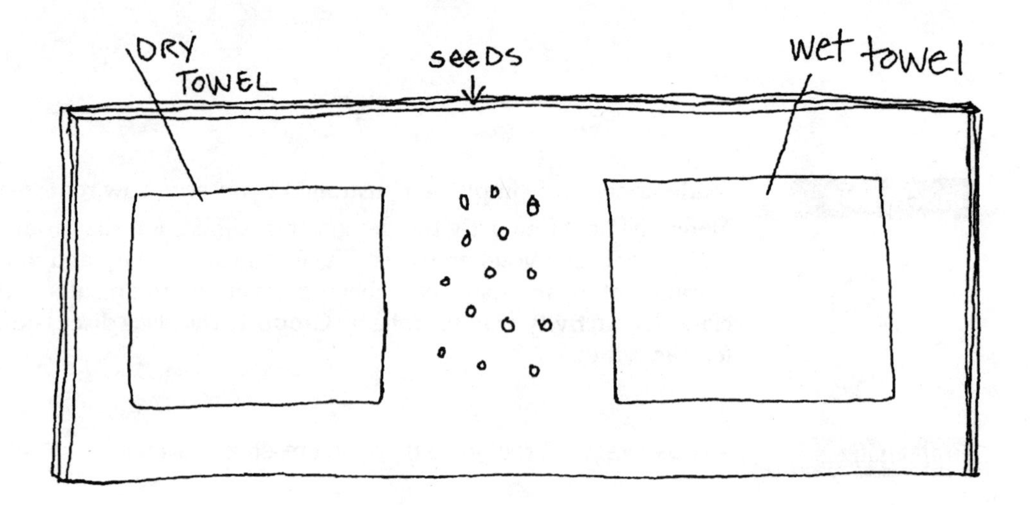

Compare the root growth between the wet and dry areas. Describe the difference. What would a plant do without water? (die) Why does a plant need water? What does water bring the plant? (nutrients) How much of the human body is water? (over 70%) What percentage of most plants is water? (over 80%)

Run, Root, Run

Description

Students will plant a seedling so that the roots are forced to grow around a block of wood. This activity may be done by Group 2.

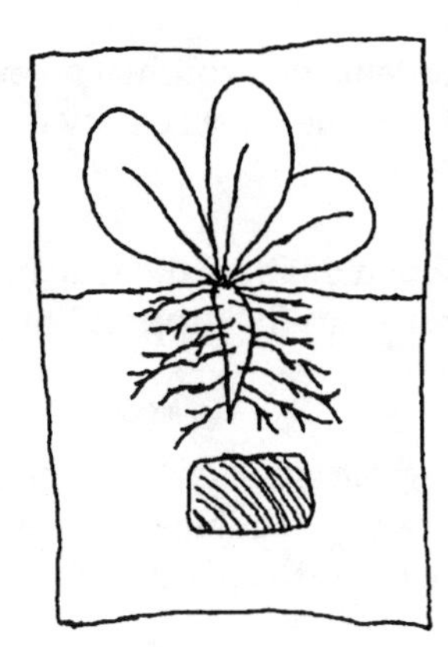

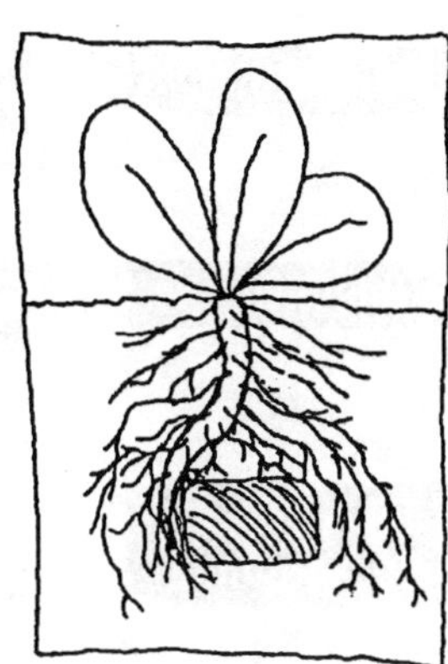

Objective

To demonstrate that roots can grow around barriers in the soil.

Materials

Seedling: marigold, pea, bean
Milk carton with clear front or clear pot, so that root growth can be observed
Potting soil
Small block of wood
Life Lab journals

See discussion on page 125.

1. Have students plant a seedling such as a young marigold in a flowerpot about 1/2″ above a piece of wood, pressing the wood down firmly so that it will make a barrier as the roots go down.
2. After one week, if roots are not visible through container, have students brush aside enough soil so that they can see the roots beginning to find their way around the wood. Replace the soil, pressing it gently, and rewater.
3. Have students check in a few more days to see if the roots have turned completely.
4. Have students make drawings of their observations and label them with the number of days since transplanting.
5. Have students report their findings to the class.

What did the roots do when they ran into the block? Why is it important for plants to be able to do this? What happens to plant roots growing in a heavy clay soil? Why?

Let's Get a Handle on This

Description

Students will conduct an experiment with climbing peas to observe the plants' use of tendrils. This activity may be done by Group 3.

Objective

To demonstrate the way in which certain plants connect themselves to outside structures for support (thigmotropism).

Teacher Background

Thigmotropism is the inherent response of certain plants to grab onto things. This is generally exhibited in vines. For example, peas have tendrils that literally grab onto the closest support in order to be able to grow off of the ground.

Materials

Two 4-inch pots
Climbing pea seeds
Small sticks
Soil
Rulers

See discussion on page 125.

1. Have students plant three pea seeds in each pot, labeling one pot "A" and the other "Control." Tell them to keep the soil moist.
2. When the seeds first sprout, have students insert a 12" stick into the soil next to the seedlings in pot A; do not have them put a stick in the control pot.
3. Have them observe the plants closely, recording the growth, number of leaves, and other observations. (After the second set of leaves appears the plant will send out tendrils that will latch onto the stick.)

4. Have students draw the two sets of plants as they grow.
5. When the plants in pot A reach the top of the stick, have them put a larger stick on the other side of the plant. The plant will then reach toward that stick and grow around it. For the next step, have them tie a string from the top of the second stick to a curtain rod or nail above a window. The vine will wind its way up the string.
6. How is the control plant growing? Have students record their observations.

Describe the parts of the pea plant. Which parts are the same as other plants? Which are different? Describe the similar ways in which the plants in pot A and the control grew. Describe the different ways in which they grew. How do tendrils help the pea plant? What types of plants need tendrils? Why was the control important to your experiment?

Which Way Did It Grow?

Description

Students conduct a simple experiment by growing radish seeds in a jar without soil to determine the direction of root and stem growth. This activity may be done by Group 4.

Objective

To demonstrate the effect of gravity on plants (geotropism).

Teacher Background

Geotropism describes the plants' response to gravity: roots grow down and stems grow up.

Materials

Two glass jars
Wet paper towels
Radish seeds

See discussion on page 125.

1. Soak the radish seeds overnight.
2. Have students line both jars with wet paper towels and place an equal number of seeds in both jars. Label one jar "A" and the other "Control."

3. Have students grow the seeds in the dark until the stems are about 1″ long. Then have them pour off the excess water from the jar.
4. When the stems are 1″ long, have students turn jar A on its side. Have them keep both jars in the dark, wait 24 hours, and then notice the growth of the stems and the roots.
5. Have students record the daily growth of the stems and roots. Have them draw pictures of the growth in the two jars.
6. Suggest that they change the direction of jar A every 24 hours but always keep the control upright.
7. Have students report their observations to the class.

What does this experiment tell you about the growth direction of the stem? What does it tell you about the growth direction of the roots? Why is this important for plant growth? Why was the control important to this experiment?

Experiment with *phototropism,* the response of plants to light. Have students cut out a 1″ square at one end of a shoe box and fill it 2/3 full with gravel and top it off with soil. Have students plant bean seeds, water to moisten, and cover the box with the lid. Do not let them peek except to water. In a week, have students observe how the new plant bends toward the patch of light at the opened end of the box.

Sugar Factories

Description

Read this short story to the class to describe a historical science experiment and to stimulate discussion on how plants grow.

Objective

To introduce the concept of photosynthesis.

Materials

None

We've learned a lot about plants and how they grow. For human beings to grow, we eat food. What do plants eat to grow? Scientists today know the answer to this question. But scientists 350 years ago did not. If you were a scientist then, what types of experiments might you have tried in order to find the answer?

Read the following story to the class:

The Tree Experiment

About 350 years ago, a man named Jan Van Helmont decided to find out how plants grow. At that time, most people thought plants ate soil. Jan wasn't sure this was true, so he set up an experiment to find out for himself. He planted a small, five-pound willow tree in a pot of dry soil weighing 200 pounds. Jan figured that if the tree ate the soil, then the weight of the soil should get less and less.

For five years Jan watered and took care of the willow. It grew very well and became a handsome 169-pound tree. Then Jan weighed the soil. He was careful to let the soil dry out so that it would be as dry as when he first planted the tree. The soil tipped the scales at 199 pounds and 14 ounces, only 2 ounces lighter than the original 200 pounds! Where did the tree get the food to grow 164 pounds? Jan

thought it all came from the water he added. Where do you think it came from? What question did Van Helmont set out to answer? What were his conclusions? Was he right?

Since Van Helmont's time we have learned that plants make their own food from the sun's energy. This process is called *photosynthesis.* Life as we know it depends on this unique ability of green plants to convert the sun's energy into food. Photosynthesis is one of the most important chemical reactions on earth. We are totally dependent on plants for our food. No other living organism can make the sun's energy available to us as chemical energy. Photosynthesis takes place within the chloroplasts of plant cells. There the raw materials, water and carbon dioxide, are combined chemically in the presence of sunlight and chlorophyll. Some of the resulting sugar is immediately transported to other parts of the plant. Some of the sugar is changed into starch and stored temporarily in the leaves. Oxygen is released into the air as a by-product of the process. We would not have any air to breathe or food to eat without green plants.

What was Van Helmont's experiment? What did the results tell him? (The willow tree did not eat soil.) If you got Van Helmont's results, what would your next experiment be? Since Van Helmont's time, what have scientists learned about what plants need in order to grow?

Sipping Through a Straw

Description

Students observe how colored water moves through a celery stalk.

Objective

To demonstrate how water moves to the leaves and is used in photosynthesis.

Teacher Background

Plants drink water through their roots in the soil. As the water evaporates from the leaves, a vacuum is created that pulls the root water upward to the leaves. The principle is the same as sipping through a straw. Water moves in the plants vessels (xylem), distributing nutrients. In the leaf, it is used in the photosynthesis process.

Materials

One fresh celery stalk with leaves per group of four
One clear glass or cup per group
Red food coloring
Water

Where do plants get water? Do other parts need water? Leaves need water to make food. How do the leaves get water? (Record responses on the chalkboard.)

1. Divide the class into groups of four and distribute materials.
2. Have each group add water and food coloring to their containers.
3. Immerse a celery stalk with bottom cut off in the water in each container.
4. Have each group label their container and set it in a visible location.
5. Make a drawing or graph of the celery plant. Daily, observe and record every few hours the height of the colored water in the stalk.
6. When the food coloring reaches the leaves, cut the stem and examine it.

What did you observe? Why is it important for water to reach the leaves? How did the water get pulled to the leaves? Can you think of a way you can make water move that is similar to the way a plant does it? (sip through a straw)

Have students make colorful daisy bouquets by placing the stems of white daisies in different food colorings mixed with water.

Plant Sweat

Description

Students conduct a simple experiment with potted plants. This experiment requires a few hours of sunshine. It should be started during the morning on a sunny day.

Objective

To determine how excess water leaves a plant.

Teacher Background

Plants recycle a great deal of water back into the atmosphere through a process called *transpiration*. Water taken in by the plant's roots is pulled up to the leaves. The excess water is released into the atmosphere through tiny openings in the leaves called *stomata*. This water is evaporated from the leaves by the heat of the sun. An apple tree can lose 15 liters of water per hour on a hot, sunny day. Thus, plants recycle a lot of water into the atmosphere when the sun evaporates water from the leaves, which in turn creates a vacuum in the leaves that pulls water up from the roots.

Materials

Three identical plant containers for each group
Soil for all of the containers
Mature plants for 2/3 of the containers
Plastic bags to fit over each container and ties or rubberbands
One measuring cup for each group
Water
Masking tape
Life Lab journals
Plastic ties for 1/3 of the containers

How does water enter the plants? (roots) How does it move throughout the plant? (pulled upward through xylem, or veins) Why do plants need water? (to make food, to keep rigid, to transport nutrients) What happens to excess water? (Record predictions.) Let's do an experiment to find out.

1. Divide the class into five groups.
2. Give each group three containers and have students prepare them by filling each with potting soil and transplanting mature plants into two of the three.
3. Each group should label the containers as follows:

Pot A—the one without a plant

Pot B—one with a plant

Pot C—one with a plant

4. Have students pour equal amounts of water into all containers so that they are well watered.
5. Have students cover Pot A and Pot B with plastic bags and seal them. Have them cover the plant in Pot C with a plastic bag tied at the point where the stem meets the soil. Be sure that the soil in Pot C is exposed to the air.

6. Ask students to hypothesize which plastic bag will collect more moisture. Share and record the hypotheses.
7. Place the covered containers in the sun for a few hours. More time will be needed on a cloudy day.
8. In the afternoon, ask the class to observe Pot A and Pot B. What do they see on the inside of the bags? Does it look like one bag is holding more moisture than the other?
9. Now have students observe Pot C. What do they see on the inside of the bag? Where did it come from? Transpiration is the plant's way of sweating, releasing excess water into the air through its leaves. The constant flow of water through the plant gives it shape and life. Water enters through the roots, carrying air and nutrients. It is pulled through the plant continuously in columns by evaporation caused by the heat of the sun.
10. Ask, Which pot produced the most moisture? Why? Have students write their results and conclusions in their Life Lab journals.

What were your results? Why do plants sweat? How is this sweat (transpiration) important to the plant? (pulls more water through plant) How is transpiration important to the atmosphere? (adds moisture to the air) What would life be like in an environment without plants?

Magical Mystery Tour

Description

Teacher conducts as a demonstration a simple experiment creating a visible chemical reaction, first with the carbon dioxide humans exhale and then with the oxygen released from an aquarium plant.

Objective

To demonstrate the production of oxygen and the utilization of carbon dioxide by plants in the course of photosynthesis.

Teacher Background

Plants have the unique capacity to make their own food. This process, called photosynthesis, requires carbon dioxide, sunlight, and water. The carbon dioxide is taken from the air, and from the chemical reaction of photosynthesis the plant gives off excess oxygen. Animals reverse the process, taking in oxygen and giving off carbon dioxide. This great exchange between animals and plants recycles the earth's limited air supply. Many scientists are currently concerned about the increase of carbons in our atmosphere from the burning of petroleum. An increase in carbon can trap more heat in the atmosphere (the greenhouse effect). Some scientists hypothesize that this could lead to dramatic changes in climate and result in the melting of glaciers. Plants help to keep the carbon ratio in balance by using carbon dioxide in photosynthesis.

Materials

Bromothymol blue, available from Let's Get Growing catalogue
Two bottles or test tubes with tight fitting stoppers that have holes for straws or pipettes
Two straws or pipettes that fit in stopper holes
Sprig of an aquarium plant (elodea or hornwart)

Preparation

Fill bottle half-full with bromothymol blue. Seal with stopper (which has straw or pipette inserted in stopper hole). Stopper needs to fit tightly in bottle. If there are gaps around the stopper or straw, seal them with tape. A clear baby bottle works well for this experiment.

Animals inhale oxygen and exhale carbon dioxide as part of their respiration systems. Plants use the carbon dioxide and release oxygen into the air during photosynthesis. This exchange can be demonstrated by the use of a chemical, bromothymol blue, which changes color when carbon dioxide amounts are increased. When you breathe into the chemical, the carbon dioxide you exhaled changes the color of the chemical from blue to yellow-green.

1. Demonstrate by breathing in a soft, steady rhythm into the straw of one of the rubber-stopped bottles.

2. Breathe into the bottles until the blue color becomes yellow-green. Explain that the yellow-green color signifies the presence of carbon dioxide.
3. Ask students to hypothesize how some of the carbon dioxide can be removed from the bromothymol blue solution. How will they be able to tell if this has happened? (The color will return to blue.)
4. Have them test the hypothesis as you remove the stopper and place a sprig of elodea in one of the bottles. Place both bottles in bright sunlight and observe changes over several days. If students develop other hypotheses, follow through on their ideas.
5. Ask students to report the results of the experiment. Discuss the importance of plants in removing carbon dioxide from the atmosphere.

What was released into the chemical to change its color? How was carbon dioxide removed from the chemical? How did the plant use the carbon dioxide? What is an important exchange that takes place between plants and animals? Why are plants important in maintaining the carbon dioxide balance in our atmosphere?

Plants Need Light Too

Description

Cork disks prevent light and air from reaching a section of leaf. The leaf is then treated and tested for starch content.

Objective

To demonstrate that plants need light to produce food.

Teacher Background

Plants need air and light. Green plants make food from water in the soil, carbon dioxide in the air, and sunlight. The carbon dioxide enters the plant through small openings in the leaves. The chlorophyll in leaves gives leaves their green color. A plant needs sunlight to keep the chlorophyll active and the leaves green. The chlorophyll reacts with the carbon dioxide and water to produce starch, the plant's food.

A simple test for the presence of starch is to place a drop of iodine on the substance to be tested. If it turns blue-black, starch is present. In this demonstration, you may want each of the four groups to set up a demonstration plant and then analyze the results together. You should be responsible for heating the alcohol. Follow the directions: *Do NOT place the alcohol directly on the flame.* Within six hours after putting iodine on the colorless leaf, the leaf should turn black except over the area that the cork was covering. The cork prevents the carbon dioxide in the air from entering the leaf. The cork also keeps sunlight from reaching the covered part of the leaf, preventing photosynthesis and food production.

Materials

Cork cut into thin disks
Alcohol burner
Pins
Rubbing alcohol
Two beakers, one larger than the other
Burner stand
One plant with large leaves per group
Iodine
Eyedropper

Preparation

(Optional) Demonstrate to the class that iodine tests for the presence of starch. Mix a few drops of iodine in water and add a few drops to water that is mixed with cornstarch. Observe the different colors. How does iodine show that starch is present? Test food samples, such as bread, milk, potato.

What do plants need in order to grow? (water, air, sunlight, nutrients) What does the plant do with these ingredients? (makes its own food) How do the air and sunlight get into the plant? (through the leaves) What do you think would happen if the leaves were blocked from getting air and sunlight? (Record predictions.)

1. Have the class work in groups of four.
2. Have groups test one leaf on each plant for starch, and record results.
3. Provide two cork disks for each plant.
4. Have students use pins to sandwich a leaf on the plant with the cork disks.
5. Place the plant in the sun for six-to-eight days.
6. Have students remove the cork disks and pick the leaves.
7. Light the alcohol burner.
8. Place each leaf in a beaker of alcohol.
9. Put the beaker of alcohol inside a larger beaker of water and place it on the burner. THE ALCOHOL BEAKER SHOULD BE HEATED IN THE WATER. ALCOHOL SHOULD NEVER BE PLACED DIRECTLY ON A FLAME.
10. After 15 minutes, the leaf color will be gone.
11. Remove each leaf from the alcohol and have students rinse them in water.
12. Have students gently blot each leaf with a paper towel.
13. Test each leaf for starch (food) by putting iodine drop by drop over the entire leaf. Iodine turns blue-black if starch is present.
14. Discuss the results.

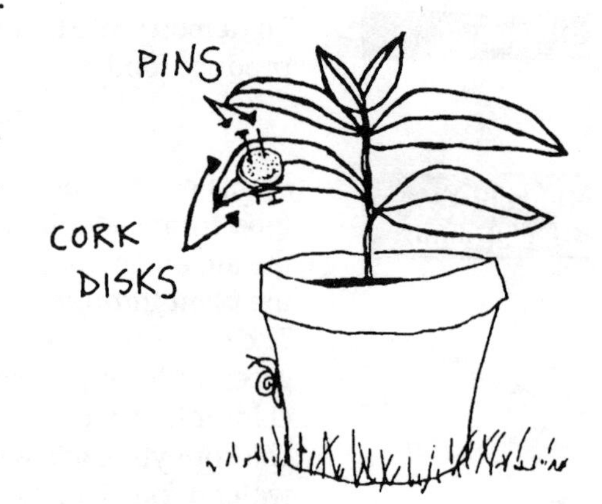

What were the results of this experiment? What can you conclude from the results? (Plant does not make food without air and sunlight.) How could you tell if food was present? Did this experiment have a control? (Yes. The part of the leaf that wasn't covered.)

Ask, Why do mushrooms grow in the dark? (They're fungi, not plants.) Where do they get their energy and nutrients?

Plant Food Magic

Description

Students role play the process of photosynthesis.

Objective

To integrate the different factors involved in plants making food.

Materials

Props (See preparation below)
Life Lab journals

Preparation

Make the following props for the skit or divide the class into three groups and have each make a set of props.

- a sun
- a large plant
- a single leaf similar to the ones on the plant
- drops of water
- signs that say CO_2 (carbon dioxide)
- signs that say O_2 (oxygen)

Have you ever thought about where plants get their food? Where do we get our food? Have you ever seen plants get their food the way we do?

The secret is that plants are able to make their own food! Green plants are the only living things on earth that can make their own food inside themselves. To do this they have a very special recipe they each follow. While I tell you about the recipe, I'm going to ask some of you to act out the story.

1. Assign students to parts for each of the props. (You may have more than one group acting out the skit at the same time.)
2. Set students with props on a stage area where they can move about. Have the plant in the center with the O_2 hiding behind it. Encourage the students to act out the story as you tell it.
3. Read the story:

   It was another beautiful day. The sun was shining brightly in the sky. It shined its light all over, and it was shining brightly on the garden at our school. Now, out in the garden in its soft dirt home sat (*variety*) plant. The plant knew it was a beautiful day and wanted to take advantage of it. It was going to be a great day for making food. First the plant would gather all the ingredients, and then it would cook them up into food that would make it grow.

First the plant stretched toward the sky and opened its leaves wide to take in as much sunlight as possible. It would need lots of light to make food. Then it took some very deep breaths. It was looking around for just the right parts of the air. Ah, yes, the carbon dioxide (CO_2) floating in the air was just perfect. CO_2 would give the plant a very important part of the recipe. The plant brought them into its leaves. Now it needed one more ingredient. For this it wiggled its roots around looking for water. Ah, the soil was nice and damp. It was easy to bring some water from the roots all the way up to the leaves where the CO_2 and sunlight waited.

Now it was time to start making food. The plant mixed the light and the CO_2 and the water all together. It was a beautiful day and it wouldn't take long to make lots of food! Presto! Pretty soon, it made enough food to add another leaf to its body. The cooking is now over, and it's time for the plant to clean up. The plant used up everything pretty well, but part of the CO_2 is left over. Oh, the plant didn't need the oxygen that was in the carbon dioxide. So it can just send the O_2 back into the air again. The plant watches it fly away.

4. Turn to the class and say: That's amazing. This means plants can make food with just light from the sun, carbon dioxide from the air, and water! Can someone tell me the story again so that I can listen this time? Select someone else to tell the story while different actors act out the skit, or have each group prepare their own skit and perform it.
5. Have students take pencils and their Life Lab journals to the garden. Have students find a green plant, draw it in their Life Lab journals, and add the ingredients that it needs to make food.

Do plants need air? What do they need air for? What part of the air do they use? What do they release back into the air after they are done making food? What are the other ingredients of their recipe? Why are plants important to all animals?

Star Food

Description

This garden experiment has students compare plant growth when aluminum foil is used to reflect more light onto some plants.

Objective

To discover whether photosynthesis can be increased by reflecting more light onto a plant.

Materials

Two short garden beds or one long one
Seedlings to transplant (larger plants such as broccoli and tomatoes are preferable)
Aluminum foil (ask students to recycle some from home)
Life Lab journals

Preparation

Prepare garden beds for planting.

What do plants need in order to make food? (air, sunlight, water) What do you think would happen if we increased the amount of sunlight plants receive? (Record predictions.) How could we do this? (Lead the class through designing the experiment.)

1. Plant each bed with the same amount of different crops. They could look like this:

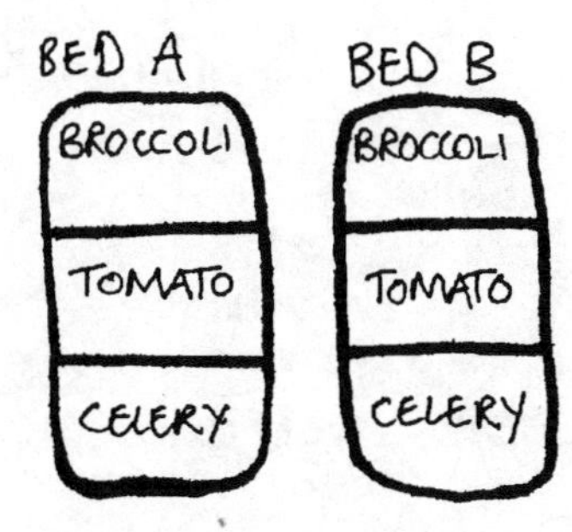

2. Place a wide collar of aluminum foil around the base of each plant in one bed and leave the other (control) bed alone. Try to have as much of the bed covered with foil as possible, but be sure to leave enough space between the plant and the foil to water. If necessary, weight down the foil with rocks. Try to treat each bed exactly the same.

3. Assign students to partner plants, one in each bed. Have them record the different growth rates for each plant by taking weekly measurements of the two.

MY TOMATO PLANT		
DATE	BED A	BED B (control)
4/15	4"	4"
4/22	6"	5"
4/29	9"	6"
5/6	13"	8"
5/13	15"	10"

4. After six weeks, have students present their results and the average height of the plant in each bed.

Did the plants of one bed grow faster or bigger? If so, why? Do you think the plants could get too much light? (Plants can sunburn.) What would happen if you did a similar experiment adding more water?

Explain that flying insects may get confused by the reflection of the sky in the aluminum foil and will avoid the plants in the reflecting bed. Do you find that fewer insects visit the plants in the foil bed? Do those plants have less insect damage?

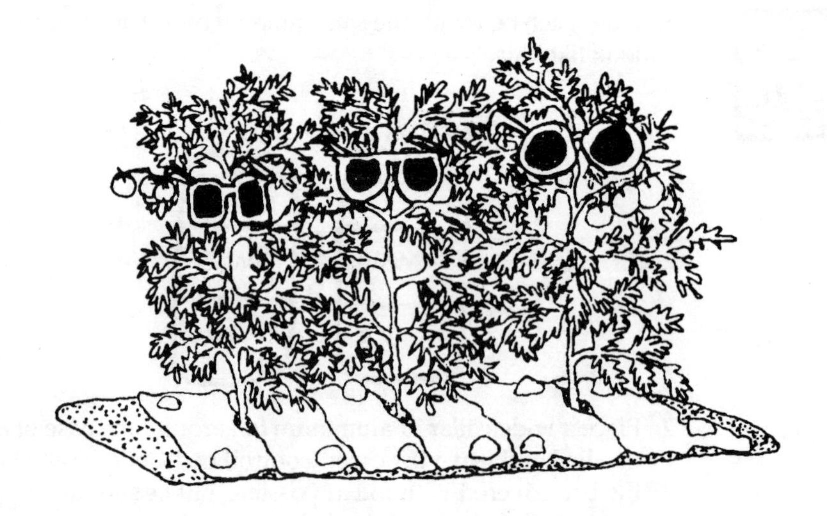

Room to Live

Description

Students plant seeds in pots at different densities and measure root and stem growth.

Objective

To discover the effects of crowding on the growth of plants.

Materials

Five 3- or 4-inch pots per group
Potting soil
62 radish seeds per group
Labels

How much room do you need in order to live? If you grew up in a room three feet high by three feet wide, do you think it would be easy to grow up to be a basketball player? Do plants need room to grow, too? What differences do you think we would find if we planted seeds in a crowded pot and in one where the seeds had plenty of room to grow? (Record predictions.) We are going to plant radish seeds in five different pots. Which pot do you think will have the plant with the longest roots? Shortest roots? Largest leaves? Smallest leaves? Longest stems? Shortest stems? Most leaves? Biggest radish? Smallest radish? (Record all predictions.)

1. If materials allow, divide the class into groups so that each group can carry out a whole experiment.
2. Fill small pots with soil and label them with numbers. Then have students sow radish seeds, equally spaced, in each pot. List the following amounts on the chalkboard:

   Pot 1: 2 seeds

   Pot 2: 4 seeds

   Pot 3: 8 seeds

   Pot 4: 16 seeds

   Pot 5: 32 seeds
3. Place pots in a sunny location and have students keep them watered as the seeds grow. The soil should be kept moist in all of the pots.
4. Have students observe the pots carefully to see whether the radishes in some pots are growing better than others, noting whether some are tall and spindly, others stunted, and so on. Have students draw pictures of their observations of each pot every few days.

5. When the radishes are full-grown, have students carefully pull them out of the pots, making sure to keep each pot separate. Have them choose the largest plant from each pot and measure the information for the chart below.
6. Compare students' results with their predictions.
7. Have a radish party with the results!

What did you test for in this experiment? Did crowding seem to affect radish growth? Why do you think some of the plants had longer stems? (to try to get to the light) Why did some plants have longer roots? (more room to spread) Bigger leaves? Bigger radishes?

Seed to Earth, Seed to Earth, Do You Read Me?

Description

Seeds are planted in five different types of soil, and the rates of germination and growth are measured.

Objective

To demonstrate the effects of different substances on plant growth.

Teacher Background

This unit examines how plants grow, and this activity ties plant growth to soil type. Water, Water, Everywhere is a related activity in the soil unit.

Materials

Five 4-inch pots per group
Radish seeds
Five different growing media (soil mix, garden soil, compost, sand, clay)
Labels
Life Lab journals

What do plants need in order to grow? (air, water, sunlight, nutrients) Where do plants get water and nutrients? (from the soil) Do you think different types of soil can affect how a seed and plant grow? (Record predictions.)

1. If there are enough materials, divide the class into groups of eight.
2. Fill five pots with five different growing media. Label each pot. Allow each group to select their own substances to test.
3. Have students pour equal amounts of water into each pot, observing and comparing the soils' water-holding ability.
4. Have students plant an equal number of seeds (approximately 10) in each container and observe differences in germination and growth. Have them keep a record in their Life Lab journals. Continue for at least three weeks.

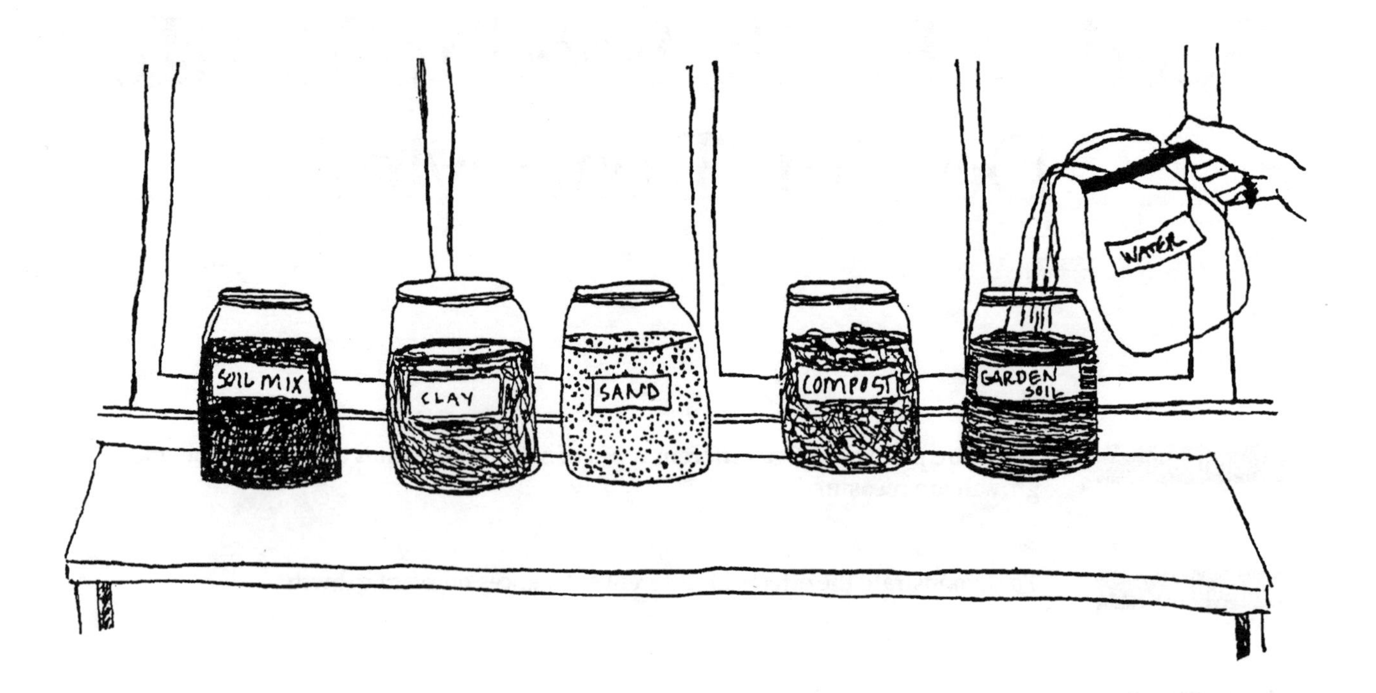

Which seeds germinated the fastest? Which plants looked the healthiest?

What does a seed need? Think about yourself for a moment. When you were born, what did you need in order to grow? What couldn't you live without? Think of the most basic things and write them on the chalkboard. Discuss and compare basic human needs with the basic seed needs.

Have students plant seeds under various conditions without soil by following the steps below.

1. Label four petri dishes:
   Dish A - Water and Light
   Dish B - Water and No Light
   Dish C - Light and No water
   Dish D - Water and Cold
2. Put a folded paper towel and two seeds in the bottom of each dish.
   Dish A - Moisten the paper towel with water and place the dish in a sunny window. Do not allow the paper to dry out.
   Dish B - Moisten the paper towel with water and place it in a dark closet or in a bag. Do not allow the paper to dry out.
   Dish C - Place the dish in a sunny window, but do not moisten the paper towel.
   Dish D - Moisten the paper towel with water and place the dish in a refrigerator.
3. Observe changes in the seeds for ten days. Make a chart and record your observations.

Stem, Root, Leaf, or Fruit?

Description

Students classify foods and spices they eat according to plant parts and make a vegetable and dip snack.

Objective

To identify and classify the parts of food we eat.

Materials

Life Lab journals
Sample of fresh spices from list of answers at end of activity
Vegetables for snack (carrots, celery, spinach, broccoli, peas, sunflower seeds)
Dip for snack (cottage cheese, onion dip, and so on)
Cutting board and knives

Preparation

Make up a picture chart of foods and spices that will be available in the classroom for this activity (see the "Food Categories" chart on the next page).

Name some plants that you eat. (List responses on the chalkboard.) Do you eat the whole plant or part of it? Let's list the different parts of plants. (root, stem, leaf, bark, flower, fruit, and seed) Do you think we eat all of these different parts? (Record predictions.) Can you name the different parts of the plants we listed that you eat? (List the part name(s) next to each plant).

1. Group students in pairs and have them refer to the picture chart of foods and spices.
2. Have students make seven category headings in their Life Lab journals: root, stem, leaf, bark, flower, fruit, and seed.
3. Tell them to write each food in one of the categories, according to what part of the plant we eat. For example, a walnut is a seed, an eggplant is a fruit, and so on.)
4. To introduce students to the wonderful world of spices, have them use their senses to explore the samples you have collected.
5. Challenge students to try classifying the spices. This tends to be a little more difficult for students, so if they cannot put the spices in categories, guide them through.
6. Now have students enjoy their new knowledge. Have them cut up vegetables and use the spices to prepare a dip. Be sure that they name the part of the plant they are eating.

What is your favorite vegetable? Which part of that plant do you eat? What is your favorite root, leaf, stem, bark, flower, fruit, and seed?

1. Have students describe their last meal in terms of plant parts. For example, a peanut butter and jelly sandwich would be ground-up seeds (peanut butter) and crushed fruit (jelly) on ground-up and baked seeds (bread).
2. Have students design a three-course meal composed only of one category. How would they enjoy such a meal?
3. Have students plant a garden bed according to the plant parts they eat, with a section for each category.

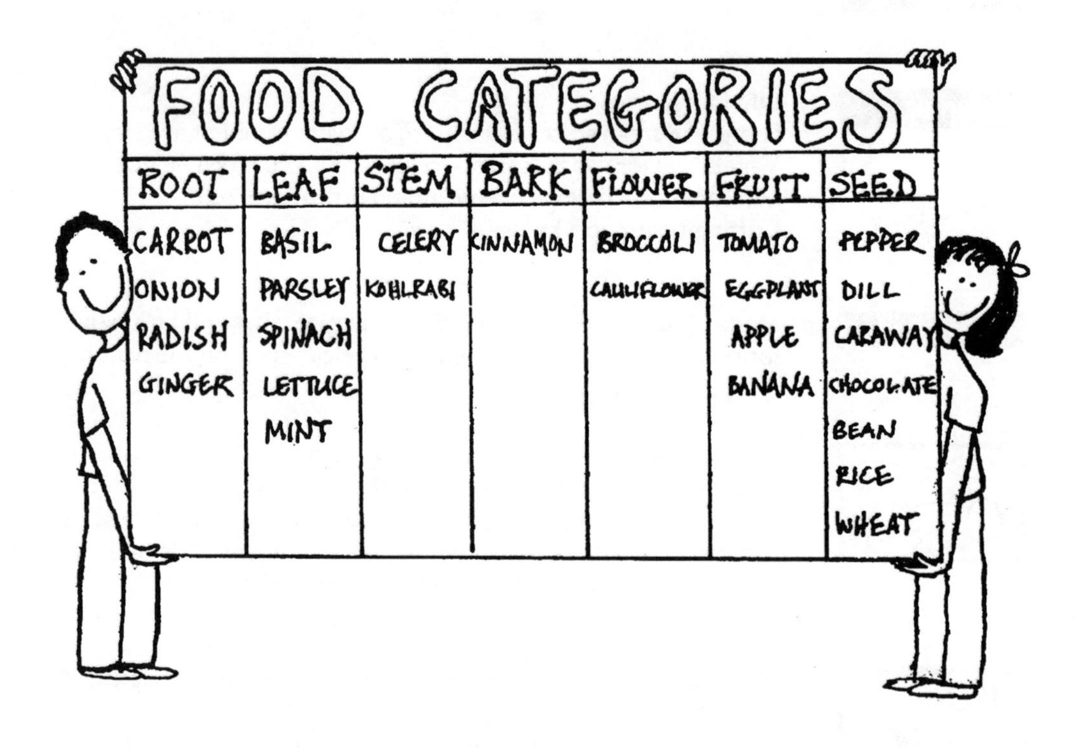

Living Laboratory

Outdoor Gardening

Vine Hill Elementary
Scotts Valley, CA

Alianza School
Watsonville, CA

Alianza School
Watsonville, CA

Alianza School
Watsonville, CA

Vine Hill Elementary School
Scotts Valley, California
and
Alianza Elementary School
Watsonville, California

Vine Hill: Our school garden started in the 1970s with federal CETA funds. We have integrated the gardening with our classroom studies of science, math, and language arts with the assistance of the Life Lab curriculum and teacher workshops.

The fenced-in garden is 50′ X 100′ and has 50 redwood raised beds that are 3′ X 10′ each. We have a 16′ X 20′ greenhouse, lath house, and tool shed. Two years ago, picnic tables and shade trees were added.

Classes are rotated on a weekly basis through the garden. There is a sequential plan for activities and all students participate in the various areas of work such as double-digging, tool clean-up, planting, transplanting, and seed collecting.

It has been a significant learning experience for the children, who eagerly look forward to fall when they'll again be involved in the Life Lab. As children become involved, they develop ownership in the garden and a protective attitude—and vandalism has decreased.

Alianza: Alianza School is a bilingual, magnet K-5 school located in Watsonville, California, a prime agricultural area on the central coast. Alianza, as its Spanish name reflects, was designed to foster an alliance between the Hispanic and Anglo segments of the community and to promote ethnic, cultural, and linguistic harmony.

Our Life Lab program began with a few planter boxes in 1986 and grew like wildflowers in the 1987-88 school year. A small group of dedicated teachers and parents acted as the catalyst for growth, and local businesses, especially nurseries and lumber yards, have generously donated or discounted materials. Parents, students, and staff formed a gardening alliance to build coldframes, a shed, planter boxes, fences, soil bins, and to prepare and plant beds of vegetables and flowers. Through our partnership with Watsonville Community Hospital, students have grown flowers to donate to patients and nursing staff as gifts. Volunteers from the local community offer to teach gardening skills on a regular basis.

Life Lab becomes a focal point for integrating different curricular areas: children are learning a second language through hands-on activities in a natural, relaxed setting, in addition to learning important concepts in science; math becomes a useful tool when applied to record keeping, graphing, and building projects. Our garden has become a resource center, open-air classroom, a picnic area, and a gathering place for friends to work, learn, and relax.

UNIT INTRODUCTION

Living Laboratory

Gardening and horticulture apply the concepts of science, weaving them into the context of our lives. In this unit, activities demonstrate basic gardening: garden planning, sowing seeds, and transplanting, watering, and maintaining plants. Students will learn these techniques and practice them over and over again. In so doing, they apply the concepts from all of the other units in a meaningful way, so that every Life Lab activity, whether taught in the classroom or in the Living Laboratory, is related to their garden system.

This unit starts with the fundamentals—soil, water, sun, and air—and proceeds to watching seeds sprout. After a few experiments to test for good ways to plant and water, students explore one of the most frustrating pests in the garden—weeds. The three experiments in this unit—"Water We Doing?," "What's In A Name?," and "Weeding, Writing, and Arithmetic"—require an extended period of time for data collection.

Whether your Living Laboratory is under grow lights in the classroom, in a planter box, or in a schoolyard garden, students will take pride in its ownership. Applying Life Lab concepts will reinforce and give relevance to their learning.

Activities	Recommended Grade Level
It's As Simple as One, Two, Three...Four *(identification of the necessities of life: air, water, sun, soil)*	2,3,4,5,6
So What? Sow Seeds! *(soil mixing and seed sowing)*	2,3,4,5,6
Inch by Inch, Row by Row *(planning and mapping garden beds)*	5,6
Transplanting, or Let's Move 'Em Out! *(technique for planting seedlings)*	2,3,4,5,6
Water We Doing? *(experimenting with the effects of watering on plant growth)*	3,4
What's In a Name? *(observation and experimentation of seed varieties)*	5,6
Weeding, Writing, and Arithmetic *(observation of a weeded and non-weeded section of garden)*	4,5,6

Sun, Soil, Water, Air

(To be chanted in a round)

Sun, soil, water, air.
Sun, soil, water, air.
Everything you eat
and everything you wear
comes from
Sun, soil, water, air.

It's As Simple As One, Two, Three...Four

Description

In this short activity, students hold soil and water in their hands.

Objective

To identify the four basic necessities of life: air, water, soil, sun.

Teacher Background

This unit of outdoor gardening activities starts by reminding students what they learned in the Growing Unit: Life depends on plants, which need soil, water, sunlight, and air in order to grow.

Materials

Water
Soil

To be good gardeners, we have to pay attention to what plants need in order to grow. If we care for them, they will grow to be healthy and give us food. So we will start by going on a mystery search for the four ingredients plants need.

1. At a water faucet in the garden, have the students cup their hands together and scoop up water and soil.
2. Tell students that they are holding in their hands the four basic ingredients necessary for all life to exist. Ask, Can you name them?

Why are these four ingredients so important to life? What would life be like without any one of the Big Four? How can you help plants in your garden get all four of them?

So What? Sow Seeds!

Description

Students sow seeds in flats.

Objective

To prepare a soil mix for flats and demonstrate seed sowing.

Teacher Background

By preparing their own soil mix, students gain a good understanding of the components of healthy garden soil. Seed sowing allows you to get a head start on spring by keeping flats in a coldframe, greenhouse, or window. Also, transplanting makes garden maintenance easier and gives the plants a better chance at survival. Please note that most root crops do not transplant well. (See Vegetable Planting Guide, Appendix pp. 452-453.)

Materials

One flat per group of five
Seeds: lettuce, broccoli, peppers, flowers, tomatoes, cabbage
Soil mix ingredients: screened garden soil, screened compost, sand
Water
Shovels
Labels
Grease pencils
Trowels

Preparation

1. To order seeds in advance from seed companies see ZIP Code Seeds, p. 116. Allow six weeks for delivery.
2. To build your own flats, see p. 447.

Have you ever seen farmers planting seeds in a field? How do you think they do it? We are going to become farmers with very small fields. As a matter of fact, our field is inside this box. (Show students the flat.) First we will prepare our own soil mix, so we are sure our seeds get all the nutrients they need. The compost provides the minerals for the new plants to grow; the garden soil helps the seedlings adapt to the soil life, including bacteria and fungi, that they will eventually live with once they are transplanted; the sand provides better drainage, preventing the seedlings from rotting. Once we fill our flats with soil, we will make rows and plant our seeds. What else will our seeds need in order to grow? (water) We will make sure the soil is always moist. We will also label our flats to know what we planted and when.

1. Prepare the soil mix using the following recipe:

2. Divide the group into teams of five and give each team one flat to fill with soil. Tell students to run their hands or a board gently over the soil surface so that it is flush with the top of the flat.
3. Have students run a finger from one end to the other in the soil to make a series of parallel shallow grooves as close together as possible. Explain that a rule of thumb for planting is that seeds should be planted to a depth that is roughly 2-3 times their size. The bigger the seed, the deeper it goes, but not too deep! It is better to be on the safe, shallow side. (Another method is to use a finger to drill or poke a shallow hole for each seed.)
4. Have students sow the seeds in the grooves so that they are evenly spaced about one inch apart. When the flat is sown, have students cover the seeds with soil or compost and water thoroughly with a fine mist.
5. Have students write the name and variety of the seed sown along with the date on a label and place it in the flat.
6. Flats should be placed in a greenhouse, coldframe, or warm area where they will get a minimum of six hours of full sun per day.
7. Have students water the flats thoroughly and be sure that they are kept moist all the time.

What were the ingredients of the soil mix? Why is each ingredient important? What are some advantages of sowing seeds in the flats?

1. Have students act out the following scene as you read it aloud:

   Make believe you are a powerful little seed. You are very tiny and sound asleep now. You are in the ground. It starts to rain. You drink a little rain water. You begin to move and start to to wake up and grow. You push and push with your little head to get through the ground and, suddenly, out pops your head. The sun shines and warms you. It makes you happy and healthy. More rain falls and you drink it. Now you really start to grow. Your arms reach out to the sun. Your legs stand firm in the soil to hold you straight and tall. The breeze gently blows you. You love the sun and the rain and the breezes. You are a healthy, happy plant.
2. Test seed germination in different soil mixes and compare the results.

Inch by Inch, Row by Row

Description

Students plan and map garden beds using information about growth requirements for each plant.

Objective

To combine several skills to create a garden design, including research, mapping, and drawing to scale.

Teacher Background

Designing the garden gives students the opportunity to practice mapping skills and a purpose to research information about specific plants. Plants should be rotated. If the same plants are always grown in the same soil, disease organisms can build up, and the soil can be depleted of certain nutrients. The roots of each plant grow to different depths, and each plant consequently has different spacing requirements. Consult the Vegetable Planting Guide (see Appendix) or seed packages for spacing information. Consult your local Agricultural Extension Office for a list of vegetable varieties that grow well in your area.

Materials

Five copies of the Vegetable Planting Guide, Appendix pp. 452-453
One copy of the Companion Planting Guide, see Appendix pp. 455-457
Five copies of the list of plants to be grown, see ZIP Code Seeds activity
One map of last year's garden or list of what was grown
Seed packets and catalogues
Graph paper
Ruler
One magnetic compass

Preparation

Make a list of all the plants to be grown. Make a list of last year's garden plants.

When plants grow, do they all look the same? What are some differences? What are some plant needs we should consider when planning our garden? (space, sunlight, root depth, height) How can we find out specific needs of each type of plant? (seed packets, seed catalogues, gardening books) To make a map of our garden plan, what other information do we need? (Be sure to agree to a common scale for mapping on graph paper.)

1. Divide the class into five groups.
2. Role play the spacing needs of plants. Have each group gather in a small space, each huddled in a ball, and ask them to stretch out slowly to their full height. How do they feel? Do they think they could each get enough food and water? Ask them to turn to the sun. Do they think they would each get enough sunlight? How could they change their spacing so that everyone is happy?
3. Distribute to each group Vegetable Planting Guides, graph paper, pencils, straight edges, and a list of plants to be grown. Explain that each group will work on one part of the problem. Later, representatives from each group will get together to compile the information to make a map.

Group 1 will draw the size and shape of the garden to scale on the graph paper, orienting it to the compass directions and showing all permanent features, such as trees and buildings.

Group 2 will create a list of the plants to be grown according to height. They should first make a bar graph, with the plant height on the y-axis and the plant name on the x-axis and use the graph to create their list. Seed packets and catalogues may be used as reference.

Group 3 will create a list of compatible plants to be grown by filling out three columns: name of plant, companion plant, antagonistic plant. The Companion Planting Guide may be used as reference.

Group 4 will list the space requirements of each type of plant using the Vegetable Planting Guide and information on seed packages. They will indicate on graph paper the space requirements by shading the number of square inches or feet needed by mature plants.

Group 5 will analyze last year's garden to ensure that plants are rotated in this year's garden. Using the list of last year's plants and the Plant Rotation information, they will construct a rough map and will recommend which plants may best be grown in each bed this year.

Compiler Group: A representative from each group will meet to create a single garden map. (See illustration.) Students should have fun advocating the needs of each plant in making decisions. Have them present their results to the rest of the class in an imaginative way.

GARDEN PLAN

thyme/dill
broccoli
basil
tomato
marigold

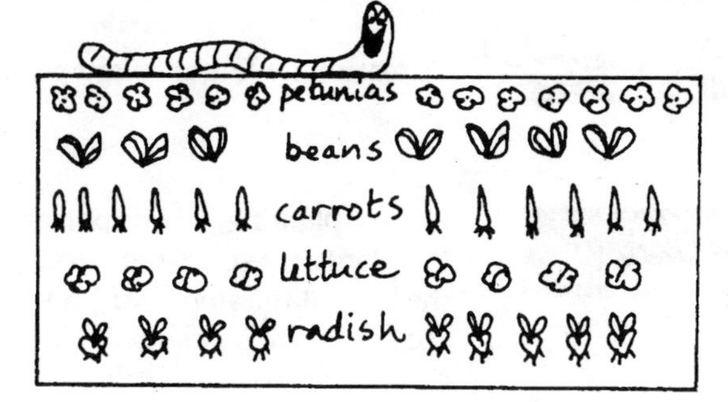

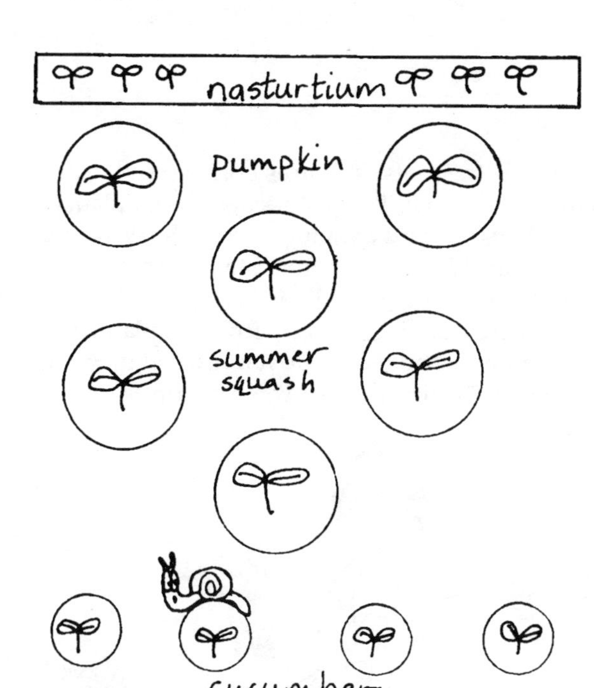

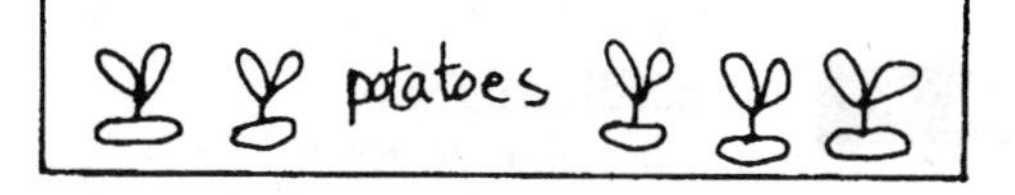

What would be the outcome if we simply scattered seeds randomly in the garden? Why bother to plan the garden? What is the difference between a garden and a natural field? What did you learn from making your garden map?

Have students make a papier-mâché or clay model of the garden.

Transplanting, or Let's Move 'Em Out!

Description

Students transplant seedlings from flats to garden.

Objective

To learn how to transplant seedlings.

Teacher Background

In transplanting, it is important to handle the roots as little as possible and to give the plants the space they need in their new home. Teach students to hold plants at the base of the stem. For younger students, you may want to dig the holes where the plant is to be planted. Use the Vegetable Planting Guide in the Appendix to determine spacing needs. The Companion Planting Guide will assist you in intercropping plants so that plants that are beneficial to each other are planted together. Seedlings need to be transplanted when they are overcrowded in the flat, have their first true leaves, or both. Note: The first leaves to emerge are called *cotyledons*. They are not the true leaves.

Materials

Seedlings ready to be transplanted from a flat
Trowels

Preparation

Prepare a garden bed for transplanting. (Note: Seedlings may be transplanted into another flat if it is too cold or wet outdoors.)

We are ready to give our seedlings a new home. What are some important things to consider? (The bed should be prepared so that the roots can grow; compost should be added to the soil to provide nutrients.) The seedling is like a baby. How do we handle it carefully? (Dig it carefully from the flat; don't touch the roots.) Picture the plants when they will be big. What kind of space will our plants need? When the plants are put back in the ground, how will we get the soil to stick to the roots? (Press the plant firmly into the soil; water it well.)

1. Mark the spaces in the bed where plants will be planted.
2. Have students dig a hole two times the size of the root ball for each plant.
3. Demonstrate how to slide the trowel gently into the soil at the edge of the flat and lift up a clump of seedlings shading their roots from the sun to prevent them from drying out.
4. Demonstrate how to separate the plants carefully. "Think them apart," trying not to break too many roots and keeping as much soil around them as possible.
5. Hand each student a seedling, making sure to hold it by the stem.
6. Help students plant each plant gently by holding it at the stem, having the roots fall straight down, gently covering the roots with dirt up to the first set of leaves, and pressing the soil firmly around the plant.
7. Water the plants well and label them.

Why was it necessary to transplant the seedlings? How did you determine how much space to leave between plants? How will you care for your transplanted plants?

Allow the seedlings in one flat to overgrow, and have students observe the effects of overcrowding on them. For an activity on the effects of overcrowding, see p. 145, Room to Live.

Water We Doing?

Description

In this two-week project, students observe the relationship between watering and plant growth and apply their experimental findings to the garden. The three activities are planned for a Monday/Monday/Friday sequence.

Objective

To control the application of water to plants.

Part One:

Materials

One package of bean or pea seeds
One 1/2-pint milk carton per pair of students and one for you
Potting soil for cartons
One large self-stick label per pair of students
One plant stake per pair of students
Several water sprayers or sprinkling cans
Several measuring cups or graduated cylinders
Newspaper
Class chart

Preparation

One week before the activity, plant a seed in each carton, first punching five identically spaced holes in the bottom of each. Water evenly as instructed on the seed packet and maintain the plants until the activity begins, at which time sprouts should just be showing.

What are the four necessities of plant life? (sun, water, air, soil) How can each of them be controlled in the garden? Which one of them can be controlled most easily? (water) How does watering plants affect their growth? (It helps them grow.) What happens to plants if they're watered too much? (They die.) What happens to them if they're not watered? (They die.) Let's set up a test to try to find out the amount of water to give a plant. First we must decide just when we are all going to water. (Decide on every day, every other day, or every third day.) We now have to agree on just how we will all do our watering. (Decide whether to pour on water or spray it.) These agreements are part of controlling a test so that we are only comparing one thing at a time. What other things must we control to make the test fair? (Elicit other factors of light and temperature.) Now I'll demonstrate various amounts of water, and each pair can decide which would be the best amount to use. (Show amounts from 25 mL to 250 mL.) Now each pair may put their names and amount on the chart. Let's begin with whoever is going to use the least water and work up from there. (Make sure that there is a diversity.) I'm going to set out a plant, too, but mine won't get any water at all. It will provide us with a point of comparison for all the others.

1. Give each pair of students a plant and a label to indicate the group and the amount of water.
2. Put plants in the designated location.
3. Have students water the plants according to the schedule for the next two weeks. Have them measure plant height, count the number of leaves, and record this information on the chart.

Which plants grew the most last week? Which grew the most leaves? Which grew the least? Which grew the fewest leaves? How did my control plant do? What pattern does the chart seem to show?

Water We Doing?

name	DAY 1		DAY 2		DAY 3		DAY 4		DAY 5		DAY
	HEIGHT	LEAVES	HEIGHT	LEAVES	HEIGHT	LEAVES	HEIGHT	LEAVES	HEIGHT	LEAVES	HEIGHT
Kevin	1½"	4	1¾"	4							
Tom	1¾"	4	1¾"	4							
Julie	1"	4	1¼"	4							
Melissa	1¼"	4	1¼"	4							

Part Two, one week later:

Materials

One set of small paper squares labeled A, B, C, D, E per student
Five 1/2-pint milk cartons

Preparation

Prepare five milk cartons of soil, each one with a different amount of water added, from 25 mL to 250 mL. Label the cartons A, B, C, D, E in random order, keeping a list of which carton received which amount of water.

What do you think is meant by the terms *dry, moist,* and *wet* soil? Let's try a test to tell whether soil has the right amount of moisture to help plants grow.

1. Pass around milk cartons of soil for dampness testing. Have each student take a pinch of each soil and place it on the paper with the corresponding letter. When all five cartons have been circulated, have each student arrange them in order from driest to dampest. Check their

results. Note: Soil that is too dry crumbles when it is squeezed; soil that is too wet oozes; soil that is just right sticks together but doesn't ooze.

2. Have students prepare soil in the garden for the eventual planting of beans.

Part Three, four days later:

One ruler per group
One sheet of newspaper per group

Repeat the questions and analysis in the discussion for Part One. What have we measured each day? (height and number of leaves) What part of the plant have we not looked at? (roots) Let's look at those roots now to see what they have been up to while we've been working elsewhere.

1. Divide the class into groups of six. Give each group one plant that represents different amounts of watering.
2. Have students gently remove the plant from its carton and lay it out on the newspaper.
3. Have students compare the extent of the plant's underground growth to its above ground growth. Was it more? Less? About the same?
4. Transplant all the seedlings outdoors, and water them.

In what ways will watering garden plants be the same as watering test plants? (Soil dampness must be just right.) How will it be different? (We can't measure so exactly, we may want to sprinkle or soak, and so on.)

Have students continue to compare the factors in watering, including method of application, frequency, and so on.

What's in a Name?

Description

Students grow different varieties of the same crop to test their suitability to the school's garden soil and climate.

Objective

To discover that there are many varieties of each vegetable.

Teacher Background

Varieties are bred by plant scientists working for seed companies and are developed for their different characteristics. Some tomatoes, for example, make a big fruit; some resist pests and diseases; some do well in certain climates and poorly in others; some grow longer, shorter, rounder, firmer, redder, bigger, faster, and so on. Many varieties are hybrids and do not reproduce the same seed from season to season. However, you can save the seeds of nonhybrid varieties and develop seeds that are especially well-suited to your garden.

Materials

A dug and fertilized garden bed
Different varieties of one particular vegetable such as carrots or lettuce.

Preparation

Prepare a garden bed for planting

Picture a lettuce plant. More than one image should come to mind, because there are many different types of lettuce! Some have dark green leaves and some have red leaves; some have curly leaves and some have long, straight leaves; some taste sweet; some are crunchy. These different types of lettuce are called *varieties*. Just as a brother and sister are closely related but different, so there are different varieties of each vegetable. Do you think some varieties may taste better than others? Would they grow better in our soil than others? Let's find out if certain vegetable varieties are better for our climate and soil than others.

1. Have students divide a garden bed or planter box into several equal sections and sow or transplant each section with one variety of one particular vegetable. Have them label each section with variety name and date planted, and water well.
2. Have students read the information on the seed packets and record their predictions about which variety will grow the best, taste the best, be the tallest, be the quickest to harvest, and so on.

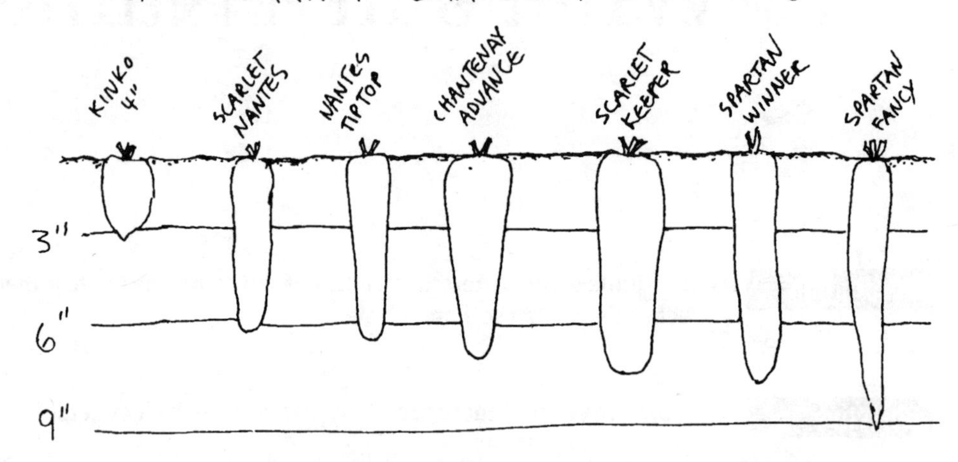

3. Have students treat all the varieties exactly the same: thin plants at the same time if necessary, water the plants the same amount, and so on.
4. Have students make weekly observations and record all the information on charts.
5. When students harvest the crop, have a tasting party.

# GERMINATED				PEST DAMAGE (SCALE 0-5)				ROOT LENGTH IN INCHES				TASTE (SCALE 0-5)			
A	B	C	D	A	B	C	D	A	B	C	D	A	B	C	D

CARROTS: "NANTES" = A "CORELESS" = C
"HALFLONG" = B "LONG" = D

Which variety produced the most? Which grew the most? Which tasted the best? Which factor is the most important? Is it better to have short, sweet carrots or long, bland ones? Which would you rather grow if you were a farmer? (Ask the class to vote on the best variety.) Would this variety necessarily be the best to grow no matter where you live? Why might another variety do better somewhere else? (different climate, different soil, shorter or longer growing season)

1. Have one student in the class call your local Agricultural Extension Agent and ask for a list of recommended vegetable varieties for your area. (Did they pick the same vegetable variety that your class did?)
2. Become plant breeders. Plant nonhybrid varieties and save seeds of the plant that grows the best. Continue from season to season until you have developed a seed especially suited to your garden.

Weeding, Writing, and Arithmetic

Description

Students maintain a weed and nonweed section of a garden bed.

Objective

To explore and observe weeds in the garden.

Materials

A dug and fertilized garden bed
String
Life Lab journals
Crop seed for one bed: bush beans or radishes

Weeds are just plants that grow wherever we don't want them to grow. They are often faster growing and heartier than the plants we plant. They compete for space, nutrients, and sunlight. But weeds can help gardeners in several ways. Some weeds, such as pigweed, have deep roots that transport nutrients from the soil. These weeds add valuable nutrients to the compost pile. Some weeds have powerful roots that break up and aerate the hard soil. Other weeds attract pests that would otherwise nibble on our crops. And some weeds attract insects that feed on still other pests. For example, the green lacewing, whose larvae devour many pests, feeds on the nectar of nettles, lamb's-quarters, and dandelions. Finally, some weeds are edible, such as pigweed, mustard, purslane, lamb's-quarters, and dandelion greens. Weeds can also be beautiful plants. The poet Walt Whitman described one familiar weed in this way: "Innocent, golden, calm as the dawn/The spring's first dandelion shows its trustful face."

Have students follow the steps below to see just what weeds are and how they affect other crops.

1. Plant seeds in straight, spaced rows so that you know where your plants are supposed to come up.
2. Divide your planted bed into two sections: weeding and nonweeding. (A small nonweeded section may easily be added to any garden, although it would be best to separate the sections by several feet.)
3. Make signs for each section: Beans With Weeds; Beans Without Weeds.

4. Draw pictures in your Life Lab journal predicting what each section will look like.
5. Keep one section weed free. In the other section, weed until the first time true leaves appear on the plant, then stop weeding and let everything grow. Water both sections the same amount.

6. In your Life Lab journal, keep a chart of the weekly growth of your crop plants and your weeds. You might also tie a string like a tight rope three inches above the ground and see whether weeds or crops reach it first.
7. Try to identify what kinds of weeds you have. This will be easier as they develop and flower. Possibilities include: lamb's-quarters, red root pigweed, dandelion, mustard, purslane, black-eyed Susans, buttercups, white clover, nettles, yarrow, sorrel, milkweed, crabgrass, cockleburr, and wild morning glory.
8. Give the number and name of any insects you see on your crops and on the weeds. Do weeds attract insects away from your crops?

When the crops are harvested and the charts are completed, the class can discuss several questions. Do weeds grow faster than other crop plants? Do weeds affect the growth of crop plants? Do weeds affect pests in the garden?

1. Some garden stores sell packets of dandelion seeds. Students might grow some of these and compare them with dandelions that pop up uninvited. These cultivated dandelions usually have larger and thicker leaves than the weed form. Tender leaves may be added to mixed green salads or thick leaves may be boiled as greens.
2. Discuss how weeds get in soil. Have students try growing the stickers in their socks after a walk through a field or overgrown empty lot. Or have them try growing the dandelion parachutes. Discuss other ways in which weed seeds might travel.

Cycles and Changes

Live Oak School, Santa Cruz, CA

Green Acres School, Santa Cruz, CA

Live Oak School
Live Oak School District
Santa Cruz, California

Our Life Lab at Live Oak School is growing. It began as a small garden project for a second grade class. Now it is being expanded to include the upper grades. Life Lab is helping to focus the community energy of a rapidly growing and diverse population, and we are building pride in our garden and school. Originally, we had a small garden between two classroom buildings. With increased teacher and parent enthusiasm and participation, we've expanded the area into a 70′ X 90′ corner of the playground. The School District fenced the area and provided funding for planter boxes, trees, and teacher participation in the Life Lab training. Life Lab is integrated throughout the curriculum to include literature, social studies, and ecological issues in lessons, as well as themes from Jack and the Beanstalk, Native Americans, rain forests, and Arbor Day. Community support in terms of time, energy, and money is tremendous. Life Lab's reputation as an effective hands-on science program had parents and kids excited and enthusiastic right from the start.

UNIT INTRODUCTION
Cycles and Changes

As we study integrated systems such as the garden, we see that change is always taking place. However, the more we observe, we see that the changes repeat in cycles. Systems—whether a human being, an individual plant, the garden, or the whole planet—conserve their limited resources by recycling them.

In this unit, cycles that are observable in nature and critical to the survival of life are studied. As change repeats itself and cycles emerge, patterns of birth, life, growth, death, and decay become evident. Composting is introduced, and offers students the chance to mimic patterns of nature. Reusing organic wastes as the source of nutrients for the garden creates a bridge between nature's cycles and those that can be part of our daily lives.

Students will collect data on changes and cycles, from seasons to oxygen-carbon dioxide exchange to making recycled paper. Advanced preparation is needed for "Let's Make a Compost Cake," and compostable materials must be collected. The experiments that involve observing changes will take longer to complete. They include "Dr. Jekyl and Mr. Hyde," "You Look Different from the Last Time I Saw You," "Bring in the Clean-Up Crew," and "Compost Bags."

As scientists, students observe and record data regarding change and cycles. As we all grasp cycles and their importance in maintaining life, we can apply this profound knowledge to our daily lives.

Activities	Recommended Grade Level
The Power of the Circle *(introduction of cycles)*	2,3,4,5,6
Adopt-a-Tree *(observation of seasonal changes)*	2,3,4,5,6
The Cycle Hunt *(study of seasons)*	2,3
Me and the Seasons *(seasonal differences)*	2,3
Collector's Corner *(categorization by seasons and creating season display)*	2,3,4,5,6
Roundabout *(water cycle)*	2,3,4,5,6
What a Deal *(observation of the exchange of oxygen and carbon dioxide between plants and animals)*	4,5,6
Dr. Jekyl and Mr. Hyde *(observation of how substances change)*	4,5,6
You Look Different from the Last Time I Saw You *(observation of decomposition and nutrient cycling of different foods)*	4,5,6
Bring in the Clean-Up Crew *(observation of rates of decay)*	4,5,6
Compost Bags *(introduces the idea of biodegradable and non-biodegradable)*	2,3,4,5,6
Teacher Background: Composting *(the how and what of compost)*	
Let's Make a Compost Cake *(making compost)*	2,3,4,5,6
The Cycle of Recycle *(personal recycling habits survey)*	3,4,5,6
A Human Paper Factory *(making recycled paper)*	3,4,5,6
Mother Earth *(Native American writings)*	4,5,6

River Song

D G D A7
It happened one day on the mountain so high

D G D A7 D
A river was born from out of the sky.

D G D A7
The rain and the snow came falling down

D G D A7 D
And they started to run as they hit the ground.

Chorus

D G D A7
Blurp ah pashosh rumbly pound,

D G D A7 D
A white rapid river makes a wonderful sound.

Over beds made of granite it swept and it rolled,
It was narrow and steep and so icy cold.
It carved out a valley and gouged out the land.
It carried small rocks and ground them to sand.

Chorus

It filled up a lake and was still for a day,
But soon the wide river went along on its way.
It rolled past rocks and banks lined with trees.
It carried small boats of fall-colored leaves.

Chorus

It wound and it wound until it wound past me,
And I knew it was happy, it was wild and free.
I knew it was happy, it was wild and free,
But I waved it goodbye as it entered the sea.

Chorus

The water in the sea soon rose to the sky
And the wind blew a cloud to the mountain so high.
The rain and the snow came falling down
And flowed to the river as they hit the ground.

Chorus

Tapes available through *Let's Get Growing.*

The Power of the Circle

Description

Students color drawings of different cycles in nature.

Objective

To introduce the concept of cycles and identify cycles in nature.

Materials

One copy of the Water, Nutrient, and Oxygen cycle poster per student (Appendix pp. 398-400)
Crayons or markers
Drawing paper

(Write the word *circle* on the chalkboard.) What is this word? What is a circle? (Write the word *cycle* directly beneath it.) What is this word? (Replace the *ir* in *circle* with a *y*.) The word *cycle* comes from an old word meaning "circle." Cycles are repeating circles.

(Draw a circle on the chalkboard and begin tracing slowly around it.) When things happen in a cycle they continue to happen in a particular order until they are back where they started, and then (retrace the circle) they start around all over again. The four seasons happen in cycles. Who can name them in order? Does it make any difference where you start? Why not? (Write the seasons around the circle.)

What other things happen in a cycle? (possible answers: clock time, phases of the moon, tides, insect life, human life, paper and glass recycling, and so on) Discuss how these things happen in a cycle that repeats itself.

Circles or cycles are very important in many cultures. Native Americans are very aware of the significance of circles. Black Elk was an Ogalala Indian Spiritual Man who lived in the late 1800s. This is something he wrote about the importance of circles or cycles:

"You have noticed that everything an Indian does is in a circle, and that is because the Power of the World always works in circles, and everything tries to be round. In the old days when we were a strong and happy people, all our power came to us from the sacred hoop of the nation, and so long as the hoop was unbroken, the people flourished. The flowering tree was the living center of the hoop, and the circle of four quarters nourished it. The east gave peace and light, the south gave warmth, the west gave rain, and the north with its cold and mighty wind gave strength and endurance. This knowledge came to us from the outer world with our religion. Everything the Power of the World does is done in a circle. The Sky is round and I have heard that the earth is round like a ball and so are all the stars. The Wind, in its greatest power, whirls. Birds make their nests in circles, for theirs is the same religion as ours. The sun comes forth and goes down again in a circle. The moon does the same, and both are round.

Even the seasons form a great circle in their changing and always come back again to where they were. The life of a man is a circle from childhood to childhood, and so it is in everything where power moves. Our teepees were round like the nests of birds and these were always set in a circle, the nation's hoop, a nest of many nests where the Great Spirit meant for us to hatch our children."

1. Have students design and draw their own Seasons poster showing the repeating cycle.
2. Use the Oxygen, Nutrient, and Water posters to introduce other cycles in nature. Discuss these cycles and their importance to our garden and ourselves:

   Water Cycle—Water is part of a very important cycle. All the water that we will ever have is already part of that cycle. Moisture evaporates from the earth and accumulates in the sky as clouds. Then it falls, returning to the land where it again evaporates. Without this constant recycling, we would quickly run out of water.

   Nutrient Cycle—When plants die and decay, the nutrients that the plants took out of the soil are put back into the soil to be used again and again. This is another cycle. Without this cycle all the nutrients in the soil would soon be completely used up and no more plants could grow.

   Oxygen Cycle—People breathe in oxygen in order to live. They breathe out carbon dioxide. Plants breathe in carbon dioxide and breathe out oxygen. We make an exchange with the plants. We need the oxygen that they produce, and they need the carbon dioxide that we and other animals produce.
3. Have students color in each of the cycle posters.

Name a cycle that you use everyday. Are there any cycles in nature that are important to you? What would happen to the water cycle if we were to use up most of the water? What would happen to the oxygen cycle if the air were polluted? How can we help keep nature's cycles healthy?

Adopt-a-Tree

Description

Students make drawings of a selected tree throughout the school year.

Objective

To observe seasonal changes in the life of a tree.

Materials

Life Lab journals

Have you ever noticed a tree on or near the school grounds that you really like? Does it look different at different times during the year? If you were to keep a diary about this tree, what could you write or draw?

1. Collectively or individually, have class members adopt a deciduous tree on or near school property. Have them observe their trees throughout the school year, keeping a log of changes and observations made every other week if possible.
2. Have students first try to get a feeling for their trees by answering the following questions: How tall is your tree? How wide? How many students does it take to hug a ring around the tree? What are the color and texture of the bark? What are the color, texture, and shape of the leaves? Do they smell? What sound does the wind make in the tree?
3. Have students draw pictures of their trees as they look now, including themselves in the picture. Then have them draw a picture of what they think the tree will look like at the end of the year and include how they will look, too.
4. Remind students that as they observe their trees throughout the year, they should be careful to note any changes, such as when leaves fall, when buds form, when birds visit, when fruit appears, whether the fruit has holes and what caused them, how the smell of the tree changes throughout the year, and so on. Suggest that students think of the tree as an apartment building and describe who lives on the ground level, in the upper stories, and any other observations of insects, lichen, moss, or other life.

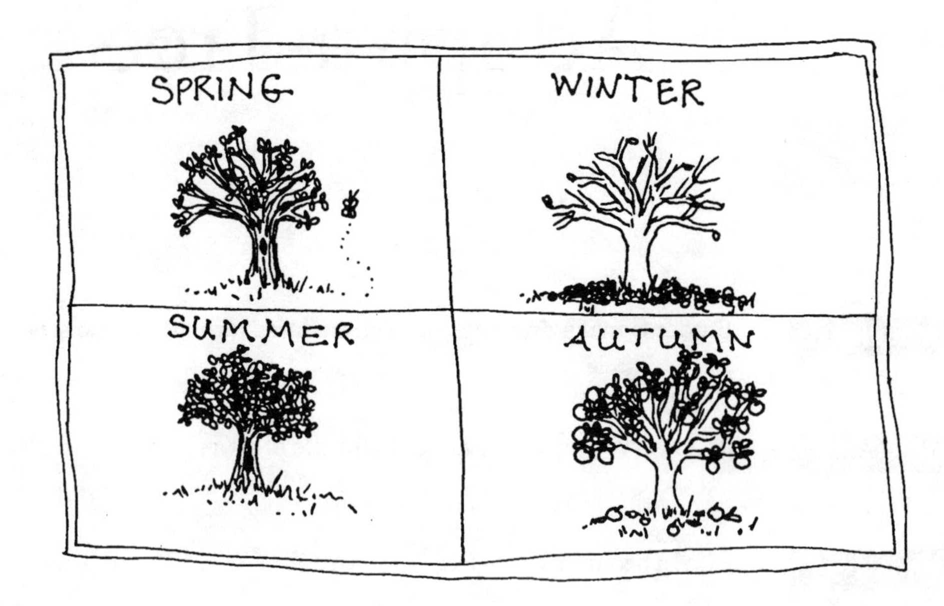

Do you think your tree will change during the year? What will it look like this time next year? Is the tree part of a cycle? Will you change during the year? What are some special things you will observe about your tree during the year? How will you keep records of your tree?

In the spring, have your class celebrate trees by planting one in your garden. A local nursery can help you select one that will grow well in your climate. Be sure students learn all about it and know how to help it grow.

The Cycle Hunt

Description

Students will participate in an outdoor scavenger hunt to learn about the four seasons.

Objective

To recognize attributes of each of the four seasons by categorizing items into the proper season.

Materials

One object or picture of an item representing each of the four seasons per student, with an equal number for each season: snow boot, bird's nest, flowers, buds on trees, pictures of snow, children swimming (if objects identified with seasonal holidays are used, be sure to include a variety of cultural holidays)
Four large sheets of butcher paper
Seasonal labels for each student

Preparation

Hide objects and pictures in the garden or other appropriate outdoor area.

(Discuss weather and weather changes for your region.) What is the current weather like? How is it expected to change through the year? Last year at this time was the weather like it is now? Can you predict what the weather will be like one year from now? Is it possible that our weather has its own cycle?

(Describe the seasons.) We divide the year into four seasons and we can tell the seasons by the weather changes that take place. (Use characteristics that are appropriate for your weather to describe the seasons: winter, December 21 - March 21; spring, March 21 - June 21; summer, June 21 - September 21; autumn, september 21-December 21. Write the names and dates on four large sheets of butcher paper, along with the names of the other months. Every year these seasons repeat themselves. Thus our seasons and our weather have a yearly cycle.

1. Divide the class into groups of four, and assign a different season to each member.
2. Give each member a label indicating the season he or she represents.
3. Have groups go into the garden or other selected area on a scavenger hunt to collect the hidden items representing the seasons. You may want to send groups out at separate times with an aide to encourage language development. Groups should follow these procedures:

- Students must stay with their groups.
- Each member of the group should find and collect one object that relates to the group's season.
- Group members may help each other.
- Each member must agree that the object found matches the season.

4. When a group has all four objects, have students return to the room and place each object on the appropriate piece of butcher paper.
5. When the groups are finished, have the entire class observe how the objects were categorized. Are there any that could be in another season?

Which season comes after spring? After summer? After fall? After winter? Describe how a fruit tree may differ in each season. What is the season cycle? How does knowing the season cycle help you plan your garden?

Make a seasonal calendar with students. Have students write the 12 months and draw a picture that symbolizes the season next to each month.

Me and the Seasons

Description

Students construct a pictorial wheel depicting seasonal differences in their activities, clothing, and environment.

Objective

To reinforce the cycle of seasons and identify different clothing and activities that correspond to the different seasons.

Materials

Two sheets of construction paper per student
One set of blackline masters for wheel parts per student, Appendix p. 401 and p. 402
One pair of scissors per student
Paste
One brad per student
Crayons or colored pens

Preparation

Make copies of the two blackline masters so that each student has both.

Ask students to name the four seasons (write them on the chalkboard). What kinds of weather are characteristic of our region during each season? Name some activities you do during each season. Discuss with class how weather for different seasons varies around the world.

1. Give each student two sheets of construction paper and a copy of each blackline master. Have them cut out the circles and paste them onto the sheets of construction paper.
2. Have students trim off excess construction paper, leaving a 1 cm border around the edge.
3. Have students cut out the window in circle 2.
4. Have students color in the pictures illustrating each season on circle 1. These pictures represent activities children in different parts of the world do during each season.
5. In the blank below the picture, have students draw an activity they do during that particular season.
6. Have students place circle 2 over circle 1 and insert the brad through the center.
7. Give students time to explore and play with their wheel.

Ask students how the weather seems to be different in each picture on the wheel. How is the clothing different? What different activities are shown? Why are the seasons in a circle?

Collector's Corner

Description

Students will gather natural objects from the garden and school yard throughout the year to create an ongoing season display in a corner of the classroom.

Objective

To be able to categorize natural objects according to the appropriate season.

Materials

Corner of the classroom with display board and table
Paper for decorative background
Lettering to label the current season
Materials gathered from the garden, school yard, or local area that are characteristic of the current season: acorns, twigs, leaves, flowers, rosehips, cast off insect exoskeletons, abandoned cocoons or nests

Preparation

Create the basic structure for the "Season Corner" by decorating the display board and labeling it for the current season. You may also want to put up a few sample objects.

What changes in weather have you observed over the past year? We divide the year into four seasons based on these kinds of changes.

Can you name some objects that you would expect to find outside during the present season? Were those objects here last year at this time? Do you expect them to be here next year? Why?

1. Accompany the students into the garden or school yard and ask each to find one object that is characteristic of the current season.
2. Return to the classroom and incorporate these objects into the "Season Corner" display.
3. You might want to use a grouping exercise to help the students organize their treasures. For example, material can be grouped according to whether it is a plant, animal, or mineral. Subgroupings could be made for the plant material, such as leaves, twigs, and seeds.
4. Suggest that students continue to observe the changes in their environment throughout the year. Encourage them to continue to bring in new objects they might find around them. Especially ask them to replace objects that have changed with new samples. For example, a closed pine cone put up in the fall will be replaced by an open or decomposing cone later in the year.

5. Be sure to have students return display objects to the environment where they were found when they are no longer appropriate in the display. Explain that these things are useful to plants and animals in the local ecosystem and shouldn't be wasted.

In what ways are the objects part of the current season? How would this object have changed if left outside? What have you found that surprised you?

1. Plan a monthly excursion into the garden or school yard to renew the display. When students bring in new objects, ask them to share information with the class about where the object was found and what kind of change it has been undergoing.
2. Invite a naturalist to class to talk about how the local seasons affect plants and animals in your area.

Roundabout

Description

Students build a terrarium to observe the water cycle and create a bulletin board to illustrate it.

Objective

To observe, describe, and illustrate the water cycle.

Materials

Large glass container: aquarium, fish bowl, large pickle jar placed on its side
Soil
Sand
Pebbles
Untreated charcoal
Small cup
Tight-fitting cover: plastic wrap, pane of glass, lid
Small plants: ferns, houseplant seedlings
Humus, moss (optional)
Spray bottle
Materials for collage: cotton for clouds, tinsel for rain, leaves and branches, small toy animals or toy people for biosphere, construction paper, and so on
Life Lab journals

What is a cycle? Name some cycles that are part of your life. (day/night, seasons, breathing, time, number cycle, nutrient cycle) Which cycles take a short time to repeat? Which take a long time? (Make a chart on the chalkboard.) Why is it that the world doesn't run out of water? Why don't the oceans and lakes dry up, as puddles do? Where does rain come from? Maybe water goes in a cycle, too, over and over again. Let's find out in this activity.

Have students help complete the following steps to build a terrarium:

1. Wash a glass container. (Two or three containers will allow for more student involvement.) Dry it thoroughly.
2. Place gravel and pebbles in the bottom.
3. Add a layer of small pieces of charcoal.
4. Put in a small cup filled with water and surround it with soil 2″ deep.
5. Give shape to the landscape by mounding the soil.
6. Plant small ferns and other plants (evergreens grow well) in the soil.

7. Decorate the terrain with moss, driftwood, twigs, rocks.
8. Moisten the soil and plants lightly using a spray bottle.
9. Cover tightly.
10. Place in indirect sunlight.

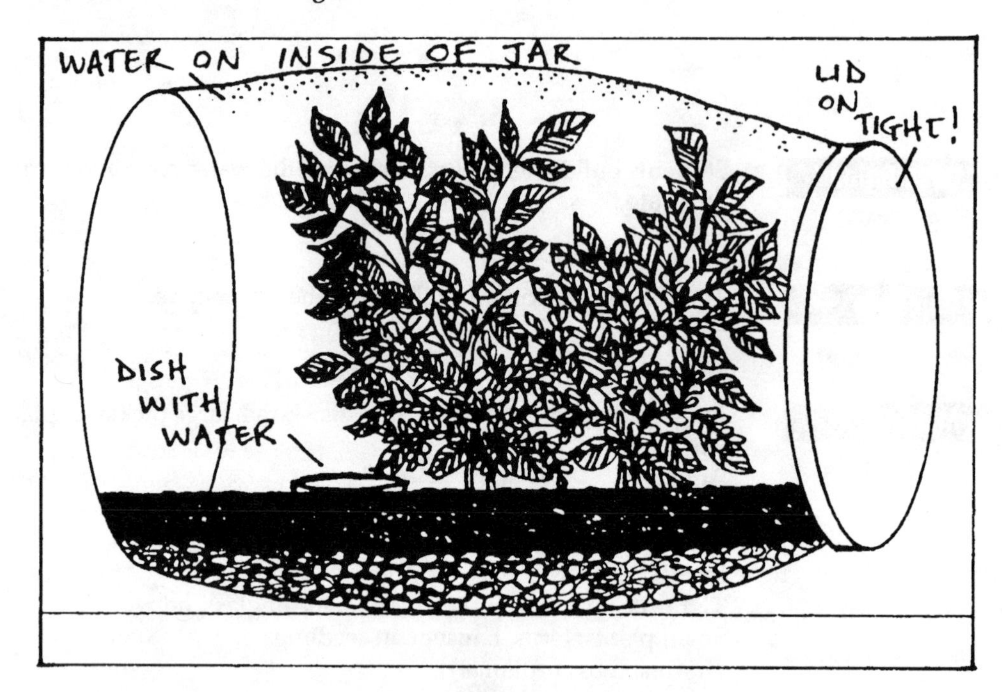

11. Gather the students around the terrarium. Ask if any water will be able to get in or out. How will the plants get the water they need in order to live? (Record predictions.)
12. Keep a student log book of observations next to the terrarium. Each day for the next several days, discuss any changes that the students observe in the terrarium. Summarize observations and relate them to concepts of the water cycle.
13. Have students create a water cycle collage for the bulletin board.

What makes the water cycle run? (sun, gravity) What would happen if the sun's energy was blocked from the earth? How is water used? Does all water return to the cycle?

1. Have the students make small, individual terrariums by taping together two clear plastic cups.
2. Host a water cycle celebration, with song, dance, and freshly collected water to drink.
3. Students can make individual water cycles by placing a seedling with soil in a small zip-lock plastic bag, adding a little water, and sealing. Use clothespins to hang the bags from a clothesline in a classroom window.

What a Deal

Description

Students will role play the oxygen-carbon dioxide cycle by riding a bicycle attached to a tree. (This activity is best done one small group at a time.)

Objective

To understand that green plants release oxygen and consume carbon dioxide while animals release carbon dioxide and consume oxygen.

Materials

Tree
Old bicycle positioned so that the rear wheel is free to rotate, as in an exercycle
Plastic tube with a cup attached to the bike and connected to tree leaves

Preparation

Set up the bicycle so that it is attached to the tree and to the plastic tube. (See illustration.)

Why do we need air? (to breathe) What is air made up of? (different gases such as oxygen, carbon dioxide, nitrogen) Is there air in outer space? (no) If our air became very polluted or if all of the oxygen were used up, would there be a way to get more air? (no) There is an important cycle that takes place between plants and animals that helps to keep a balance between oxygen and carbon dioxide in the air. Isn't it amazing that while we need oxygen from the air, plants give oxygen to the air; and while plants need carbon dioxide from the air, all animals, including us, put carbon dioxide in the air?

(Draw this cycle on the chalkboard.) What does this cycle indicate? (As long as we have plants and animals, there will be enough carbon dioxide and oxygen in the air; that the gases get used over and over again.)

1. Take part of the class to the bicycle tree. Ask, Who would like to play "What a Deal?" Point out the special breathing tube and cup that connects the rider to the tree leaves.
2. Have one student begin to pedal the bike, explaining that he or she must breathe into the tube connected to the tree. The student is exhaling carbon dioxide to the tree and receiving oxygen back from the leaves.

3. Ask the students why this connection with the tree is so important. What do we give to the plants? (carbon dioxide) What do the plants give us? (oxygen) Could we survive without plants? (no) Explain that this exchange is the greatest deal on earth.
4. Continue to have students ride the bike. Ask each rider a question relating to the oxygen-carbon dioxide cycle: What are you breathing out? What are you breathing in? Where does the oxygen come from? What does the plant make with your carbon dioxide? What does the plant release?

Why are plants important to us? Why are animals important to plants? How do we keep enough oxygen and carbon dioxide in the air? How do we change the number of plants on earth? (logging, cities, roads) If we run out of air, can we make it in a factory?

1. Have students read *The Giving Tree* by Shel Silverstein and *The Lorax* by Dr. Seuss.
2. See related activity, Magical Mystery Tour, p. 137.

Dr. Jekyl and Mr. Hyde

Description

Students observe various substances to determine whether they change over time.

Objective

To observe how different substances change.

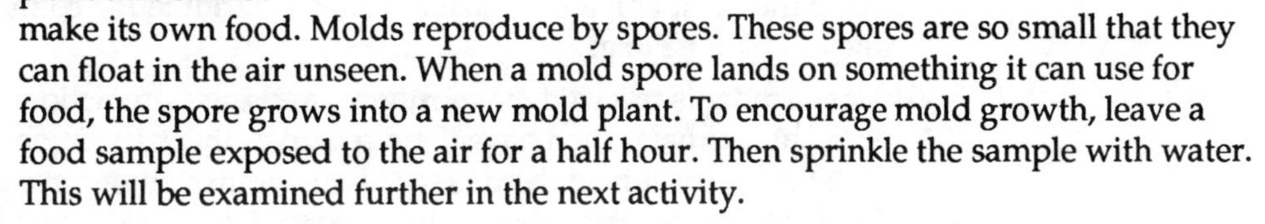

Teacher Background

Students will observe change in organic and inorganic substances. Molds are certain to grow on some of the foods. Mold is a nongreen plant that cannot make its own food. Molds reproduce by spores. These spores are so small that they can float in the air unseen. When a mold spore lands on something it can use for food, the spore grows into a new mold plant. To encourage mold growth, leave a food sample exposed to the air for a half hour. Then sprinkle the sample with water. This will be examined further in the next activity.

Materials

Clear containers with lids
Clear plastic bags
Butcher paper
Markers
Life Lab journals

Preparation

After discussing the topic and making predictions, students will collect substances from home to test in the activity.

What does it mean when something changes? How do you know something has changed? Can you give examples of changes occurring around you? (Examples: It was sunny this morning, now it is cloudy; I'm two inches taller than I was last year; A candle burns down to a puddle of wax; An ice cube melts.) Can you think of something that never changes?

1. Label a sheet of butcher paper: PREDICTIONS. Have students make three lists on the butcher paper: **1.** THINGS THAT WILL CHANGE **2.** THINGS THAT WILL NEVER CHANGE **3.** THINGS THAT MAY OR MAY NOT CHANGE. Possible responses are shown below.

1.	**2.**	**3.**
cereal	sand	pepper
orange peel	glass	salt
leaves	rubber	paper clips
wax	paper	mustard
butter	steel wool	dried apple
bread	pencil	
	leather	

2. Record students' predictions. Do not correct their placement of an item. Let them discover what will happen through observation. Post the list in the classroom.
3. Ask students to bring in a variety of substances from home, including those they listed, along with liquid and solid food, hardware, seeds, cloth, and anything else they think of and can easily obtain. Encourage them to bring in things they are not sure about. Leftover items from their school lunches would make good samples.
4. As students bring in samples, place them in separate transparent containers and label them with name of substance and date. If the substance is not on the list, have students add it to the column where they think it belongs.
5. Have students record their substances and their predictions in their Life Lab journals. Then have them record daily observations for each substance. Encourage them to try to distinguish among the different kinds of changes they observe in order to determine some of the possible causes and to relate certain kinds of changes to the kinds of substances.
6. Continue the activity until interest begins to wane. Then have students draw conclusions regarding their samples and compare their observations with their original predictions.

Which of the substances changed in the way you predicted? Which ones surprised you? Did certain types of substances change while others did not? Do you have any ideas about why the food substances changed? (part of the decomposition in the nutrient cycle)

You Look Different from the Last Time I Saw You

Description

Students examine mold and decomposition of food.

Objective

To demonstrate decomposition and nutrient cycling.

Teacher Background

Decomposition of plants and animals is the work of decomposers, countless billions of small animals and plants that live in the soil, air, and water. Many of them are so small that they can be seen only with a powerful microscope. One type, called *bacteria*, is so tiny that one spoonful of soil can contain more of these creatures than there are people on earth. Fungi and molds are another type of decomposer. They are much larger than bacteria and sometimes can be seen with the naked eye.

Decomposers use dead plants and animals for food. When something dies, the bacteria, fungi, and molds that happen to be present start eating it. They eat and grow and multiply so rapidly that in a very short time millions of them are working on the dead plant or animal. It is their eating that causes what we call *decay* or *decomposition*.

When things decay, they form a material called *humus*. Humus is rich in minerals and other nutrients. The humus becomes part of the soil.

Materials

One petri dish with cover per group of four
Soil
Small pieces of fruit, vegetables, bread (1/2 cubic inch)
Water
Masking tape
Life Lab journals
Hand lenses
Microscope (optional)

Have you ever noticed what happens to leaves when they fall off the trees or to the bodies of birds and animals when they die in the forest? Have you ever taken a walk in the woods and come across an old tree stretched across the path, with moss and mushrooms growing on it and hundreds of spiders and bugs making their home in it? Then you have seen the beginnings of the way in which an important part of the soil is made. This part, which comes from dead plants and animals, is called *organic matter*. How do you think these dead plants and animals become soil? (Record ideas.) What do you think organic matter adds to the soil? (Record ideas.) Let's set up some demonstrations and see what we observe.

1. Give one petri dish to each group of four students.
2. Have students place a thin layer of soil on the bottom of the dish.
3. Have them put three pieces of the same food on top of the soil in each dish. Moisten the food and the soil with water.
4. Have students cover the dish.
5. Be sure students label each dish with food type, the date, and group name or number.
6. Have students remove the cover for a few seconds every couple of days to let in some air.
7. Have students observe the dish daily, using a hand lens or microscope or both for closer observations.
8. Make a class chart as illustrated and have students record detailed observations in their journals.

DECOMPOSERS CHART

NAME	FOOD	STARTED	FIRST CHANGE	CAN'T RECOGNIZE	DISAPPEARS
Carla	bread	2/3	2/7	2/14	2/23
Alex	lettuce	2/3	2/5	2/10	2/20
Erin	apple	2/3	2/4	2/15	

Which food started to change the quickest? Which food had the most visible decomposers? Why do you think decomposed organic matter is valuable to soil? Next time you see mold on food, how will you react?

Bring in the Clean-Up Crew

Description

Students bury various objects in the ground and dig them up weekly to observe changes.

Objective

To observe the rate of decay of various materials.

Materials

Various decomposing and nondecomposing materials: metal, glass, plastic, rubber, vegetables, bone, wood, paper, rope, leather, feathers, and so on
Shovel
Markers: sticks, stones

Preparation

Gather a variety of objects that will and will not decompose in the ground. Display them during the discussion.

Have you ever found an object that was buried in the ground? Could you recognize what it was? Do you think the soil made it change faster? Let's list these objects according to which will change the most in the ground to those that will not change at all. Now guess what we will do with the objects.

1. Have students take the variety of different materials you have gathered and bury each item in a different hole, all at the same depth. Have them mark each spot with a stick or a stone.
2. Once a week, have students dig up the items and record how fast and in what ways each item is decaying. (If you think students might damage items when they dig or might have difficulty finding each one, put a screen over the material before putting the soil in the holes).
3. Expand the activity by placing the same or similar items on the surface of the soil and have students compare rates of decomposition.

What factors affected the speed of decay of the various objects? How did the results compare with your predictions? What do you think happens to materials at the garbage dump?

Compost Bags

Description

Students manage natural decomposition by creating their own compost bag.

Objective

To introduce composting as a way of managing the nutrient cycle.

Teacher Background

This project will stimulate students' curiosity and serve as a good motivator for making garden compost. Prepare the bag with them in Part One without telling them what it is for. Let the students speculate. Start this project one month prior to making compost in the garden. This will allow the organic substances in the bag time to decompose.

Note: Before closing the bag, breathe air into it. Decaying matter smells if it does not have enough air and if the anaerobic bacteria that create the smell are active.

Materials

Part One
One large plastic garbage bag with closing tie
One gallon of wet soil
Three grapes
Five pieces of plastic cup
One handful of grass clippings
Three leaves of lettuce
Two nails
One slice of white bread
Three squares of wet toilet paper
One slice of whole wheat bread
Life Lab journals

Part Two
Sifting screen
Walnut shell
Apple

I have the ingredients for a very special recipe. When you mix ingredients do you always know what the results will be? Let's see if you can guess while we put these ingredients together.

Part One:

1. Take the large plastic bag and put a gallon of wet soil in it.
2. Add the rest of the materials.
3. Mix all the ingredients well into the wet soil, so that they are distributed throughout the bag.
4. Blow the bag up with air, and then close it tightly.
5. Record on a card what was put in the bag and the date it was sealed; tape it to the bag.
6. Hang a sign on the bag: What's going on in here?
7. Put the bag in an out-of-the-way place in the room.
8. Ask the students to write in their journals predictions about what will happen, or draw what they think the contents will look like in a month.

Part Two:

One month later, complete the following steps outdoors:

1. Ask students to read the list of materials that were put in the bag.
2. Ingredient by ingredient, ask students to hypothesize as to what happened to each.
3. Open the compost bag in the garden.
4. Pass the contents through a screen.
5. Have students observe the condition of the ingredients.
6. Ask students to refer to their original predictions made in the classroom and draw conclusions. What changes occurred?
7. Discuss decomposition, decay, and nutrient recycling.
8. Introduce composting as a way to put nutrients back into the soil using natural decomposition. Explain that composting takes nature's process of recycling nutrients and accelerates it. A well-made compost heap creates an environment in which decay-causing bacteria can live and reproduce at the highest rate of activity. As a result, fresh manure, food scraps, leaves, seeds, wood ashes, sawdust, and other compost materials are converted into dark humus.

Hold up a walnut shell and an apple. Ask, How could this walnut shell become an apple and become you? (The walnut shell is dropped under an apple tree. The walnut shell decomposes and adds nutrients to the soil. The roots of the apple tree absorb the nutrients, some of which go into the fruit of the tree. You come along and pick and eat the apple.) How could the apple core become part of you?

TEACHER BACKGROUND
Composting

"Behold this compost! Behold it well. . . . It grows such sweet things out of such corruptions. . . ."

Walt Whitman

Composting Defined

Compost is a mixture of decomposed vegetation that is used to improve soil structure and provide plants with necessary nutrients for growth and development. Composting is the art of ecologically reusing waste. When making a compost pile, we are mimicking the nutrient cycle in nature. We are promoting the biological decomposition of organic matter under controlled conditions and demonstrating the concept of cycles and changes.

Green Acres School, Santa Cruz, CA

What Occurs

Decomposition is the result of the efforts of billions of microorganisms, mainly bacteria and fungi, which eat the highly organized matter and in so doing break it into smaller, simpler molecules that become available as nutrients for plants. As they eat their way through the compost, they give off heat. This heat speeds the decomposition process and can be felt and measured by the students.

Green Acres School, Santa Cruz, CA

Why Compost?

Compost allows students to see that what we call waste may be nutrients in disguise. Organic matter (food wastes, weeds, leaves, manure) can be added to the compost pile and retrieved a few months later as valuable fertilizer. Composting gives students an opportunity to experience the nutrient cycle and create abundant fertilizer that builds the garden soil—and is free!

Composting Methods

There are two basic methods for making compost, the fast method and the slow method. The fast method produces compost in 2-3 weeks. The pile is built and then turned every few days, so that the outside of the old pile becomes the inside of the new pile. This method speeds up the decomposition process. The fast method is labor-intensive and is not advisable for Life Lab school gardens. We recommend the slow method, which takes 2-4 months to produce a useable product. In this method, a layered pile of organic matter is built, and is left to decompose until it is ready for use in the garden.

Collecting Materials

Traditionally, compost is made in the fall when there is an abundant supply of dried materials. However, building compost piles throughout the year will provide your Life Lab with compost year-round. The more materials you gather, the more compost you can make. You may want to schedule a school-wide composting day and gather materials in advance. Most decomposable materials are useable. Good examples include kitchen vegetable scraps; crop wastes; and straw, chicken, goat, horse, and cow manure. DO NOT INCLUDE human, dog, or cat feces; animal parts; noxious weeds; plants with resins; meat or fish; greasy foods; toxic materials. There are all kinds of compost materials around just going to waste! Check stables for manure and straw and produce companies and grocery stores, restaurants, and the school cafeteria for food wastes. Check to see whether there are any composting guidelines or restrictions in your community before you begin.

Materials for the compost pile may be divided into three categories:

Carbon Dried Matter	**Nitrogen** Fresh Matter	**Soil/Minerals**
dried leaves	kitchen scraps	soil (introduces the micro-organisms)
straw	manure	minerals (adds nutrients)
dried grass	lawn clippings	
branches	leaves	
	weeds	

Be sure to include plenty of materials from all three categories.

Building Your Compost Pile

Carbon + Nitrogen + Soil/Minerals + Air + Water = Compost

Compost is made by layering carbon and nitrogen materials and soil in alternating 4″-6″ layers.

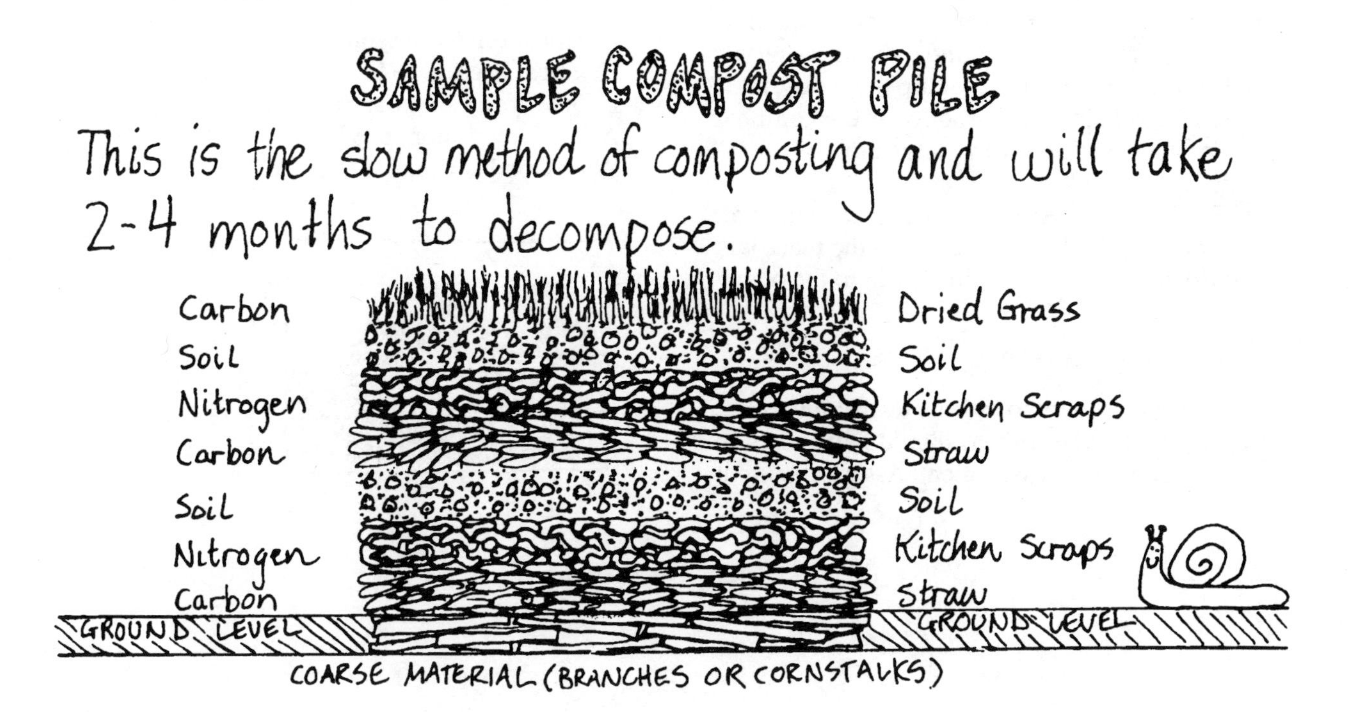

SIZE OF THE PILE The pile should be a minimum of 4′ X 4′. This size is essential for optimum decomposition and adequate heat retention. The maximum height should be five feet; any taller than this, the pile becomes too compressed and is deprived of air. It can be as long as you want to make it. When the pile is properly built, it will heat up to 160° F and destroy any pathogens.

SHAPE The pile should be rectangular. This is assured by constantly forking materials out to the corners and edges with each added layer. Otherwise, the pile will become a giant mound or pyramid, which will allow heat, moisture, and nitrogen to escape.

AERATION The process of composting is most efficient when aerobic decomposition is taking place. This is decomposition in the presence of air. If the pile is too dense or wet, anaerobic decomposition will occur and produce a strong smell. Incorporate coarse and bulky materials to help air circulation. If anaerobic decomposition develops, turn the pile and add coarse material.

MOISTURE Moisture goes hand in hand with aeration. The pile should be the consistency of a squeezed-out sponge. A soggy pile will encourage anaerobic decomposition. If the pile gets too wet, add straw and porous materials and mix them in well to soak up the moisture. If rain is a problem, put a roof over the compost area or cover the pile with a large tarp.

SIZE OF MATERIALS The smaller the material, the faster it will decompose. Chop the materials up with a spade while adding them to the pile.

Involving Students

Try having students work in teams: Nitrogen, Carbon, Soil. Each student should have a tool in hand, and have an assignment. Assign a student to oversee the watering of each layer. Remind the students that the pile needs to have very straight walls, and the shape needs to be rectangular. Some students can be responsible for shape. Team students can always be chopping the materials so that they decompose faster. (See activity, Let's Make a Compost Cake, p. 199.)

Ready-to-Use Compost

In general, the pile has finished decomposing when the compost

- is dark brown and looks like soil
- is not made up of easily recognizable ingredients
- has an earthy, humus-like odor

The exterior of the pile will not fully decompose. Check at least six inches into the interior of the pile to observe its characteristics. The entire process will take 2-4 months.

Once the pile is decomposed it should be sifted through a heavy gauge wire screen. It is then ready to be used in soil mixes, garden beds, or used as a fertilizer around trees, shrubs, and perennial borders. Be generous. An adage recommends that the less your soil looks like compost, the more compost you need to add!

Compost related lessons:

Compost Bags, p. 194
Let's Make a Compost Cake, p. 199
A Soil Prescription, p. 88
What Good Is Compost?, p. 91

Let's Make a Compost Cake

Description

Students build a compost pile.

Objective

To experience the process of decomposition and the nutrient cycle.

Teacher Background

See detailed description of composting, p. 196. Be sure students wash hands well when done with this activity.

Materials

Compost materials
Shovels and spading forks
Wheelbarrow
Water access and hose with fan spray nozzle
Meter stick
Soil thermometer
Life Lab journals

Preparation

1. Select a permanent compost area for the garden. The ideal location is close to the garden for easy hauling as well as easy access.
2. Collect composting materials.

What types of materials decompose? (materials that have been alive) Why is it important for these materials to decompose? (they become nutrients for other plants) Is this a cycle? What are the parts of this cycle? (living plant or animal grows, dies, decomposes, provides nutrients for another living plant or animal to grow) What is the cycle called? (nutrient cycle) Do you think we can create a nutrient cycle in our garden? (Record predictions.)

1. Demonstrate building a miniature compost cake with samples of carbon, nitrogen, and soil prior to building the actual pile. Discuss the different ingredients that can be used in the pile. Stress the importance of the size, ingredients, and moisture level.

2. Divide groups of up to ten students at a time into teams of Carbon, Nitrogen, Soil. Assign one student to be waterer. Equip students with shovels, spading forks, and a wheelbarrow. Rotate groups after one round of layering carbon, nitrogen, and soil or let each group build their own pile. (You can never have too much compost!)
3. Using their spading forks, have students loosen the ground where the pile will be. The area should be a minimum of 4′ X 4′.
4. Have students begin layering materials in the following order, lightly watering each layer:
   - Stalky material first (for drainage)
   - Straw or dried plant matter (carbon layer)
   - Manure, weeds, or kitchen scraps (nitrogen layer)
   - Topsoil or old compost (bacterial activators)
   - Carbon layer
   - Nitrogen layer
   - Soil layer
   - Repeat layers until the rectangle is three feet high
5. Be sure students maintain the rectangular shape of the pile and keep the corners square. Like the foundation of a house, each layer becomes the base for new layers, and if they're not square, the pile will collapse and the heat needed for decomposition will be lost.
6. Have students measure and record the dimensions of the compost pile.
7. Have students make a hole toward the center of the pile and take the temperature.
8. Have students draw the compost cake in their journals, recording layers, measurements, and temperature.

What are the ingredients of a compost cake? What will happen to the organic matter? What will the pile look like in a few months? How will the compost be useful after it is decomposed? What materials could you use at home to make compost?

1. Record the temperature of the compost cake each day for the next week and put the readings on a class graph. The pile will heat up to approximately 160° F and then start to cool down. Let the students feel the heat from the pile. Discuss how the heat is being produced through the biological activity of the microorganisms.
2. In a month, measure the dimensions of the pile again. Ask, How has it changed? What layers can you identify?
3. Have students observe chunks of compost through a microscope and record what is seen.

The Cycle of Recycle

Description

Students complete a self-survey to determine how much they recycle.

Objective

To illustrate how the garbage explosion is a serious problem, and how our daily habits help or hurt the use, misuse, and reuse of materials.

Materials

A Short History of Trash poster, p. 405
Self-survey for each student, p. 403

Preparation

Prepare a copy of the poster and the self-survey for each student.

Imagine that your life is part of a cycle. Describe how this is possible. If everything you use during a day is part of this cycle, too, can things really be thrown away? What happens to something that is put in the trash? (Record ideas.) Every day we use materials—paper, metal, glass, plastic, wood—and every day we make decisions about whether to throw away or recycle those materials. Often we don't even think about how we are using our resources; we take them for granted and just throw them away out of habit.

1. Have students complete the self-survey to find out what their personal habits are in the area of garbage and recycling.
2. When the surveys are completed, add up the points and then read the Self-Survey Results for more information. Discuss the questions and students' responses.
3. Have students redo the survey using a different colored pen or pencil to circle the responses they would give if they could change their habits. Ask, How could you be convinced to recycle more and throw away less?
4. Have students study the poster, A Short History of Trash. Ask, Why has the amount of garbage people make increased so much in the last 30 years? List responses on the chalkboard. Explain that the average American throws away over one ton of trash each year. In the state of California, people throw away 146 million tons of trash every year. That much trash would make a solid block of trash ten feet tall and fill two freeway lanes all the way from Oregon to Mexico. And that's only one year's worth.

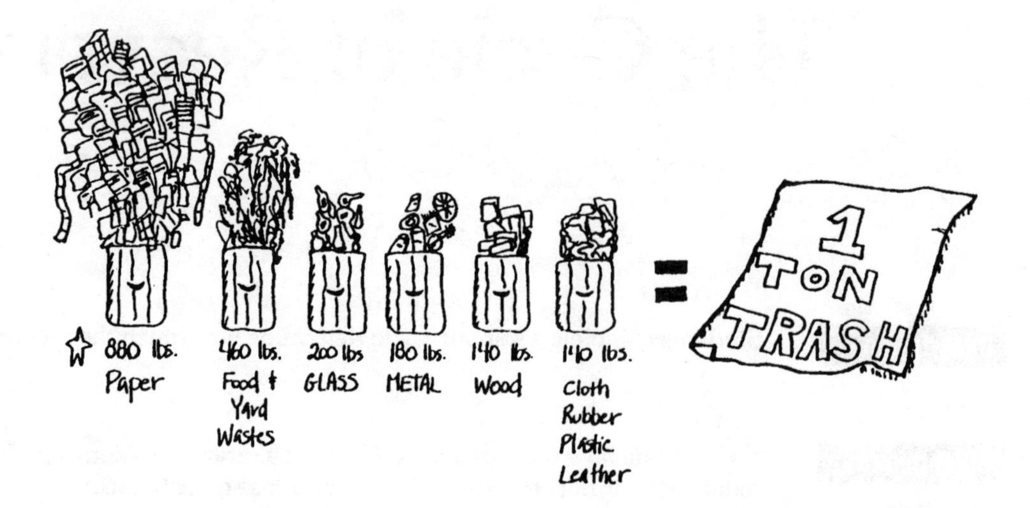

Where does all that garbage go? What kinds of problems does garbage make? How could more recycling make a difference? Where is the dump site in our city? Are we running out of room? How does it feel to throw things away? How does it feel to recycle?

1. Set up a recycling system in your classroom. Set up a give-away box for discarded but still useable things. Set up another box for extra snack and lunch foods. Set up a paper recycle bin and find out where you can recycle it locally.
2. Invite a speaker to class who is involved in recycling in your community.

A Human Paper Factory

Description

Students make their own recycled paper.

Objective

To demonstrate how new paper is made from used paper.

Teacher Background

This simple art project is well worth the gathering of materials! In making their own paper, students get a full understanding of what happens to recycled paper. Add a few colored sheets to the blender mix and see what happens! You'll soon develop your own designer paper.

Materials

Bucket or plastic tub
Blender
Sheets of used white writing paper
Large pan at least 2" deep
Fine mesh screen cut to fit inside the pan
Piece of cloth or old bedsheet
Sponge
Sunshine

Paper that has been used once can be recycled and made into new paper. How is this done at a factory? Guess the steps involved in remaking paper. (Write students' ideas on the chalkboard for later comparison.)

Have students make paper according to the following directions.

1. Tear paper into very small pieces.
2. Fill a bucket or tub with warm-hot water and add paper to it. The ratio should be one part paper to one and a half parts water.
3. Soak the paper until the pieces are soft.
4. Fill blender 2/3 full of the water and paper mixture.
5. Blend until the paper dissolves into a thin pulpy texture.
6. Fill the pan with 1" of water and set screen inside it.

7. Pour the watery pulp onto the screen.
8. Swish-shake the screen UNDER the water to spread the pulp evenly. This is the most important step. If the layer of pulp on the screen is too thick you will end up with cardboard instead of paper.
9. CAREFULLY, SLOWLY lift the screen out of the water. Let water drip through.

10. Set the screen on a flat surface and place a cloth over the paper pulp on the screen. Press a damp sponge on top of the cloth to remove as much excess water as possible.
11. Flip the whole thing over and remove screen.
12. Place the paper on the cloth in the sun to dry out.
13. When dry, peel the paper off the cloth. It is ready to use!

How do you think recycled paper is used? (paper towels, cardboard, newspaper) Why is it important to recycle paper? (save trees and trash) Do you know how much paper each of us uses in one year? (about 880 pounds)

Mother Earth

Description

The following passages are from Native American writings or speeches from the mid-1800s, and are offered here as a resource for your use.

Objective

To have students relate to the Native Americans' understanding of cycles and interdependence with the earth.

Native Americans considered trees and other living things to be their brothers and sisters. Trees were sacred members of the natural world and deserved reverence and respect. When Native Americans needed to cut down a tree, they would offer prayers to its spirit both before and after harming it. They would thank the tree for giving its life so that they might be able to carry on their own life.

Reading I:

Read and discuss the following passage by Walking Buffalo of the Stony Tribe, Alberta, Canada.

> "Hills are always more beautiful than stone buildings, you know. Living in a city is an artificial existence. Lots of people hardly ever feel real soil under their feet, see plants grow except in flower pots, or get far enough beyond the

street light to catch the enchantment of a night sky studded with stars. When people live far from scenes of the Great Spirit's making, it's easy to forget his laws.

Did you know that trees talk? Well, they do. They talk to each other, and if you listen they'll talk to you. I have learned a lot from trees: sometimes about the weather, sometimes about animals, sometimes about the Great Spirit."

Reading II:

Native American Indians were very aware of the significance of circles. They not only observed the cycles of nature that surrounded them and that they were dependent upon, but they also understood that they were part of the circle of life.

I HAVE KILLED THE DEER
Taos Pueblo Tribe

I have killed the deer.
I have crushed the grasshopper
And the plants he feeds upon.
I have cut through the heart
Of trees growing old and straight.
I have taken fish from water
And birds from the sky.
In my life I have needed death
So that my life can be.
When I die I must give life
To what has nourished me.
The earth receives my body
And gives it to the plants
And to the caterpillars
To the birds
And to the coyotes
Each in its own time so that
The circle of Life is never broken.

Reading III:

The following represents excerpts from a letter, written in 1855, that was sent to President Franklin Pierce by Chief Sealth of the Duwamish Tribe of the State of Washington. It concerned the proposed purchase of the tribe's land. Seattle, a variation of the chief's name, is built in the heart of Duwamish land. The letter excerpts are printed courtesy of Dale Jones of the Seattle office of Friends of the Earth.

THIS EARTH IS SACRED

"The Great Chief in Washington sends word that he wishes to buy our land. The Great Chief also sends us words of friendship and good will...we will consider your offer. What Chief Sealth says, the Great Chief in Washington can count on as truly as the return of the seasons. My words are like the stars—they do not set.

How can you buy or sell the sky or the warmth of the land? The idea is strange to us. Yet we do not own the freshness of the air or the sparkle of the water. How can you buy them from us? Every part of this earth is sacred to my people. Every shining pine needle, every sandy shore, every mist in the dark woods, every clearing and humming insect is holy in the memory and experience of my people.

There's no quiet place in the big cities...no place to hear the leaves of spring or the rustle of insect's wings. But perhaps because I am a savage and do not understand, the clatter only seems to insult the ears. And what is there to life if a man cannot hear the lovely cry of a whippoorwill or the arguments of the frogs around a pond at night? The Indian prefers the soft sound of the wind darting over the face of the pond and the smell of the wind itself cleansed by a mid-day rain, or scented with a pinon pine. The air is precious to the Indian. For all things share the same breath—the beasts, the trees, the man.

If I decide to accept, I will make one condition. All people must treat the beasts of this land as their brothers. I am a savage and I do not understand any other way. What are humans without the beasts? If all the beasts were gone, men would die from great loneliness of spirit, for whatever happens to the beast also happens to man. All things are connected. Whatever befalls the earth befalls the sons and daughters of the earth.

When the last Indian has vanished from the earth and the memory is only the shadow of a cloud moving across the prairie, these shores and forest will still hold the spirits of my people, for they love this earth as the newborn loves its mother's heartbeat. If we sell you our land, love it as we've loved it. Care for it, as we've cared for it. Hold in your mind the memory of the land, as it is when you take it. And with all your strength, with all your might, and with all your heart, preserve it for your children. This earth is precious."

What do these readings tell you about Native American culture? What do you think was happening to Native American land at the time of these writings? How was the cycle represented in Native American life? Describe how Native Americans saw themselves as part of the earth's cycle.

Have students develop their own oral stories about how cycles are important in their lives. These could be written down and put in a "Class Cycle Book," which, of course, would be in the shape of a circle!

LIFE LAB SCIENCE PROGRAM
408-459-2001

If you wish to order additional copies of *The Growing Classroom,* contact your local bookseller, or call:
Addison-Wesley
Publishing Company
Jacob Way
Reading, MA 01867
(800) 447-2226

Life Lab offers various materials and inservice training for teachers and schools interested in implementing an indoor/outdoor garden-based science program.

Please send more information about

___ inservice training ___ video ___ awareness presentation

Name ____________________ Position ____________________
School Name ____
Address ____
____
Phone _(____)_______________________________________

LET'S GET GROWING

A Catalogue of Supplies for the Life Lab Curriculum and Other School Garden Projects.

INCLUDES: Selected Garden Tools and Seeds; Indoor Growing Supplies; Soil Testing equipment; Lab and Science Supplies; Weather Instruments; Books; and More!

PLEASE, SEND A COMPLIMENTARY COPY TO:

NAME ____
ADDRESS ____
____

PLACE
STAMP
HERE

LIFE LAB SCIENCE PROGRAM
1156 High Street
Santa Cruz, CA 95064

PLACE
STAMP
HERE

LET'S GET GROWING
C/O General Feed and Seed Co.
1900 B. Commercial Way
Santa Cruz, CA 95065

Interdependence

Hall Elementary
Watsonville, CA

Hall Elementary
Watsonville, CA

Hall Elementary School
Pajaro Valley Unified School District
Watsonville, California

For many children at Hall School, recess no longer means four square or hopscotch. Instead, flocks of these elementary school students can be found digging, hoeing, watering, and checking their compost piles. The Life Lab program has taken off in a big way, and green thumbs are becoming contagious!

Set in agricultural Watsonville, the 75′ x 100′ garden is an ideal place for students to learn about science. The garden is composed of communal flower areas and raised beds, but almost daily another class wants to get in on the action and goes out to claim some ground and dig their own. This year the students designed and built a solar greenhouse to extend the growing season and get plants off to a healthy start.

The garden is used by individual teachers as part of the school's whole language program (and for English as a second language), by the "recess gardener," the school science program, and the school garden club. Hall's motto: "I DIG GARDENING."

UNIT INTRODUCTION

Interdependence

Connected (to be chanted in a round):

Connected, connected,
Everything's connected.
Connected, connected,
To everything else.

"When you pick a flower, you shake a star."
—John Muir

All living organisms depend on energy from food, creating a web of interdependence with a very delicate balance. The sun provides food energy for plants, and on this all other life depends. Humans inhabit only a small part of the earth, and we have been here only a short time. However, we have made ourselves an integral part of the survival of most food webs. A change at one end of a food web—use of a pesticide, for example—can affect interdependence in unexpected ways. Science has the potential to investigate the impact of these changes and predict the effects of different technologies on earth. Scientists, through the study of interdependence and energy transfer, can help preserve the resources on which all living organisms depend.

This unit looks at interdependence in the human and animal communities, with a focus on food chains. Developing an awareness of the earth's limited resources, students role play food chains and the effect of changing one link in the chain. From simple animal food chains to complex webs, students gain an understanding of interdependence and apply the concept to their school and community. Most of the activities can be done in the classroom and all are short-term.

Activities	Recommended Grade Level
Eat the Earth *(demonstration of the earth's finite resources)*	4,5,6
I Eat the Sun *(demonstration of a food chain)*	3,4,5,6
The Hungry Bear *(demonstration of a food chain)*	4,5,6
You Are What You Eat *(demonstrates the interdependence of a food web)*	5,6
Lunch Bag Ecology Part One *(our dependency on plant and animal life)*	2,3,4
Lunch Bag Ecology Part Two *(our dependency on plant and animal life)*	4,5,6
DDT Chew *(role play of bioaccumulation through food chain)*	5,6
The Day They Parachuted Cats into Borneo *(demonstrates ecosystemic interdependency and the impact of change)*	5,6
We're Just Babes in the Woods *(demonstrates the short time humans have been on earth)*	4,5,6
Caught in the Web of Life *(examines human dependencies in their communities)*	4,5,6

Eat the Earth

Description

An apple used to represent the earth is sliced into portions representing the oceans, land, farmland, and potable water.

Objective

To illustrate the percentage of the earth that is capable of supporting human needs.

Materials

A few apples
Knife

The earth has many natural resources that are very important to us. Can you name some of them? (soil, water, air, minerals) How are these resources important to us? (We could not live without them.) Can they be replaced ? (no) Let's imagine that this apple is the planet earth. We know that the earth is made up of water and land. How much of the earth do you think is water? (Record guesses.) How much is land? (Record guesses.) (Cut one or two apples according to students' guesses.) Let's make an accurate representation. Our goal is to find out how much soil there is for us to grow our food in and how much water there is to drink.

1. Cut an apple into quarters and set three of them aside. The remaining quarter represents the part of the earth's surface not under salt water.
2. Next, cut this quarter in half and set one piece aside. Explain that the remaining piece represents the part of the earth that is suitable for human habitation. The other part is too cold, too dry, too mountainous, or too hot.
3. Now cut the part on which humans can live into four equal slices. Rather thin, aren't they? Just one of these four small slices represents the part of the earth that supplies most of our food and clothing, the small part that is presently tilled. It is not too wet, not too cold, and not occupied by cities, factories, or highways.
4. Cut a very small piece from the remaining slice. This represents the 3/100 of 1% of the earth's vast surface that contains potable water.

What does cutting the apples into portions tell you? Do we need our natural resources? Does it seem as if we have enough land to farm and enough water to drink? Who is responsible for taking care of these resources? How do we want to treat the remaining healthy portion?

I Eat the Sun

Description

Students are given labels so that they can group themselves into a food chain.

Objective

To illustrate the concept of a food chain.

Teacher Background

All living organisms depend on energy from food. A chain is created as one organism consumes another to get its food. Every organism is part of a chain. Every chain begins with the energy from the sun, which is needed by plants to make their own food. The first two links in a food chain will always be the sun and a plant. The food chain is actually a cycle. Consumers at the top of the chain, such as coyotes, will provide energy for decomposers who break down organic matter, putting nutrients back into the soil to nourish plants.

Materials

Scraps of paper with the following labels (adjust the total to your class size): 1—sun; 14—plant; 18—microorganism decomposer; 4—snail; 2—chicken; 1—coyote

Name one of your favorite foods. (Write a healthy response on the chalkboard.) Why is this food important to your body? What was this before you ate it? (plant or animal) Did it need energy to grow? Where did it come from? (Trace examples of the food chain back to a plant and the sun.) Everything is linked together in a food chain. (Try other examples if desired.)

1. Pass out one labeled paper to each student. Tell them to group themselves with the other students who have the same label.
2. Ask the groups to put themselves in order according to who feeds whom.
3. After students are in order, have each group beginning with the sun say who they are and whom they feed. For example, "I am the sun, and I feed the plants." Who does the coyote feed?

What do all food chains start with? Why? Why are there more snails than chickens? How can the chain be broken? How is the chain a cycle? Can you name a food chain that we are part of?

The Hungry Bear

Description

Students create a food pyramid represented with paper plates in the form of a triangle.

Objective

To illustrate the concept of a food pyramid.

Teacher Background

The sun's energy flows through many links in the plant and animal world. These interrelationships form food chains and complex food webs on our planet. The plants and animals that depend on each other for food each form a link in the food chain.

Each chain is actually a pyramid. The size of the consumer increases at each link of the chain; each consumer needs to eat more of the previous level in order to obtain its needed energy. For example, one bear must consume more than one salmon to furnish needed energy. Thus, food chains are more accurately represented as pyramids.

Materials

21 paper plates
Felt tip pen

Preparation

Place the plates in a pyramid shape on a table or the ground. Do not label the plates.

Imagine that you are a bear living in the forest. Summer is ending, and you are enjoying wandering among the trees. You hear the sound of a nearby stream, and you suddenly realize how hungry you are. You follow the sound and edge up to the side of a wide, deep-running stream. Your paw is raised in the air as you carefully observe the passing water.

1. Write BEAR on the top plate. Ask, What is this hungry bear trying to do? (Catch fish.)
2. Have students decide upon a type of fish the bear will eat such as salmon. Write the name of the fish on one of the plates in the next level of the chain. Ask, Is one fish enough for the bear? Will the bear need more? When the class agrees, write the name of the fish on the second plate at this level. Ask, Why does the bear need two salmon?

3. Continue the discussion for the rest of the food chain, having students identify what and how many each organism eats and writing the name on a number of plates.
4. Emphasize in discussion that the food chain forms a pyramid: it is necessary to have more organisms at the base of the food chain. Also stress that the sun is the source of energy for the plants, which in turn make their own food.
5. Tell the students that a big storm comes through the forest and disturbs the stream. It destroys almost all of the aquatic insects (substitute your organism at a middle level). Turn the plates over at this level. Have the class infer what would happen to the consumers higher on the food chain. What would happen to the producers lower on the chain?

List the organisms in this food pyramid that would be affected by death of the algae? How would they be affected? What would happen to the bear if this stream dried up?

What is the source of all food pyramids? Why are food chains really pyramids?

You Are What You Eat

Description

Students form a food web by connecting links in the food chain with string.

Objective

To illustrate interdependence within a food web.

Teacher Background

Food chains are not isolated; they are interrelated, forming a complex food web. This interdependence allows for diversity in food choices to optimize chances of survival and at the same time helps to keep populations in balance with the food sources. This activity will give students the opportunity to visualize how the complexity of the food web develops from a simple food chain.

Materials

String
One piece of paper per student
Tape
Reference books (optional)

Preparation

At the start of the activity each student will be assigned a different animal or plant. You may want to assign them in advance and have students research their plants and animals to learn where they live and what they eat and who might eat them.

What is a food chain? Do animals always eat the same thing? Can the same type of animal be part of more than one food chain? What happens as these different food chains are brought together?

1. Assign students an animal or plant from the chart. Have them make signs showing the names of their animal. Tape the signs to the front of each desk and arrange the desks in one large circle.
2. Pick any animal that is *not* a plant or decomposer and hand that student the end of a ball of string. Then ask him or her to pick something from the circle that his or her animal would like to eat. Run the string from the first student to the second. Then ask the second student to pick something to eat. Continue until all students have at least one part of the string in their hands. You may have to help them understand that plants eat the products that decomposers break down. Also, decomposers eat dead things. Encourage the decomposers to

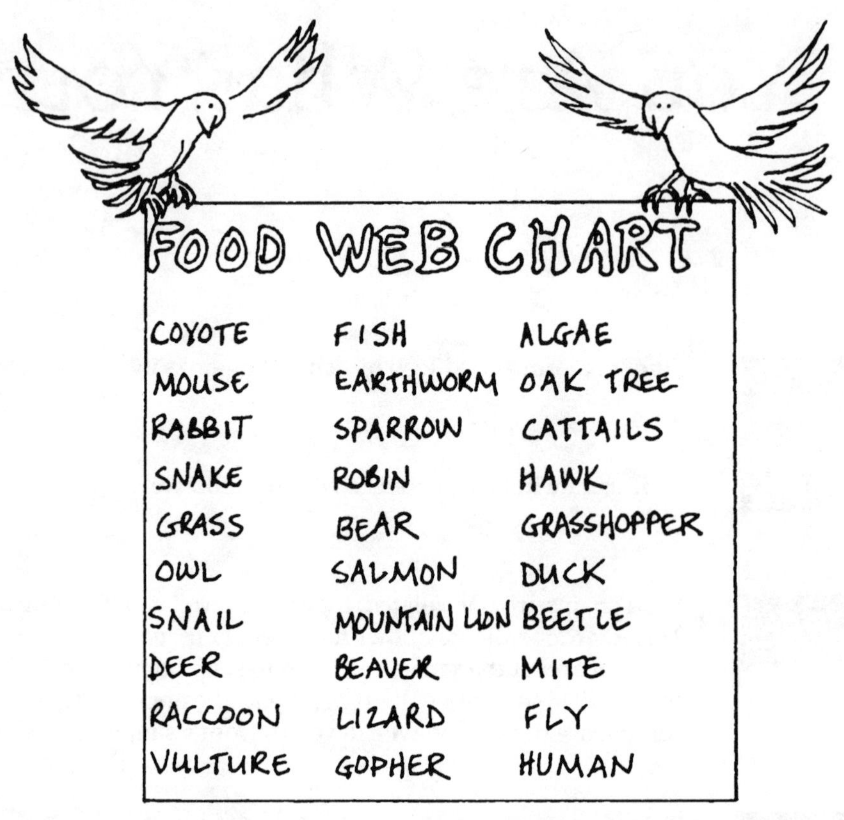

choose to eat animals such as mountain lions and vultures that are eaten by nothing else.

3. Continue until all students are connected to the food web and it is too complex to take any further.
4. Ask students what would happen to the food web if: air pollution kills all of the trees; new birds migrate to the area; frost wipes out the insects; the area is stripmined; trees are planted; flooding inundates the area. Have them think of additional factors that might alter the food web. The students' roles directly affected by a given change can drop the string, or the string can be cut. Discuss the collapse or extension of the web.

John Muir, the famous naturalist and writer said, "When we try to pick out anything by itself we find it hitched to everything else in the universe." Discuss what this means with the class. Ask, How are you part of the food web? What was the result of making a change in the web? Identify ways in which people are causing changes to the earth. Name an action you have taken that resulted in an unexpected change.

Lunch Bag Ecology
Part One

Description

Students analyze their lunch to discover their food's origin.

Objective

To illustrate our interdependence with plants and animals.

Teacher Background

Children often don't understand food origins. This activity quickly clarifies our interdependence with other forms of life.

Materials

Students' lunches

Do you ever think about where your food comes from? What are you having for lunch today? Are you eating any plants? Any food from animals? Are you part of a food chain? Let's find out!

1. Have students list what they are having for lunch. Then ask the following questions:
   Who is having roots? Who is having leaves? (carrots, lettuce) Who is having something from a tree? (apple, orange) Who is having ground-up grass seeds? (bread) Who is having bird, cow, or pig for lunch? (chicken, cheese, ham)
2. To develop broader concepts of food chains, ask:
   Who is having water? (raw fruit or vegetable) Who is having sun? (light trapped in green leaves) Who is having soil? (minerals in foods; emphasize that cheese was produced from a cow that ate grass growing in soil) Who is having compost? (used as fertilizer on your vegetables)

How many food chains were you part of at lunch today? Could you live without being part of a food chain? Why is it important to understand our need for plants, animals, sun, water, soil, air?

Lunch Bag Ecology Part Two

Description

Students analyze how their lunch was transported, stored, prepared.

Objective

To explore the food supply system.

Materials

Students' lunches

Do you ever think about where your food came from? What plants are you having for lunch today? What food comes from animals? Did you ever wonder how this food got to your lunch? We have a very complex food system. Our food can travel thousands of miles to get to us. But in many other countries, people eat food that is grown or raised near where they live. Let's examine our lunches and see how the foods got to us.

Have each student select a lunch item and explore the following:

1. In what season was it probably harvested or butchered? If your food was not picked recently, how was it stored? (dried, canned, bottled, frozen, pickled, bagged, boxed) If we were unable to preserve and store foods after seasons of great harvest (winter, spring), what would happen to our population after lean harvests?
2. Transportation has helped us transcend local seasonal limitations. Who has lunch food grown in another climate and transported here? Where is the food from? How far has it traveled? In what season was the food raised? Who has a lunch that traveled over 1,000 miles? (examples: Danish crackers, Florida oranges, New Zealand lamb, Guatemalan bananas)
3. Who has a lunch that has traveled less than 500 miles? What would your lunch consist of if you ate only fresh foods? Could you eat bread in the winter? Apples in the spring? What do you think people who lived here 200 years ago had for lunch? Have students make a menu.
4. What farming techniques manipulate climate to increase plant food production? Find a food in your lunch that is affected. For instance, fruit and vegetable crops are irrigated during the hot, dry California summer; smudge pots and fans prevent winter frost from killing citrus crops; greenhouses create summer in winter for tomatoes and flowers.

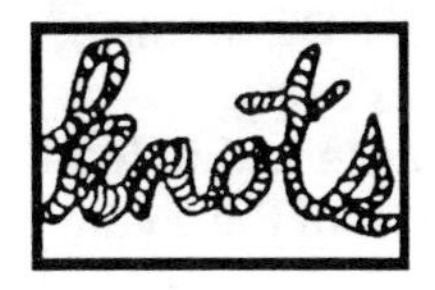

Prepare a menu of local foods. Would this menu need to change throughout the year? What are some advantages of eating foods produced locally? (fresher, lower cost, energy efficient, support of local agriculture and economy) What are some advantages of eating foods from around the world? (more choices) How can you tell where your food comes from? (Look on labels of processed foods, ask store clerks, shop at local farmers' markets.)

DDT Chew

Description

Students will role play the use of a pesticide and its transfer up the food chain.

Objective

To illustrate bioaccumulation through the food chain, or the persistence of certain substances and their effects on human health, agriculture, and the environment.

Teacher Background

This lesson will use the historically significant pesticide DDT to exemplify bioaccumulation. Recognized in 1948 for its effectiveness in killing mosquitos in Third World countries, DDT helped to control malaria. Also effective for the control of numerous agricultural pests, DDT was used in the U.S. until 1973. At that time DDT was cited as an environmental hazard due to its slow breakdown and its accumulation in the food chain. DDT was very effective in its goal of eliminating certain pests; however, its long term effects on the environment were not determined until it was too late. DDT is one of a class of pesticides that is stored in body fat and passed up the food chain as each predator becomes prey. This has a dramatic effect on birds such as pelicans. DDT has also been found in human breast milk. (Note: Not all pesticides are accumulated in the food chain.) The Environmental Protection Agency issued a federal ban on the use of DDT in this country; however, DDT is still present in our environment and poses many problems. A significant point of this activity relates to interdependence within food chains: When we try technologically to change something at one level of the chain, it may have an unknown impact at another level.

Materials

Eight jackets or sweaters

Have you ever seen an apple with a little hole in it? What does that hole probably tell you? (that an insect larva was in the apple) If you were a farmer and every year you had these holes in your apples, would you be concerned? What are some options you might consider in order to deal with the larvae? (Read information to see if the insect's life cycle can be interrupted; talk to a pest control company; talk to a local farm advisor.) Let's imagine that we are visiting an apple orchard that has an infestation of codling moths. These moths are threatening the apple harvest, so the owner of this orchard has decided to spray the trees with DDT pesticide.

1. Select eight students and set them in a row facing the class.
2. Introduce yourself as Farmer Jones. Explain the serious problem you have had with the codling moth larvae. These larvae are threatening

the apple harvest, and you have decided to spray the apple trees with a pesticide called DDT.

3. Give each of the eight students a jacket. The students you have selected are the larvae and the jackets represent the pesticide with which the larvae have been sprayed.
4. Select four more students to represent mice. These four will enter the orchard looking for food and each will eat two larvae. As they eat the larvae, they remove the jackets and put them on to signify the passage of the pesticide from the insects to the mice. Each mouse is now wearing two jackets, signifying two units of spray.
5. Select two more students to enter the orchard. These two will be snakes. Each snake will eat two mice. As the mice are eaten, their jackets are removed and placed upon the snakes, again signifying the passage of the spray from one organism to another. Each snake now wears four jackets.
6. Select one student to represent a hawk. The hawk will eat the two snakes, acquiring all eight jackets.

What do the hawk's eight jackets represent? How did the hawk acquire the jackets? (the spray was inside its prey) Explain to the students that DDT is no longer used in the U.S. because of its long-lasting properties, and that its accumulation in the food chain threatened human health, agriculture, and the environment. Most pesticides do not bioaccumulate. How can a change in one part of a food chain affect another part? Do we always know the impact of our actions?

1. See "The Garden Puzzle," p. 240, for a role-playing activity about alternative pest control.
2. Invite a farmer to class. Discuss alternative methods of pest control.
3. Have students research DDT and its effects on your community.

The Day They Parachuted Cats into Borneo

Description

Students discuss this story of how DDT affected villages in Borneo.

Objective

To demonstrate the ramifications of making a single change in an ecosystem.

Teacher Background

See Teacher Background, p. 222. Explain why DDT was banned in the U.S.

Materials

None

When we learn about a new topic, it is helpful to look at the subject's history and find out about its past, and how it has affected people and the environment.

Read aloud the following true story to your students and discuss it. (From Laycock, *Let the Wild Ones Stay Home.*)

Some time ago, the World Health Organization sent supplies of DDT to Borneo to fight mosquitos that spread malaria among the people. The mosquitos were quickly wiped out. But billions of roaches lived in the villages and they simply stored the DDT in their bodies. One kind of animal that fed on the roaches was a small lizard. When these lizards ate the roaches, they also ate a lot of DDT. Instead of killing them, DDT only slowed them down. This made it easier for cats to catch the lizards, one of their favorite foods. About the same time, people also found that hoards of caterpillars had moved in to feed on the roofing materials of their homes. They realized the lizards that previously had kept the caterpillar population under control had been eaten by the cats. And now, all over North Borneo, cats that ate the lizards died from DDT poisoning.

Then rats moved in because there were no cats to control their population. With the rats came a new danger: plague. Officials sent out emergency calls for cats. Cats were sent in by airplane and dropped by parachute.

One simple change in the ecosystem had set off a whole chain reaction.

Have students identify actions they or others have taken that caused unexpected changes or outcomes. What happens when we change just one thing? What does it mean to be interdependent?

We're Just Babes in the Woods

Description

A 300-foot path is marked outdoors to create a walking timeline of the earth's history.

Objective

To illustrate the relative time that humans have been present on earth.

Teacher Background

We have a sense that the earth was always as it appears now and that humans were always part of the system. The timeline in this activity puts the tremendous changes in the earth's history into perspective.

Materials

Two stakes
12 colored flags or markers
Tape measure

Our planet has a natural history that is billions of years old. It is very difficult for us to imagine one billion, but what do you think it was like long, long ago on our planet? Were there the same mountains and streams that there are today? Were there oceans? Were there always animals? Were there always human beings? We're going to mark a line outside that is 300 feet long. This line will represent the history of the earth, from its beginning until now. Then we'll mark events along the line to give us an idea of when they happened. This kind of graph is called a *timeline.*

1. Have students put stakes at both ends of a 300-foot path.
2. Start at the beginning and walk the class through events in the earth's history. Place a colored flag at each event's location on the timeline or have students write the event on paper and attach it to a stake at the location.
   - Beginning our walk through time, we see no life for 79 feet.
   - At 80 feet, single-celled life appears.
   - We move on to 125 feet, at which time the first vertebrates are seen.
   - The first land plants appear at 160 feet.
   - The first land vertebrates, the amphibians, develop at 190 feet.
   - Dinosaurs march onto the scene at 215 feet and remain until 270 feet.
   - In the meantime, the first birds and mammals appear.

- The Grand Canyon is formed at 290 feet. With only 10 feet remaining, humans have not yet appeared.
- Finally at 299 feet, the first human-like creature appears.
- At 299 feet 8 inches, stone-age people appear.
- At 299 feet 11 inches, the calendar begins and civilization as we know it begins.
- Within the last inch, Columbus arrives in America, and the Declaration of Independence is signed.

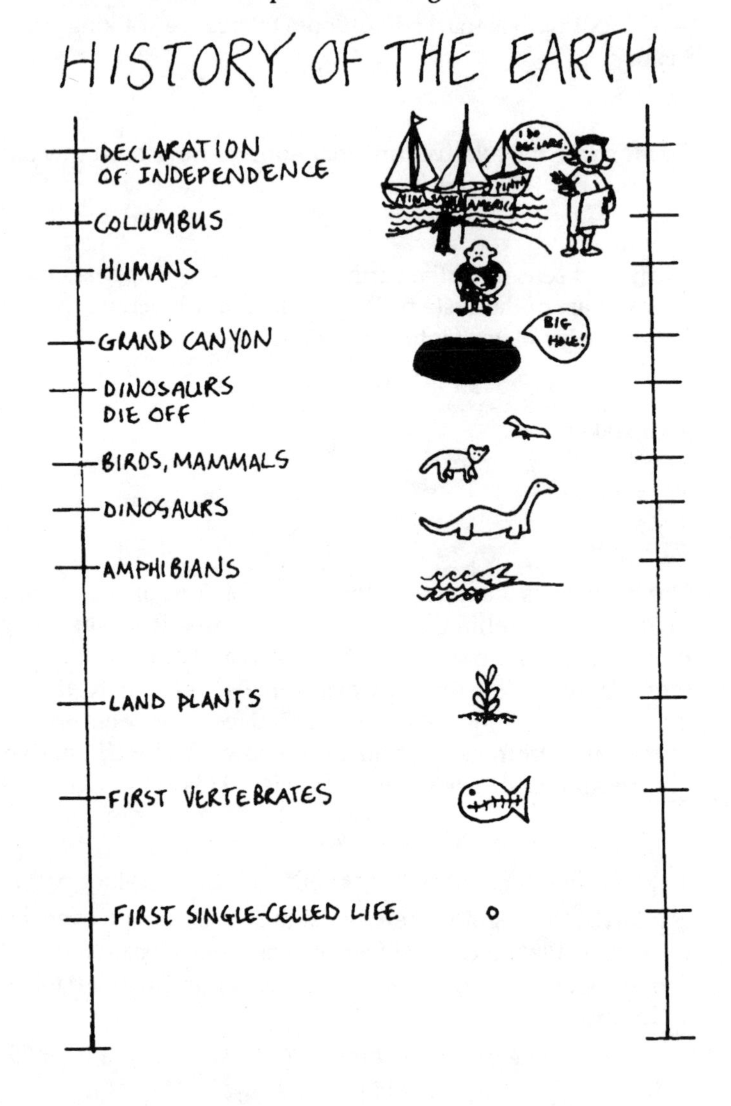

Life has existed on earth for some 200 feet of our walk; humans have been here for less than 12 inches. The dinosaurs dominated the walk for 55 feet; we have dominated it for less than two inches. Yet sometimes we look upon the dinosaurs as unsuccessful animals that could not adapt. If the path were to be continued into the future, would we be here 55 feet from now? What have we done to the earth in our short time here? What can we do to ensure that both humans and the earth will continue to survive?

Caught in the Web of Life

Description

Students make a web showing the interdependence of people at school and in the community.

Objective

To examine interdependence i

Materials

None

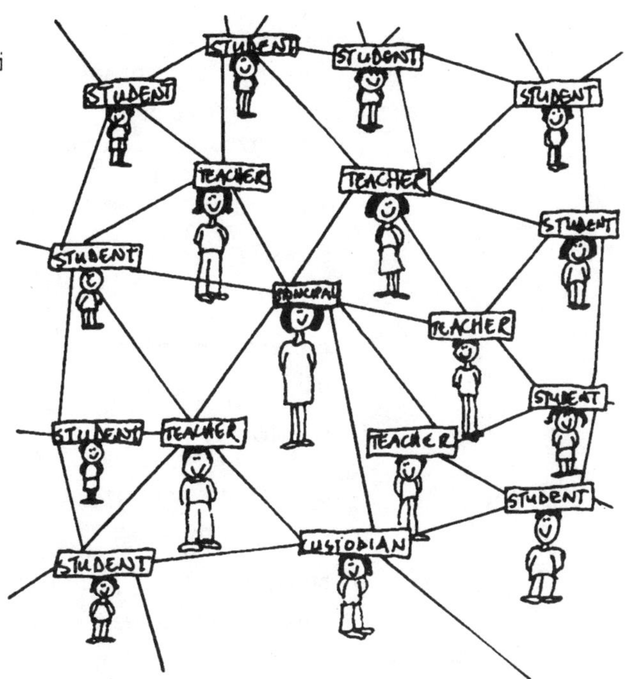

What is a food web? What does it tell us about our relationship to plants and animals? (We are all interdependent.) Do you think if we looked at the people who are at our school we would find interdependencies there? Do we need each other to make our school operate well?

1. Have students think of all the different people who contribute to the school and write them in a circle on the chalkboard as they are suggested.
2. Starting with one, ask if this person needs or depends on any other person listed on the chalkboard. If there is a dependence established, draw a line connecting the two.
3. Continue around the circle. Ask, What is our result? What if one of the people in the web was removed? What would be the effect?

4. Repeat, constructing a web of community occupations. Write each occupation on the chalkboard in a circle and again draw the lines that connect one with another as an interdependency is established. Try taking one occupation away, such as garbage collectors. What happens?
5. Explain that any community, natural or human, is composed of producers, consumers, and decomposers.

   Producers—green plants that convert the sun's energy to food.

   Consumers—animals that eat the green plants or eat other animals.

   Decomposers—plants or animals that eat dead producers and consumers, breaking down and recycling them in the process.

   Introduce these concepts and give examples of each. Review some of the examples from natural communities.
6. Using the web of the human community, list the human producers, consumers, and decomposers and their needs. For example:

Human Producers	*Basic Needs*
Farmers	Plants, soil, water, sun, air, fertilizer, livestock
Human Consumers	
People in general	Food, shelter, clothes, sun, air, water, fuel
Car drivers	Water, gas, maintenance
Human Decomposers	
Garbage collectors	Gas, water, electricity, soil, water
Sewage treatment plant users	Air, sun, animal decomposers
Landfill/dump users	

7. Divide the class into pairs and have them draw interdependencies of their school or town. Have students share their drawings with the class.

List the effects of a garbage collectors' strike on a community. Identify the kind of workers who would lose a job if everyone stopped using cars tomorrow. What would happen in the natural world if there were no decomposers? How does knowing you are interdependent affect your attitude toward people, the school, the community, and the environment?

Garden Ecology

Bayview Elementary,
Santa Cruz, CA

Bayview Elementary
Santa Cruz, CA

Bayview Elementary School
Santa Cruz, California
and
Redding Elementary School
Redding, Connecticut

The Life Lab program at Bayview started in 1983 between two classroom wings with an 11′ X 60′ strip of ground, with garden beds for five Life Lab classes K-6. After one year, the program grew into an unused area near the playing field, approximately 30′ X 150′, plus another small area of garden beds for three Kindergarten classes. About 280 children regularly participate in the Life Lab classes, and another 40 per week come for lunchtime gardening. Other projects, such as bulb planting, have involved as many as 450 students.

The garden has survived a near drowning during the first two years, and again when a trench was bulldozed through it. It struggled for two years with the effects of a drought. The community has been involved, and we are constantly planning future projects.

Much is learned in the Life Lab Garden. Some things are the foundation of future learning and ways of learning: wondering why an insect was attracted to a specific plant, and learning how to pursue the answer. Some things have been sheer enjoyment: the crisp taste of carrots just pulled, the color of poppies in the sunshine, or the feel of Lamb's Ears on a cheek. Some things have encouraged the conquering of fears: touching a first earthworm or a rolly-polly bug. Many things grow in Life Lab: the size of the garden, the number of involved classes, as well as the plants themselves. The most important things that grow, however, are the children as they learn new skills and challenge themselves.

The Redding Elementary School houses 585 children, grades Kindergarten through 5. We have a Living Laboratory on our grounds (started in 1987 after a Life Lab inservice workshop) that provides a beautifying effect for all to see and enjoy. In addition, our children have learned cooperative group methods during the course of weeding, turning over the soil, and planting. Science has come alive in the classroom through experimentation, observation, and recording of data in the indoor and outdoor growing projects.

Our guidance counselor was instrumental in introducing this enriching science program to our curriculum that now incorporates all elementary grade levels and all subject areas.

Life Lab at Redding Elementary School has become a "living" laboratory. Children are learning and growing in mind as well as body. Everyone and everything is flourishing.

UNIT INTRODUCTION

Garden Ecology

In our Living Laboratory we continually try to replicate nature. We copy the nutrient cycle by making compost; we mulch to keep in moisture and avoid soil erosion as in the forest; we water to mimic a spring rain. And in our Living Laboratory we are able to observe the cycles from seed back to seed, changes from germination to decay, and the interdependence of pollinator and pollen that are the basic processes of nature. This unit introduces us to the ecology of our Living Laboratories, and in so doing gives us an opportunity to apply many of the concepts learned in "Cycles and Changes" and "Interdependence."

By understanding ecological principles we can manage our garden system to promote the health of the crops. From providing flower food for pollinators to shading out weeds, this unit will have students observe and experiment with various ecological processes. Students focus on the garden habitat to explore flower parts and pollinators, plant-insect interactions, and effects of plants on each other. They experiment with intercropping, companion planting, allelopathy (ability of one plant to give off chemicals that inhibit the growth of another plant), and plant spacing. In the final activity, "Under the Big Top," students observe and role play these interactions.

This unit is best taught in the spring when flowers and pollinators are in abundance. Most of the activities involve the outdoors. "Companion Planting," "Plant Architects," "I Need My Space," and "Poison Darts" require extended time for data collection.

The magic and science of the Living Laboratory are highlighted in the study of interactions. It is like watching a circus performance in which everything works together as precisely as trapeze artists flying into each other's grip high above the ground.

Activities	**Recommended Grade Level**
Flower Power Part One *(flower part identification)*	4,5,6
Flower Power Part Two *(relationships between pollinators and flowers)*	4,5,6
Magic Spots *(observation of habitats)*	3,4,5,6
Garden Puzzle *(demonstration of companion planting by role playing)*	4,5,6
Companion Planting *(companion planting garden experiment)*	5,6
Plant Architects *(designing plants and observing their impact in an "imaginary" garden)*	4,5,6
I Need My Space *(garden experiment on the effect of seed density on weeds)*	4,5,6
Natural Defense *(experimentation of the effects of "plant solutions" on crops)*	5,6
Under the Big Top *(observations of interactions in the garden)*	4,5,6

I'm A Tree

G **C** **G**
I'm a tree, I'm a tree and I'm growing very tall.

G **D**
I'm a tree, can't you see me shake my leaves?*

G **C** **G**
I'm a tree, I'm a tree and when the breezes blow in fall,

G **D** **G**
I give a voice to the wind.*

I'm a tree, I'm a tree and I'm growing very tall.
I'm a tree, can't you see my branches spread?
I'm a tree, I'm a tree, can you hear the robin call?
I give a home to the birds.*
I give a home to the wind.*

I'm a tree, I'm a tree and I'm growing very tall.
I'm a tree, can't you see my acorns grow?*
I'm a tree, I'm a tree and when the bushy tails
crawl,
I give food to the squirrels.*
I give a home to the birds.*
I give a voice to the wind.*

I'm a tree, I'm a tree and I'm growing very tall.
I'm a tree, can't you see my green leaves?*
I'm a tree, I'm a tree and when the sun it shines
its all,
I give shade to the world.*
I give food to the squirrels.*
I give a home to the birds.*
I give a voice to the wind.*

In Unison:
I give shade to the people of the world.
I give shade to the people of the world.
(Repeat.)

* These are the action parts of the song. The group may be divided into four sections, with each performing on cue. With smaller groups, act out all of the parts together. "Shake my leaves" is the cue for the first group to stand and shake their leaves. At "voice to the wind," they make wind sounds. The second group stands on one leg and spread their arms out like branches when they hear "my branches spread." At "homes to the birds," they make bird sounds. The third group stands and makes a hat with their arms and hands, like that of an acorn, when they hear "my acorns grow" and they make squirrel sounds at "food to the squirrels." The last group shows their leaves at "my green leaves," and they wipe their foreheads and give a sound of relief at "shade to the world."

Flower Power Part One

Description

Students dissect and draw flowers to learn about their parts.

Objective

To learn the structure and function of flower parts.

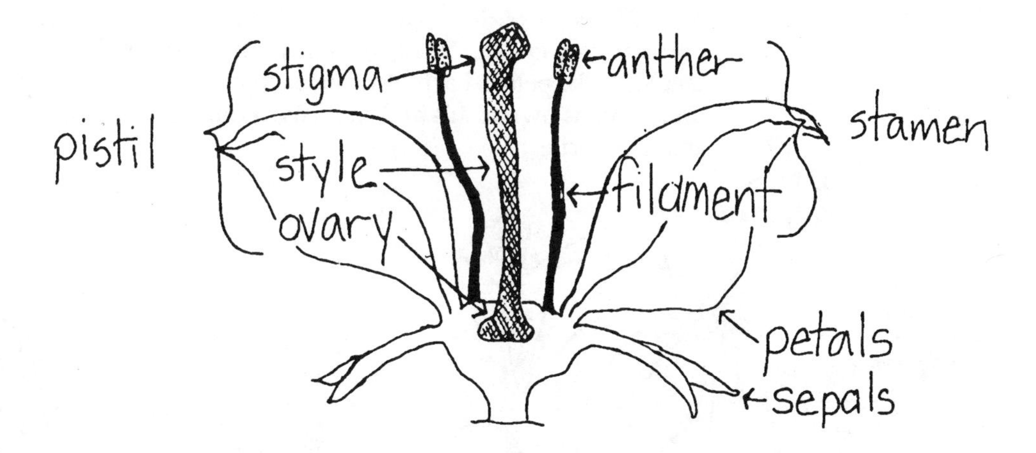

Teacher Background

Evolutionally advanced plants (angiosperms) produce flowers, where the sex cells are contained for the plant's reproduction. The *stamen* is the male organ for reproduction and is composed of the *anther* and *filament* (or stalk). At its tip is the anther, the organ that produces the pollen. *Pollen* is composed of fine powder-like grains that contain the male sex cells. The *pistil* is the female organ; its parts are composed of the *stigma, style,* and *ovary*. During pollination, male pollen lands on the stigma, travels down the style, and fertilizes the ovary. This fertilized egg develops into the seed. *Sepals* are the leaf-like parts under the petals. They are usually green and photosynthetic (able to produce food from the sun). *Petals* can be all colors and shapes and smells and serve to attract pollinators.

Materials

Life Lab journals
Drawing and coloring materials

Each flower seems to be unique, with its own special beauty. But all flowers are composed of the same parts. You and your friends are all unique, but you all have the same parts too: eyes, ears, noses, fingers, and so on.

1. Divide the class into small groups.
2. Ask each group to go into the garden and carefully collect one flower, preferably one that no other group has. Simple flowers with easily identifiable parts are foxglove, sweet pea, tomato, potato, bean, mustard, poppy, lily. Avoid composite flowers such as daisies and sunflowers.
3. When groups return, ask them to look carefully at their flower; then have them spend some time drawing a colored picture of it.
4. Ask the students to gently take their flowers apart and draw each part. Use the drawings as a guide to flower parts and discuss the function of each part.
5. Students should examine, draw, and label the sepals, petals, pistil, and stamens.

   Note: Members of the daisy family (Compositae or Asteraceae) have *composite* flowers: the "disk" flowers in the center and the "ray" flowers that look like petals. Each disk and each ray is a separate reproductive unit, with its own pistils and stamens, though many of them are actually sterile.

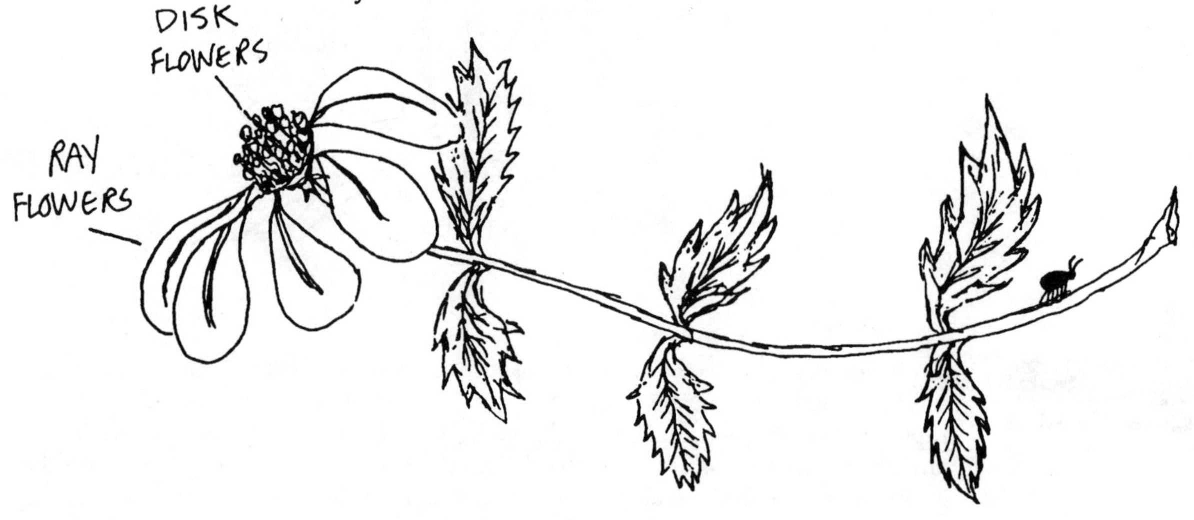

What is the name of the pollen-bearing, male part of the flower? What is the female part? What part of the flower swells to become the fruit and seeds? How does pollen get to the pistil? List things that would change if there were no more flowers.

Flower Power Part Two

Description

Students role play flowers and pollinators and try to find their perfect match!

Objective

To learn about pollinators and their relationship to flowers.

Teacher Background

Nature has provided various ways for pollen to reach the stigma in flowering plants. Some flowers are wind-pollinated and have very light pollen grains that are blown from plant to plant in fine clouds. Other plants have male and female parts that develop at different times. Insect-pollinated plants produce nectar that many insects collect for food. As the insect enters the flower to get the nectar, part of the insect is dusted with pollen. When the insect enters the next flower some of the pollen brushes off onto the stigma. Some flowers produce smells that attract insects while others have colors that are attractive to insects. By impersonating flowers and pollinators, students learn that there is a great variety of pollinators and that each has a special relationship to certain kinds of flowers.

Materials

Pins to attach tags to students
Construction paper
Marking pens

Unlike animals, plants can't move from place to place to find their mates. How then does the pollen from one flower get to the pistil, or female part, of another flower?

That's where pollinators come in. A pollinator is anything that helps spread flower pollen. There are all kinds of pollinators: birds, bats, bees, bugs, and more! Even the wind is an important pollinator. Pollinators may drink nectar from the flowers, and some, such as honey bees, collect and eat the pollen, too. In the process they spread pollen from flower to flower without even trying. Once the pollen fertilizes the egg in the flower ovary, the plant will go on to produce fruit and seeds. So we have pollinators to thank for all our fruits and nuts and lots of our vegetables, too.

1. Write the following list on the board:

Pollinator	*Type of Flower Preferred*
Beetle	White or dull-colored, fruity or spicy fragrance
Honey bee	Showy, bright petals, often blue or yellow
Mosquito	Small flower, often white or green
Butterfly	Red, orange, blue, or yellow flowers
Bat	Large flower with fruity fragrance and lots of nectar
Hummingbird	Red flower, little or no fragrance
Moth	White or yellow flowers with heavy fragrance
Wind	Small, odorless, colorless flowers

Grasses, corn, and so on tend to be wind pollinated. Since they rely on the wind, they don't have to produce showy or scented flowers to attract pollinators.

2. Divide the class into two groups. One group will be Pollinators, the other Flowers.
3. Assign each member of the flower group to a flower type.
4. Have members of the flower group write on construction paper a short description of what type of flower they are (bright red, no scent; white, very sweet-smelling) and pin the descriptions to their shirts.
5. Now take the pollinator group aside and whisper an identity (honey bee, wind, bat) to each member.
6. Then have the two groups mingle silently. Have each pollinator refer to the list on the chalkboard in order to find his or her right flower. Remind the class that there can be more than one pollinator to a flower because different pollinators may like the same type.

7. Since the pollinators have no identifying tags, have each flower guess in turn the identity of his or her pollinator ("I'm a bright red flower, so you're probably a hummingbird.")
8. Now go outdoors and have the pollinators find a *real* flower they like! Then ask the flowers to look around the garden for their *real* pollinators. Can they find any bees, hummingbirds, beetles, or wind?

When you look at insects near flowers now, what will you try to observe?

Most scientists believe that flowers and their pollinators *coevolved*. That means that they changed over time to suit each other; they *adapted* to each other. How does this coevolution benefit the flower? How does it benefit the pollinator? During this activity you learned that often several pollinators like the same flower. For example, bees and butterflies often visit the same type of flower. How would more than one pollinator be an advantage for the flower?

Go outdoors with students and sit quietly near some flowers. Watch carefully. What pollinators do you observe? How long does a pollinator stay on each flower?

Magic Spots

Description

Choosing one location in the garden to observe for several weeks, each student will record the habits of insects and changes that occur in this habitat.

Objective

To demonstrate the variety of living organisms and their interrelationships in an undisturbed environment.

Teacher Background

A *habitat* is the environment that is necessary in order for a living organism to survive. Insects and plants in the habitat will provide shelter and food for each other. To observe a habitat, students should get as close to the area as possible, be very still and quiet, and disturb the area as little as possible.

Materials

Garden or natural environment
Drawing boards
Life Lab journals
Pencils
Insect reference books (optional)
Hand lenses (optional)
Bug boxes (optional)

What do you need to do to be a good observer? (List ideas on chalkboard.) If you were to pick a spot in the garden to observe, what would you be sure to do? Imagine one of your favorite spots in the garden, your very own magic spot. If you were to go there and sit very quietly for a while, what do you predict you would see? (Have students record ideas in Life Lab journals and list some of the predictions on the chalkboard.) What might an insect be doing in a flower? Under a leaf? In the soil? Do you think if you are good observers you would be able to see what the insects are doing?

1. Have each student choose his or her own magic spot in the garden or in a natural environment. The spot should contain only one or two mature plants.
2. Have students observe in their special spots until they discover at least three different types of insects. They can look carefully under leaves, inside the plant, and in the soil around the plant.
3. Have students draw their special habitat including the plants and the insects' locations in relation to the plant.

4. Have students draw and describe in their Life Lab journals at least one of the insects: What does it look like? (winged, legs, mouth) Where does it live? (under the leaf, in the ground) What does it eat? (the plant, aphids, flying insects) Note: To assist students in this, you may want to use bug boxes that allow insects to be caged and magnified for a short time. Be sure to return the insect carefully to its habitat.
5. Repeat this activity at least once a week for several weeks. Have students record the changes they observe in their habitat. Have students share observations.

Which habitat had the largest variety of insects? What kinds of insects were found in more than one habitat? What does the plant provide for the insects living around it? What do the insects provide for the plant? How could you control the insects eating your plant without destroying other insects in your habitat? Categorize insects and plants in your habitat as helpful or harmful to your plants. Explain your reasons for putting the insects in the different categories.

1. Have students use insect guides to research the name and characteristics of the insects in the habitat.
2. Have students draw illustrations of each insect, labeling the drawing with what the insect eats, where it lives, and its relationship to humans. Collect the illustrations to make a class Garden Insect Book.
3. Have students write stories about their Magic Spots.

Garden Puzzle

Description

Students role play plants in a garden to show how different plants have different space requirements, how different-shaped plants can be grown together to use root and canopy shapes efficiently, and how efficient use of garden space will leave little room for weeds.

Objective

To demonstrate the principles of companion planting (intercropping).

Teacher Preparation

Traditional farming techniques all over the world have developed intricate planting systems for intercropping plants that can be beneficial to each other. These systems have evolved over thousands of years. One of the most successful is the Mexican intercropping of corn, beans, and squash. The corn acts as a strong stake for the pole bean vine to grow on, the bean as a legume fixes nitrogen into the soil (see The Matchmaker, p. 93), and the squash, a low-growing plant with large leaves, shades the ground and inhibits weed growth.

Materials

Pictures of corn, bean, and squash plants
Companion Planting Guide, Appendix pp. 455-457

What makes a living organism a plant? (ability to make its own food; has roots, leaves, and so on) Do all plants look alike? What are some of the differences? (shape, growth rate, fruit production, type of roots, and so on) When we plan our garden we are faced with a puzzle of different shapes, sizes, and types of plants. It has been learned through thousands of years of farming that some plants actually help each other grow.

If some of you will help me act this out, we can show how this works.

1. Create boundaries for an imaginary garden large enough for 3/4 of the class. Then narrate the action in the following steps as students act out each role.
2. Act I. Scene 1. Create a small corn field. Have half of the students stand with their hands above their heads to represent tall, slender corn plants. Have these corn stalks step into the garden and space themselves so that with their hands outstretched horizontally, their fingertips just touch. This represents a common corn monocropping system.
3. Scene 2. Now we need some weeds. Weeds are persistent, unwanted plants that take advantage of any space left by the corn plants. They can be tall and thin, like the many varieties of wild grasses, or they can be short and squat or even prostrate, like field bindweed or pigweed. Choose some students to be tall weeds; they will look similar to the corn plants, with their hands up over their heads. Other weeds can be represented by students stooping over. Others will crawl and grab the ankles of any other plant in the garden. Gradually have the weeds enter the garden until all the available room is gone. The crawling weeds can be really disruptive and may even slowly pull other plants down. Ask the corn how it feels competing for space.

4. Act II. Scene 1. In Act II we will plant other kinds of plants between the corn plants. A new corn crop should assemble as before in the garden. Now, to keep those pesty weeds out maybe we could find some plants that eliminate the extra spaces in between the corn. Hmmm. . . should we use tomatoes? Well they might fit, but I heard the corn talking about tomatoes and they don't seem to like being very close to them. They take up too much of their root space and also seem to make the corn a little sick. Well, we'd better see if we can find a plant that gets along better with the corn. (Refer to Companion Planting Guide, p. 455-457.) American Indians used to plant pole beans alongside the corn. Farmers in Mexico still do this. The pole beans use the corn stalk for support. The beans don't strangle the corn, and they actually bring nitrogen from the air into the ground to help feed the corn plants. Well corn, what do you think? Should we let these beans lean on you for a little support if they come through with the snacks? The beans enter the garden. Only one bean plant to one corn plant, please!

5. Scene 2. Now we still have a problem. It looks like there is still room for the weeds. (Enter weeds again.)
6. Act III. Scene 1. We'll have to find another companion for the corn and bean dynamic duos. Let's find a plant that is low to the ground, doesn't need too much light and one that won't put down deep roots, like the corn and beans. We've got to fill in pretty irregular spaces. . . . What about squash plants? They have broad leaves, don't need too much light and the Indians and Mexicans planted squash along with corn and beans, too. (Have students emulate squash plants by squatting down and filling in the empty spaces in the garden.)
7. Scene 2. OK, weeds. Ready for the last try? All right, but where will you go? There is no extra space in that garden, and its sure hard for weeds to grow above the squash to get sunlight. Even a tenacious bindweed would have trouble finding a home in there.

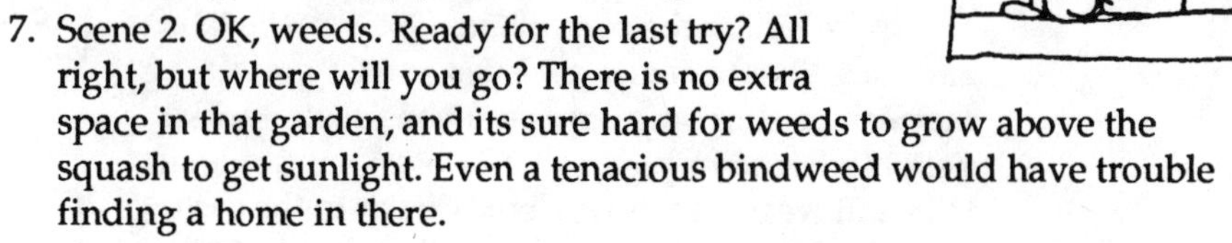

8. And thus, the principles of companion planting are revealed. The garden has become a place where three crops can live together comfortably, support one another, and even help each other to eat. Discuss with the class how these plants are beneficial to each other.

How can companion planting prevent weed growth? What are three things to consider when choosing companion plants? Why can weeds grow easily in a field of corn? How can this be prevented?

See Companion Planting, p. 243, to try some of your own garden experiments with companion plants.

Companion Planting

Description

Students plan and conduct a garden experiment on the effects of intercropping (companion planting) selected crops.

Objective

To determine if plants influence one another's growth.

Materials

Prepared garden beds
Seeds or seedlings
Companion Planting Guide, Appendix pp. 455-457
Life Lab journals
Meter stick
Garden markers and grease pencils

Preparation

1. Set the stage for this activity with the Garden Puzzle, p. 240.
2. Prepare garden beds for planting

Do you think plants influence one another? Many scientists are doing research now to determine how plants affect each other. It seems that some plants have chemical reactions with each other. One plant might repel a bad insect from another plant if they are side by side. A bad insect is one that damages the plants in a garden. One plant may attract an insect that helps or hurts another plant. Roots of certain plants may exude, or give off, chemicals that bad insects don't like. Plants also have physical effects on each other. One plant may shade another, protect it from wind, and so on. What other ways could plants affect each other? (Let students be imaginative. They don't have to be realistic—the concept is the important thing.)

1. Prepare three garden beds (or divide one bed into three equal parts) in exactly the same way.
2. Choose a plant you want to test and refer to the Companion Planting Guide, pp. 455-457, for one plant it likes and one it dislikes.
3. Have students plant Bed A with the selected plant and its friend, Bed B with your plant and the one it dislikes, and Bed C with the plant by itself. Bed C will serve as the control for the experiment. Be sure to plant the same number of the selected plant in each plot. For example, if you picked beans to test, Bed A could be beans and carrots (a friend); Bed B: beans and onions (an enemy); and Bed C: just beans.

4. Discuss with the group how to tell if the beans are growing differently in each plot. Some ideas include measuring and taking the average height of ten plants in each bed weekly; selecting 20 leaves from each bed and comparing insect damage; counting how many beans are harvested from each bed; measuring and averaging the length of ten beans from each bed; comparing the taste of beans from each bed.
5. Have students design a graph to record their data. (Use our sample as a guide.) Have them record their predictions as to which bed will grow the best beans.

DATE	HEIGHT OF BEAN PLANT IN INCHES			AMOUNT OF INSECT DAMAGE (SCALE 0-5)			DATE OF FIRST FLOWERING			LENGTH OF MATURE BEAN IN INCHES			TASTE (SCALE 0-5)		
	BED			BED			BED			BED			BED		
	A	B	C	A	B	C	A	B	C	A	B	C	A	B	C
1/1															
1/8															
1/15															
1/22															
1/29															
2/5															
2/12															

6. Have students record weekly observations in their Life Lab journals. In which bed did the beans sprout first? Grow fastest? Have less damage from insects and disease? Flower first? Taste the best?
7. After the harvest is over, analyze students' results.

What was the purpose of the control in this experiment? Did your results match your predictions? What did you learn from this experiment? How will this information help you plan next year's garden?

Try other plant combinations. The list of companions should provide lots of ideas.

Plant Architects

Description

Students design their own crop plant, construct it, then place it in the garden to determine the effect of its shape on weed growth.

Objective

To illustrate how the shape of some plants can affect the growth of other plants, especially weeds.

Materials

Construction paper
Scissors
Glue
Cardboard tubes
Coat hangers or sticks for stems
Planter box or bed

One way that plants protect their food is by their shape. Some, like squash, grow big broad leaves that make so much shade weeds don't get enough sun to grow. Another plant, like parsley, is short and skinny, but it grows so fast and close together that it leaves no room for weeds. We want to find out how the shape of a plant can affect the growth of weeds in the area around it. If you were to design a plant whose shape would not let weeds grow around it, what would it look like?

1. Divide the class into groups of four.
2. Have each group design and construct an imaginary crop plant. Encourage a variety of designs—tall skinny ones, short fat ones, and so on. They need not be models of real plants, and the students can make up their own names for them. In their designs, have them consider how the plant will get light, and how its shape will protect it.
3. Next have the students prepare a bed of soil in a planter box or a small bed outside, but do not have them plant any crops. They should weed the bed completely.
4. Have the groups plant their constructed plants in 3/4 of the prepared bed. Encourage the groups to work together in developing their planting pattern. Leave 1/4 of the area unplanted as a control to see the number of weeds that grow without the plants. Make sure there are no weeds already growing in either area.
5. Leave the garden for two weeks, then return and have the groups count the number of weed plants in the area immediately surrounding their own constructed plant. Compare this area to the control area.
6. Have the students replant their constructed plants in a pattern they think would be the best for keeping weeds from growing, and repeat the experiment.

Did some plants have more weeds surrounding them than others? Did some have none at all? What were the shapes of the plants that had the fewest weeds around them? What prevented the weeds from growing?

I Need My Space

Description

Students plant three experimental plots with identical seeds at varied intervals to discover a way to keep weeds from growing big and strong.

Objective

To demonstrate the effect that spacing has on controlling weed growth.

Materials

Radish seeds
Three 2′ X 2′ plots in a garden
Life Lab journals
Picture of a radish plant (See illustration below.)

Preparation

Prepare three 2′ X 2′ garden plots for planting.

All plants need light. Without light, a plant cannot grow strong and tall. It would be weak and withered. We will find out if we can plant radish seeds so that they have enough room and light but at the same time shade the weeds so that they will not grow. What are some things we should consider when deciding how far apart to plant our seeds? (how much space the radish needs to develop, what planting directions the package gives)

Have students complete the following steps:

1. In plot #1, plant the radish seeds six inches apart in rows that are six inches apart.
2. In plot #2, plant the seeds three inches apart in rows that are three inches apart.
3. In plot #3, plant the seeds one inch apart in rows that are one inch apart.

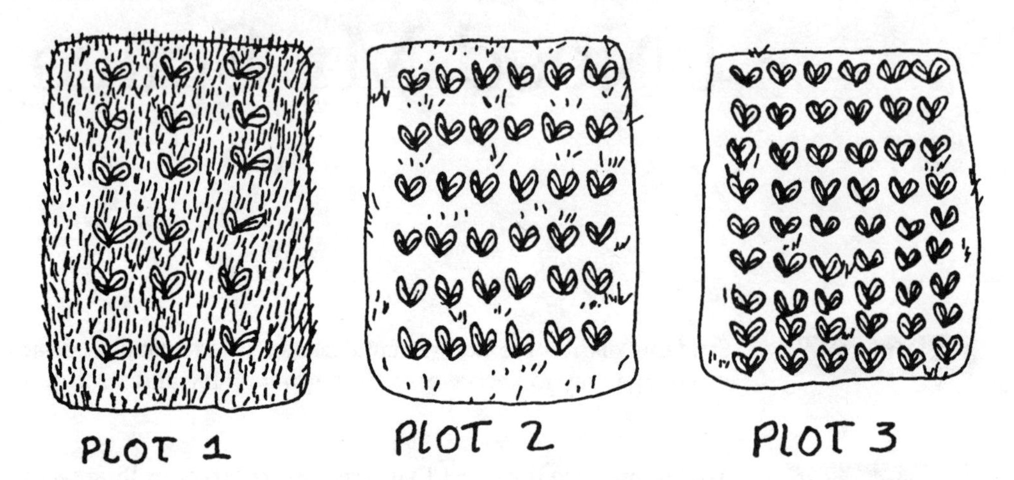

4. Water each plot equally, as needed.
5. Weed the plots completely after ten days.
6. Wait two weeks and then count the number of weeds in each plot. Keep a count of the number of weeds over one inch tall in each plot. Also keep records of the space between the canopy of the plants until the radishes are ready to harvest.
7. Explain to students that this activity demonstrates the principle of interference. All garden systems are designed to eliminate extra space for unwanted plants or weeds. The experiment has the radishes planted so close together that when they get larger, or more mature, they will be crowded. To continue the experiment and to determine the optimum spacing for mature plants, have students thin the radishes so that by the time the plants mature, the leaves just touch.

Which plots have fewer weeds? Why? Name one reason why it is good to plant things close together. Name one reason why it is bad. How can you determine the best spacing between plants?

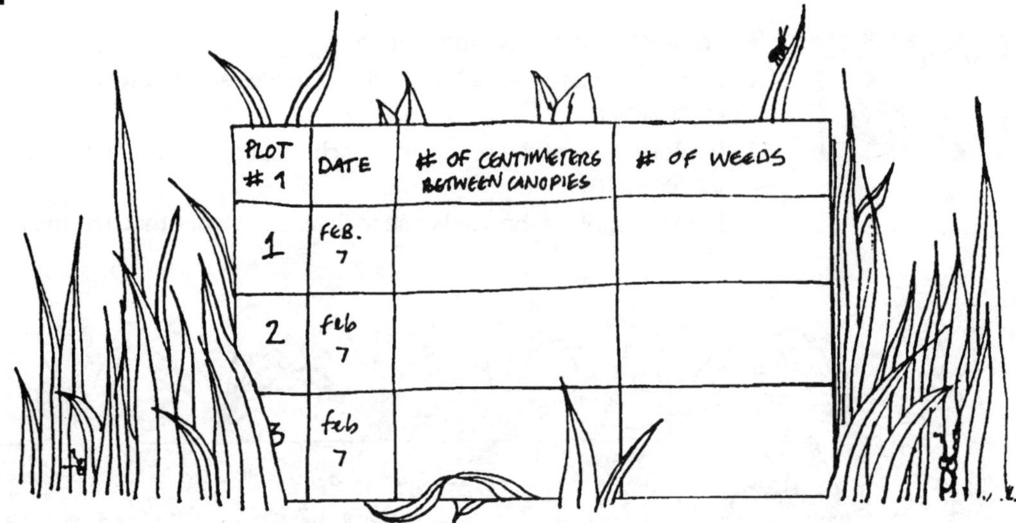

PLOT # 1	DATE	# OF CENTIMETERS BETWEEN CANOPIES	# OF WEEDS
1	FEB. 7		
2	feb 7		
3	feb 7		

Natural Defense

Description

Students conduct a weed growth experiment using leaves from certain plants.

Objective

To illustrate how a plant can emit natural poisons that inhibit the growth of neighboring plants.

Teacher Background

Have you ever noticed how some trees and plants, such as eucalyptus, pine, redwood, and tomatoes, don't have many weeds growing near them? Researchers theorize that leaves and other parts of these plants have natural chemical inhibitors that reduce another plant's growth. Rain washes these chemicals into the soil and can stunt or even prevent plants from growing, thus protecting some plants' soil nutrients, water, and light. In this activity, students will conduct an experiment to explore this concept.

Materials

Four one-gallon plant containers for each group
Water
Garden soil
Masking tape
Marker
Life Lab journals
Leaves from some of the following plants: eucalyptus, bay, pine, tomato, walnut

Preparation

1. Divide the class into groups of five.
2. Have each group collect three types of leaves from the above list, put them in bags to dry, and label the bags. Have them collect as many different types as possible.

Have you ever noticed how animals sometimes protect their food, such as a dog growling when a cat comes near at dinner time? What do you suppose a plant might do if it wants to protect its "food" (the soil, sun, air, and water)? Since they can't growl or chase or bite, some plants have found ways to reduce competition with other plants for necessary resources. Sometimes a weed can do this by preventing crops from growing next to it. But other times, a crop plant is the one that gives off the inhibitor so weeds can't grow underneath it. In this activity, we're going to see if we can find plants that protect their food by producing their own weed inhibitors to keep weeds from growing near them.

1. Give each group four one-gallon containers.
2. Have each group fill their containers with regular garden topsoil dug from the same area.
3. Have them label one soil container "control" and the remaining soil containers with the name of a collected leaf.
4. Have students crumble the dried leaves from one type of plant and mix them in with the garden soil from the corresponding container. The leaves should comprise about 5 percent of the mix. The control container will not have any leaves mixed in.
5. For two or three weeks, have students water each soil container with the same amount of water. Be sure they keep the soil moist so weeds can grow.
6. After two weeks, have students count the number of weeds that have germinated in each container, recording the results. Have them continue to observe and water for three weeks.
7. Did any group have a container where no weeds sprouted? If so, it shows that the leaves added to the soil may have some chemical in them that limits other plants' growth.
8. Have the students look around the rest of the school grounds to see if they can find any plants that have nothing growing beneath them. Explain how the rain washes the chemical inhibitor from the leaves into the soil.

What did this experiment tell you? Which leaves reduced weed growth? Why? Why is this an important adaptation for some plants? How can this help us in the garden?

1. Have students use leaves with chemical inhibitors as a mulch around perennial plants or on paths.
2. Have students make a leaf extract of the inhibitors by putting fresh leaves in water (one part leaves to ten parts water), and letting it steep for two hours. Pour the extract on weeds, and observe and record the results.

Under the Big Top

Description

Students observe the ecological interactions of the garden by viewing garden activity as a performance of nature.

Objective

To recognize the dynamic interconnection of soil, water, air, sun, plants, and animals in the garden; to reinforce concepts learned in this unit while strengthening observation skills.

Materials

None

Let's list different parts of the garden. (plants, insects, soil, roots, flowers, and so on) How do these parts help each other? What do the roots give the soil? What does the soil provide for the roots? What do spiders provide? How do broad-leafed plants affect garden life?

Note how the removal of some garden elements would change the garden. What might the garden be like if you took away the worms? What if it were located in a place that had more shade?

We're going to watch the garden put on a performance—a big show or circus. It has been preparing for this performance for months and now it is time for us to enter quietly and observe the show. This performance requires all the garden actors to work together, supporting, protecting and assisting each other. Many things are happening at the same time in this show, some very quickly and others very slowly. How is the stage constructed? (The stage may be raised or flat. It is the soil and roots that form the foundation of the garden.) How is it secured and decorated? (The roots secure the stage. There may be a backdrop of trellises and poles.) Where is the spotlight? Does it move? (The spotlight is the sun, which moves across the sky; the sunlight is broken up by leaves into many small spotlights.) How do the performers respond to the spotlight? Do any hide from the spotlight? (Plants respond to sunlight and shade, some stretching up, some moving sideways. Some plants and animals prefer shady areas.) How do the performers move? Which ones move quickly and which move slowly? (Some movement can be recognized only over a long time, such as flowers opening and closing to the sun.) How do the actors help each other to perform their acts? (You may review interactions between garden actors that you have observed throughout the unit.)

1. Give the students about ten minutes in which to witness the performance from different spots in the garden. Have them each look for at least three performers that are interacting. To provide a different perspective, you may ask students to imagine that they are very small and are witnessing the show from a leaf or from the back of an ant.
2. Call intermission and ask the garden watchers to discuss with each other what they witnessed in the garden show. Note the variety of interactions described by students.
3. Have students return to the performance in groups of three, each group with its own observation point. After five or ten minutes of quiet observation, have each group act out the interactions of three performers that they observed.

Did anyone observe something that they hadn't seen before in the garden? How is the garden show different from the show happening just outside of the garden bed? What role do we play in the garden show? What do we give to and take from the garden? If the garden performers came to watch the show put on by the students and teacher in the classroom, how would they describe that performance?

1. Students may write a review of the garden performance, commenting on the performers, the stage, and the special effects as if they had seen a circus show. Students may want to refer to a newspaper review of a circus or other show for a review format.
2. As a class or in small groups, students may draw the performance, including what's happening underground and above the plant tops. This drawing could take the form of a large mural to be displayed in the classroom.

Garden Creatures

Robert F. Kennedy School
San Jose, CA

Happy Valley School
Santa Cruz, CA

Freedom Elementary
Freedom, CA

Freedom Elementary School
Pajaro Valley Unified School District
Watsonville, California

"The Life Lab has added so much to our program. It's a hands-on total involvement from the compost under their feet to the dirt under their nails, an experiential kind of program. Children learn best from direct participation in the learning experience and Life Lab totally fulfills this goal."

Bill Staver

Started by a small group of teachers in 1985, eighteen classrooms now participate in Life Lab with teacher and garden aide direction. There are thirty garden beds, including individual class gardens and several cooperative gardens where herbs and flowers grow. Within the area, there is a tool shed, greenhouse, worm bin, coldframe, and an outdoor classroom area. Many of the older students volunteer during their recess to help work with the younger children.

Life Lab offers a different teaching environment. The garden becomes the classroom and learning becomes experiential. As one student summed up Life Lab, "The garden is where we get to learn about real stuff like how plants grow and how they help us. It's different from having been in the classroom all day. You get to learn by doing stuff!" Whether working in a garden bed, studying a worm, or taking soil samples, Life Lab encourages a positive learning experience.

Children take pride in their gardens, projects, and other work. They have grown protective of their Life Lab. They learn respect for the outdoor environment as well as for one another.

UNIT INTRODUCTION

Garden Creatures

As we learned in "Interdependence," various types of predators may feed on the same prey. This is especially evident in the Living Laboratory, where sometimes the most plentiful inhabitants aren't vegetables at all but snails and 12-spotted cucumber beetles. The understanding of food chains and life cycles is critical to the study of the creatures that live in your Living Laboratory. Ladybird beetles, spiders, and earthworms can mean a healthy garden while snails, slugs, and aphids can mean trouble. Understanding the relationship of predator-prey and their varied habitats will give you the scientific background to develop a sound pest management plan. However, don't let those pests escape without careful examination. Insects traveling on a thin film, having a skeleton on the outside of their bodies, changing through three completely different forms in a life cycle provide interesting opportunities for observation and a greater understanding of the processes, forms, and functions of nature.

This unit introduces the anatomy and life cycles of some insects as well as the discovery of their preferred habitats. The application of the food chain concept makes it possible to categorize insects as helpful or harmful. Special studies are made also of butterflies, snails, spiders, and earthworms. "The Butterfly Flutter By" and "The Great and Powerful Earthworm" will require extended observation time. This unit is recommended for spring.

Get your hand lens, terrarium, and Life Lab journal ready. It's time to observe and record your garden creatures. The more you watch, the more you see.

Activities	**Recommended Grade Level**
Earth, Planet of the Insects *(effects of insects on earth)*	3,4,5,6
Insect Anatomy, or Make No Bones About It! *(identification of insect parts)*	3,4,5,6
The Butterfly Flutter By *(observation of insect life cycle)*	2,3,4,5
Who Lives Here? *(observation and collecting of insects)*	3,4,5,6
Buggy Diner *(feeding preferences of insects)*	5,6
Ladybug, Ladybug, Fly Away Home *(demonstration of beneficial garden insects)*	4,5,6
Slimy Characters on Trial *(pest and pest damage identification)*	4,5,6
My Friend Tummy-Foot *(snail anatomy and behavior)*	4,5,6
Following the Thread *(garden observation of spiders)*	3,4,5,6
The Great and Powerful Earthworm *(experiment showing earthworms as soil tillers and decomposers)*	2,3,4,5,6

Banana Slug

C F
Sticky as peanut butter, shade of yellow

G C
Look like banana and oh so mellow

C F
Good life givers, living on the ground

G C
Chewing on leaves when they fall down.

Chorus

C F
Banana Slug, banana slug

G C
I like them, they're beautiful.

C F
Banana slug, banana slug,

G C
They're part of the circle.

On the side of their head there's an all-purpose hole
They've got one foot and plenty of soil
All day long they work and toil
Munching down duff to renew the soil.

Chorus

They make soil to grow new trees
Trees make air for us to breathe
They're part of the circle that lets things grow
So be kind to banana slugs and let things flow.

Chorus

Banana slug, slug, slug, slug, slug, slug
Banana slug, slug, slug, slug, slug, slug
Banana slug, slug, slug, slug, slug, banana slug!

(Banana slugs live in the redwood forest. They are bright yellow and can grow to about eight inches.)

Earth, Planet of the Insects

Description

Students observe insects in the garden, categorize insects as being harmful or helpful, and write an insect story.

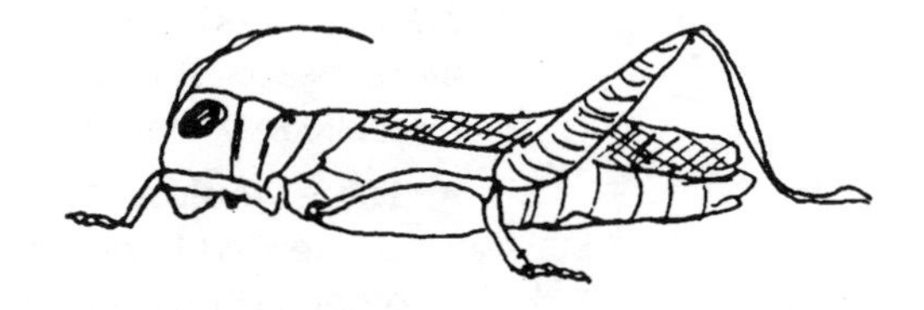

Objective

To explore the many ways that insects affect all life on earth.

Teacher Background

Insects exist in enormous numbers. The average insect population per square mile is estimated to be equal to the total world population of people. Less than one percent of the nearly one million insect species are pests to humans, our domestic animals, and useful plants. Of the 100,000 insect species in the United States, only about 600 are considered to be serious pests. Damage done by insect pests is easy to appraise; the value of beneficial insects is harder to estimate. Bees, wasps, flies, butterflies, and other insects pollinate flowers that provide us with fruits and vegetables. Honey, wax, and silk are important insect products. Some insects are vital links in the food chains of fish, birds, and other animals. Other insects are parasites or predators of damaging pests.

Materials

Blackline masters of Beneficial Insects and Common Pests, pp. 407-408
Life Lab journals

There are more insects than any other creature on earth. They have been around for at least 400 million years, whereas people have been on earth for only about 100 thousand years. And we are still discovering new kinds of insects! What are some things we already know about insects? (List responses on the chalkboard.) What are some things you would like to find out about insects? (List responses.)

1. Go into the garden. Have students sit in pairs quietly for several minutes in different spots in the garden and observe. Ask, What kinds of insects do you see? What are they doing? Are they eating? Are they being eaten? Have students go around and look under rocks or logs. Ask, What insects do you see there? What are they doing?
2. Back in the classroom, ask students to help make a list of the ways in which insects are harmful or helpful to us. Write responses on the chalkboard. The following list gives some examples.

Insect Activities That Harm Us

- Eat our crops (caterpillars, beetles)
- Eat wood (termites)
- Eat clothes (clothes moths)
- Sting or suck blood (wasps, mosquitoes)
- Transmit diseases (mosquitos, fleas)

Insect Activities That Help Us

- Many are pollinators (bees, butterflies)
- Bees produce honey
- Silkworms make silk
- Some are predators and parasites of insects that harm us
- Many are food sources for animals (birds, frogs)

3. Now ask students to pretend that, starting tomorrow, there will be no more insects. Ask them to write about what life would be like without insects. How would things change? Possible answers include:
   - There wouldn't be so many fruits and vegetables. Bees, for example, are needed to pollinate apples, cherries, cucumbers, and many others. Without pollinators, there would be few of these fruits.
   - Many animals would die. Animals such as frogs, birds, and anteaters depend on insects for food.
   - There would be no honey or silk.
   - There would be no butterfly wings, cricket calls, lights from fireflies.
   - There would be no mosquito bites!

   And much more—insects are really very important to life on earth!

Name three things you observed about insects in our garden. What are some ways insects are important to you? Are humans important to insects? What might a harmful garden insect be doing? What might a beneficial garden insect be doing?

Have students use the blackline masters to identify insects they observe in the garden. What makes them harmful or beneficial to us? What are they doing in the garden? Have them consider how harmful insects might be controlled without harming the beneficial ones.

Insect Anatomy, or Make No Bones About It!

Description

Students examine insect anatomy and draw garden insects.

Objective

To identify the parts of insects.

Teacher Background

All insects, even though they appear to be as different as a moth and an ant, have similar body parts: head, thorax, abdomen. Attached to the thorax are three pairs of legs and, in most species, two pairs of wings. The head has one pair of antennae. This combination of features distinguishes insects from closely related animals such as spiders, ticks, mites, centipedes, and millipedes.

Materials

One Insect Anatomy blackline master per student, p. 406
One bug box per group of three students (order from Let's Get Growing, p. 480)
Life Lab journals
Drawing materials

How many different insects can you name? (List responses on the chalkboard.) They look very different, don't they? What makes each of these animals an insect? (Let students generate as many ideas as possible.) Let's find out what these insects have in common.

1. Distribute the Insect Anatomy drawing and give students a few minutes to study it.
2. Point out to students that virtually all insects share the following characteristics:
   - Six jointed legs (Note that spiders, which have eight legs, are not insects.)
   - A body divided into three main parts: head, thorax, and abdomen
   - A pair of antennae
   - Most have wings—usually two pairs
   - A hard outer covering called an exoskeleton

3. Discuss and identify each of the parts on the drawing. Point out that one big difference between insects and other animals is that insects have an exoskeleton, a skeleton on the outside of their bodies. Discuss the human skeleton and its function. (It supports the tissues and protects the internal organs.) What is the main advantage of a hard exoskeleton? (It acts as a coat of armor that protects the insects.)

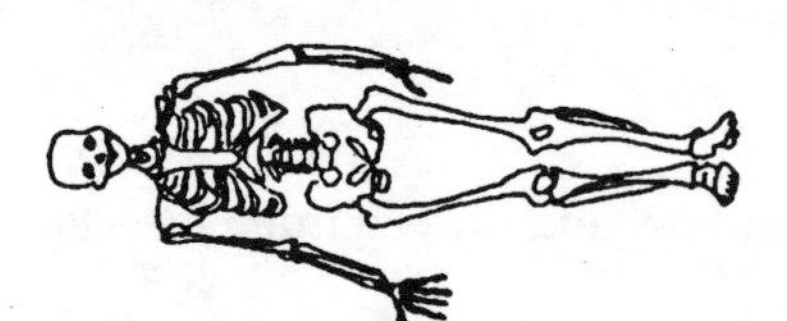

4. (Optional) Give students time to color in the insect anatomy.
5. Give each group of three one bug box. Have them go to the garden with drawing materials.
6. Each group should carefully collect one insect in the bug box, study it, and identify its parts. Then they can draw it and label the parts on the drawing. Be sure to release the insects carefully.
7. Have groups share their drawings, comparing ways in which the insects differ and ways in which they are alike.

A moth and an ant have many characteristics in common. What are they? Why do insects have an exoskeleton? How do you think insects grow if they have an exoskeleton? (They shed it.) What would it be like if you had an exoskeleton?

Identify the insects that were drawn. Use an insect reference guide or the blackline masters, p. 407 and p. 408.

The Butterfly Flutter By

Description

Students collect caterpillars and observe and raise them in the classroom until they are butterflies or moths.

Objective

To observe the metamorphosis of insects in their life cycle stages: egg, larva, pupa, adult.

Teacher Background

In their life cycles, insects go through one of two types of metamorphoses. A *simple metamorphosis* has three stages: egg, nymph, adult. Nymphs are young insects that as soon as they hatch resemble miniature adults. The nymph will shed its skin, or molt, as it grows. Nymphs are active. They commonly live in the same place and have the same feeding habits as adults. Grasshoppers, cockroaches, and lice go through a simple metamorphosis. More types of insects go through a *complete metamorphosis*, changing in appearance and habits through four distinct stages: egg, larva, pupa, and adult. The egg hatches into a larva, an active, feeding stage. The mature larva forms a pupa, a resting, nonfeeding stage from which the adult emerges. Caterpillars are the crawling larvae of butterflies and moths. They have chewing mouthparts and can be major garden pests. The adults have sucking, or nectar-feeding, mouthparts and never do damage.

Materials

One clear jar with a lid per group of five
Leaves and sticks
Art materials
Insect reference guides
Life Lab journals

Preparation

Put air holes in the jar lids.

Have you ever watched an insect grow up? How would you like to adopt a caterpillar from the garden and watch it grow? What do you need in order to take care of the caterpillar? (place to keep it, food) What does a caterpillar eat? (Accept all ideas.) How will you find a caterpillar?

1. Divide the class into groups of five. Give each group a jar with lid with air holes.
2. Each group plans a caterpillar home by
   - discussing together how to set it up (insect reference books can give them information on food supply and so on)
   - decorating the home by supplying it with food (green leaves) and a stick for the caterpillar to climb on
   - recording predictions in their journals as to how big the caterpillar will get, what it will look like as an adult, how long it will take to become an adult
   - designing an observation chart to be kept next to the home
3. When the whole group is ready and you have checked the home to make sure it is prepared, have each group collect one caterpillar from the garden. (Note: A caterpillar has six legs toward the front end; a worm has none!)
4. To welcome the caterpillar have each group measure its length, draw it, label its parts, and, of course, name it!

5. Have students observe the caterpillars regularly, making sure they have fresh leaves, and record any changes. The pupa stage will be a chrysalis or cocoon. The adult stage will be a moth or a butterfly. Have students draw each stage including as many observations as possible.
6. When the adult emerges from the pupa, have students observe it and then take it to the garden and release it. Have them observe it for as long as possible and spend time observing other butterflies in the garden. What do they eat? How is it different from what caterpillars eat? Explain that the butterfly will find a spot to lay eggs from which more caterpillars will emerge.
7. Discuss and draw all stages of the butterfly's life cycle. How is it a cycle?

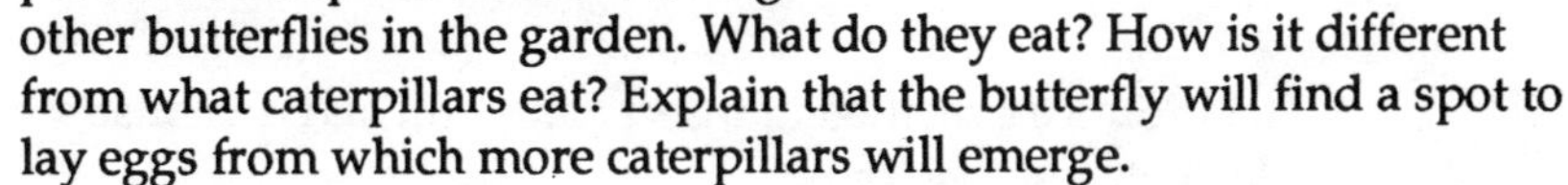

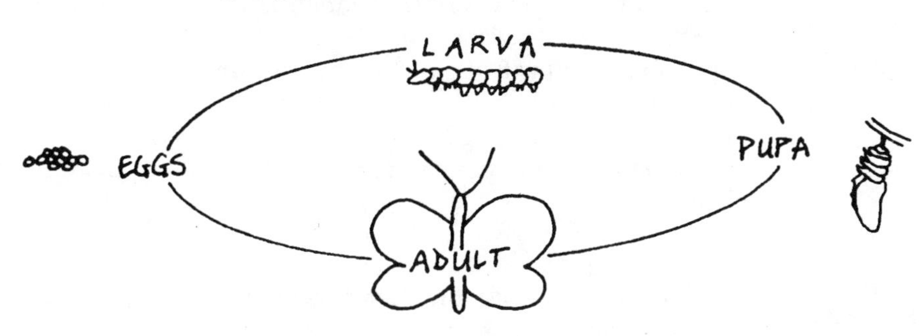

How do you know that the butterfly is an insect? What are the stages of the insect life cycle? Describe three interesting things you learned about caterpillars and butterflies. Why might it be difficult to identify an insect even if you know what the adult looks like?

Have students identify different insect stages using an insect reference guide. See if they can find insects in the garden at different stages. If an insect is a beneficial insect, at which stage do you think it eats the most harmful insects?

Who Lives Here?

Description

This lesson is divided in two sections. In Part One (optional), students make insect collecting nets. This can be accomplished in small groups, resulting in one net for each group of five students. In Part Two, students collect and identify insects.

Objective

To introduce a method of insect collection and demonstrate the variety of insects in a habitat.

Teacher Background

Insect nets are fun and easy to construct, as shown in Part One. However, if you don't have time to make them, they are readily available and inexpensive. (See the Let's Get Growing catalogue, p. 480.) Part Two is important in demonstrating how different habitats support different insects. Try to select as many different habitats as possible (garden, landscaped area, field, and so on). In comparing insects within one habitat, you will often find they consume different food resources and avoid competing with each other. Thus some will have chewing mouthparts, others sucking, and so on. Bug boxes are great for studying insects because one side is magnified for closer observation. After observation, insects can be released to their habitat.

Materials

One set of Insect Collecting Net directions per group, p. 450
Net materials for each group (see direction sheet)
One jar with air holes per group or several bug boxes per group
Insect identification books
Magnifying lenses
Life Lab journals

Preparation

Precut net materials for each group.

Where do insects live? (everywhere) What determines where they live? (availability of food, water, light, shade, shelter) If you were to look under a log or a rock, what insects and other animals would you expect to find? (earwigs, snails, slugs) Would you find the same insects in a sunny spot? (no) Insects, like other animals, live in habitats that provide them with the food, water, and shelter that they need. Let's list some habitats we can explore around school to find insects.

Part One:

1. Determine how many insect collecting nets you want to make. Divide the class into small groups and have each group make their own net, following the directions on the Insect Collecting Net handout.

Part Two:

1. Have each group choose a different habitat for collecting insects (garden, orchard, field, rocks, baseball field, and so on).
2. Students should spend approximately 20 minutes collecting. The net should be brushed through weeds, bushes, and branches of trees, sweeping insects off the plants. After a few sweeps, students may flip the end of the net over the rim to trap the insects, flip it again to force the insects to the bottom of the net, and examine the catch.
3. Have students trap the insects they wish to keep and transfer them to the jar or bug box. Discuss gentle handling of insects so that wings and legs are not damaged.
4. Have each group record the habitat explored and describe the characteristics of each insect. Are the wings leathery or transparent? Does the insect have sucking or chewing mouthparts? Is the abdomen exposed or covered by the wings? Does it have two or four wings? Have students compare one insect to another.
5. Have students use the insect identification book to identify the insects.
6. Have students determine each insect's food requirements from books or observation of mouthparts. What characteristics do insects that eat the same food have in common?
7. Have each group draw a picture of the habitat, showing the plants, insects, and other organisms they find. How does each insect live in this habitat?
8. Have each group share its discoveries with the whole class.
9. Release the insects in their habitats.

Do different insects live in different habitats? Do different insects live in the same habitat? What are some characteristics that insects share? How do they differ? Name insects in your habitat that had the same food resource. How could you tell? Give an example of how one of the insects you collected depended on its habitat. How were any of the insects you found beneficial to their habitat?

1. Have students choose one insect and write a story about it. Have them imagine why the insect developed its characteristics (wings, chewing mouth, color, and so on) to live in its environment.
2. Have students make an insect habitat mural, showing insects in their varied environments.

Buggy Diner

Description

Students collect common garden insects and set up a simple food preference test for them.

Objective

To demonstrate the feeding preferences of some common garden insects.

Materials

Ten plastic petri dishes with covers
Garden insects
Variety of plant leaves
Plant and insect identification guides (optional)
Life Lab journals

Preparation

The day before this experiment, collect ten garden insects and place them out of direct sunlight in a jar with air holes. Do *not* feed them.

Did you ever wonder what insects eat? Not all insects eat the same foods. Some like to eat the leafy parts of vegetables and some like the roots. Some insects prefer other insects! In this experiment we will construct an insect diner and serve food samples to the invited insects.

1. Divide the class into ten groups. Have each group collect six different leaves from the garden. These will be food samples to determine the insects' preferences. Be sure the students know what type of plant each leaf comes from.
2. Have students record observations about the leaves in their Life Lab journals. What plants do they come from? What do they look like? Does it look like they provide food for insects or other garden animals? How can you tell?
3. Have students cut similar-sized circles from each plant leaf. It is important that they be the same size in order to get accurate results.
4. Have students label the bottom of each dish with the leaf names. Place six leaf samples around the inside edge of each dish.

5. The diner is now ready to open for lunch. Who will you invite? Each group can choose one of the insects from the jar. For example, one group may invite a common and voracious eater, the spotted cucumber beetle named Spot. First make a prediction as to what you think Spot will order to eat. Place Spot in the prepared petri dish and cover it.
6. Let Spot stay in the diner for 24 hours. Observe Spot every two or three hours to see what he is munching.
7. To verify results, some groups may introduce insects of the same variety.
8. Estimate the amount of each leaf that was eaten. Record the results in your Life Lab journal. You can use a chart similar to the "Diner Check." If Spot eats the same plants that we eat (cabbage, lettuce, and so on), would we want Spot in the garden?

DINER CHECK

DATE: January 2

DINER'S NAME: cucumber beetle

FOOD CHOICES (CHECK PREFERENCES)

1 cabbage leaf	
2 cucumber leaf	✓
3 bean	✓
4 grass	
5 tomato leaf	
6 lettuce	

CONCLUSION: PREFERS... #2, #3

Which insects preferred to eat foods that we eat? Do we want them in the garden? How could you use a plant they did not like to protect the food we eat? Which insects ate foods we do not want in the garden? How can we encourage them to stay in the garden?

INSECT	FOODS THEY LIKE . . .
SPIDER MITES	BERRIES
APHIDS	CABBAGE, BEANS
CABBAGE LOOPERS	CABBAGE, LETTUCE
SNAILS AND SLUGS	LETTUCE, CABBAGE

Have students spend time observing in the garden. Ask, How many insects can you find eating? Can you find insect damage (holes in leaves and so on) but no sign of insects? When do they eat? If you have a garden at home, take a flashlight at night and go out into the garden to observe.

Ladybug, Ladybug, Fly Away Home

Description

Students design a terrarium home for ladybugs and observe their habits.

Objective

To illustrate how a garden can become a home for beneficial insects.

Teacher Background

There are many insects that are beneficial to us in the garden, because their food source is other insects that eat our crops. The best known of these insects is the ladybug, also known as the ladybird beetle or lady beetle. These predators go through a complete metamorphosis, and their larva stage looks very different from the adult. The common food of both the larva and adult is the aphid, a major crop pest that sucks liquids from leaves. In the early 1900s, aphids were causing major damage to California's fruit trees, and ladybugs were released to control the aphids in orchards throughout the state.

Ladybugs are readily available for purchase. Just be sure they are not about to swarm (and fly off). These beetles hibernate over winter and then fly away to a new home. If they were collected during hibernation, they will swarm before they settle in a new home!

Materials

One terrarium with a tight mesh cover
A variety of seedlings, including flowers and weeds
Soil mix
Ladybugs (to order, see Additional Resources, p. 480)
Life Lab journals

Sometimes it's good for gardens to have plants that you cannot eat. Even weeds can sometimes help certain crops. In this activity we will examine one way in which weeds can reduce crop damage from insect pests. Not all insects are pests in our gardens. For example, ladybugs help by eating insects that damage crops. They eat aphids that eat many vegetables, so we want to encourage ladybugs to live in our garden. One way we can convince ladybugs to do that is to make sure our garden has plenty of attractive ladybug habitat and enough food. So in this activity we're going to try to find out where ladybugs like to live.

1. Show the class the variety of seedlings and weed plants to choose from. Design the terrarium so that the ladybugs will have as much variety as possible to choose among. The class can plant whatever they want, as long as they plant some crop plants and some weeds. Transplant the seedlings into the terrarium with three inches of soil on the bottom. Let the plants grow for a few weeks.
2. Ask the students to guess what plants the ladybugs will choose.
3. Before you introduce the ladybugs, gather leaves with aphids for ladybug food. Scatter these around the terrarium so that they are not in just one location.
4. Release a busload of ladybug tourists in the terrarium. Return the next day and have the students write in their journals which plants the ladybugs have made their homes.
5. Have students observe the ladybugs for as long as possible. Keep a log by the terrarium to record observations and changes. Be sure to keep the ladybugs well fed with aphids.
6. When ready, take the terrarium to the garden and release the ladybugs. Have the students try to observe the new homes they choose.

Do ladybugs seem to like some crops better than others? Why? Do they seem to like weeds even better than crops? Why? Is there a food source on the plants where they made their home? If you want ladybugs to live in your garden to eat insect pests, which plants would you grow to attract them?

1. Have students go into the garden with their Life Lab journals and pencils and tally the number of ladybugs they find on different plants. Did specific plants have more ladybugs than others?
2. Have students put a ladybug and an aphid under a microscope and watch a predator-prey relationship up close.

Slimy Characters on Trial

Description

After determining the numbers and locations of a certain pest in the garden, students hold a trial to decide if the pest is in fact guilty of eating their plants.

Objective

To identify pests and pest damage.

Teacher Background

This activity provides students with an excellent opportunity to use careful observation and evidence collection to support a hypothesis. Sometimes an animal that is considered to be a pest is just resting in the hole made by another pest. We recommend that the class choose an easy-to-identify pest from your garden.

Materials

Life Lab journals

We share our garden with lots of critters. Some we see and some we don't. They can live under the ground, on the ground, or in the air. Crawling, climbing, flying, jumping, slithering—they are always moving, probing, searching for food. Who lives in our garden? Where do they go for food? Let's be detectives and find out. What do detectives do? (collect evidence)

First, we will select an animal we think is a major pest in our garden. It needs to be one we can easily find. We will divide into teams. Some of the teams will look for evidence of damage by the suspected pest; other teams will look for evidence of innocence. Then we will present the evidence at a trial.

1. Appoint a judge, a prosecutor against the pest, and a defense lawyer for the pest. Divide the rest of the class into two teams, one team to work with each attorney.
2. Assist the judge in developing the rules for collecting the evidence and in determining when the trial will start.
3. Advise the attorneys to meet with their groups and to divide them into teams of three to go into the garden with pencil and paper and find as many pests as they can. Look into overgrown places and on and around the crop plants, especially in between leaves where they like to hide.

4. Each team should count and record how many pests there are on any plant and shrub and then make a map.

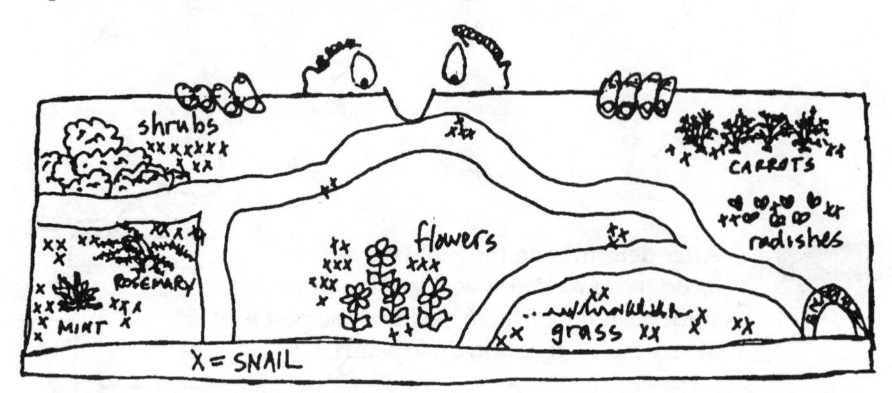

5. Ask, Does the pest seem to have a favorite hiding place or favorite food? How do you know they were eating it? What signs can you find?
6. The prosecutor's team should gather and record any evidence of damage such as damaged leaves, descriptions of evidence leading to damaged plants, and the number of pests on plants nearby, and so on.
7. The defense attorney's teams should gather evidence indicating that the damage was caused by pests other than their clients. (Which ones? Slugs? Beetles? How can you prove it?) They should look to discover other pests that feed on leaves and describe and map them.
8. Now return to the classroom and hold a mock trial. Everyone except the attorneys and judge sits as jury. Remind the jurors to be impartial. The prosecutor should present the evidence to convict the pests of plant-eating, and the defense attorney must try to prove their innocence by presenting evidence that the damage could have been done by other insects. Attorneys may call on members of their teams as witnesses. The defense attorney could suggest that the simple presence of the pests near the scene of the crime does not prove their guilt.
9. Now let the jury decide—remember, innocent until proven guilty! If guilty, the judge must decide the sentence.

Is it always easy to tell what pest has been eating a plant? Why or why not? Did your pest seem to have a favorite food in the garden? Would it die if you didn't grow that crop any more? Are all insects pests?

My Friend Tummy-Foot

Description

Students adopt pet snails, gathering basic data and observing their anatomy and physiology.

Objective

To describe the basic structures and functions of a land snail, devise and test a simple hypothesis, and communicate results of testing.

Teacher Background

The garden snail is, in many ways, an ideal classroom animal: It is anatomically relatively simple, slow but active, easy to handle, tough except for the shell, and fascinating to watch. Snails can be collected readily in nearly any garden from early spring to late fall. If they must be purchased, they are inexpensive and available from biological suppliers.

Snails are close cousins of abalone and clams. We call ours Tummy-Foot because that's literally what *gastropod*, the name of the class of animals, means. They have blood, but it has no hemoglobin and is clear. They possess a well-developed nervous system and even have what might generously be called a protobrain. They breathe through an opening near the top of the shell; they defecate and urinate through openings at the side. The genital opening, barely visible, is behind the upper antennae on the right. These antennae have eyes at the tips; the lowest pair act as feelers and odor sensors. Snails produce lots of sticky mucus in glands around the foot. The mouth contains an upper tooth-like "jaw" and a lower muscular "lip" behind which lies a leathery "tongue," the radula, covered with tiny teeth that are important in breaking down food. The radula both scrapes and carries food like a conveyor belt.

Materials

One terrarium with a lid
Three or four lettuce leaves (replenish as needed)
A jar lid filled with water
One garden snail per pair of students (see Preparation for Part One)
One clear plastic tray per pair of students
One hand lens per pair of students if possible
5"X7" file cards and a box ("Snail Registration Cards")
One small bottle of fingernail polish
One small jar lid or 35mm film canister lid per pair of students
Two or three tablespoons margarine or butter
1/4 of an apple per student
Balance with 1, 2, 5, and 10 g masses
One ruler per student
Newspaper for lining desks
One set of Snail Registration Cards per pair (see Preparation for Part One)

Part One

Preparation

1. A few days ahead of time, set up terrarium with lettuce and water inside. Direct students to begin collecting snails from the garden.
2. Prepare 5″X 7″ cards (see sample).

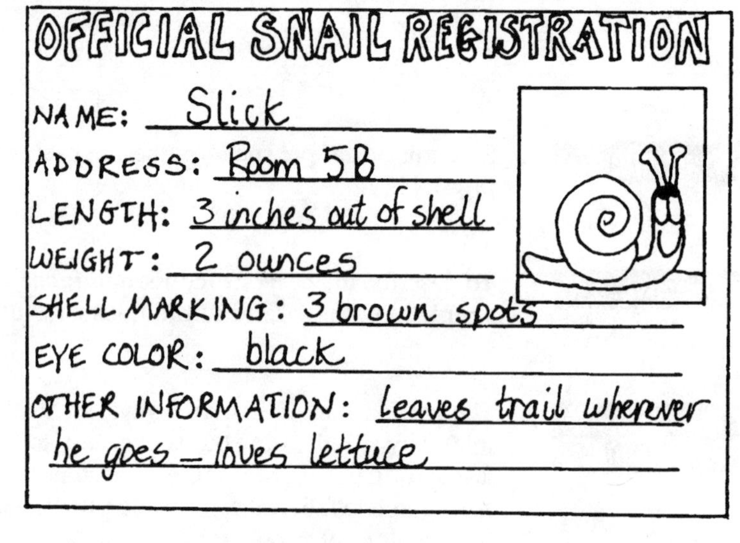
OFFICIAL SNAIL REGISTRATION

NAME: Slick

ADDRESS: Room 5B

LENGTH: 3 inches out of shell

WEIGHT: 2 ounces

SHELL MARKING: 3 brown spots

EYE COLOR: black

OTHER INFORMATION: leaves trail wherever he goes — loves lettuce

What do we know about snails? We've collected a lot of them. Do you think if we spend time observing them we can learn more? We'll be working in pairs, and each pair gets a snail to adopt. I call my snail "Tummy-Foot." Why is that name appropriate? First we'll prepare an Official Registration Card for our snails. Here's the information required on my driver's license. (Read items.) So, what information can we register about our snails? (List responses on the chalkboard: name, length, weight, shell whorls and markings, eye color, pod print, mug shot, drawing, and so on.) In order to adopt a snail, there are two rules we have to agree to:

- TOUCH ONLY YOUR OWN SNAIL.
- KEEP YOUR SNAIL SAFE FROM ALL HARM.

1. Design the Registration Card with the whole class.
2. Distribute the snails in trays (with newspaper) for observation, one snail for each pair of students.
3. Student pairs must first mark their snail on its shell with fingernail polish so that there will be no confusion later.
4. Next have them draw their snail's mug shot on the Registration Card.
5. Have students begin measuring and analyzing their snail to fill out the rest of the card.

What color is your snail? How long is it? How much does it weigh? Describe the head, tail, foot, shell. How many openings does its body have? Why do I call mine Tummy-Foot?

Part Two

Let's read over our Official Snail Registration Cards. Do they tell everything about our snails? (No, they only consider anatomy.) What else is there we ought to investigate about our snails? (How they work, what they do, why, and so on.) Today, let's concentrate on how snails function. What are some things we should focus on? (Write headings on the chalkboard: Movement, Sensing, Feeding, Communicating, and so on.) Let's list our discoveries as we go along. Whenever you're pretty sure about an observation, come up and write it under the proper heading. If you need new headings, you can make them. If you see something on the board you don't agree with, investigate it further and then we'll discuss it.

1. Distribute the snails again.
2. As earlier, the approach is guided discovery, with the necessary framework given in the information on the chalkboard. The activity should take 40 to 50 minutes.

Discuss the findings recorded on the chalkboard and attempt to resolve any conflicting findings. What was the most interesting function you discovered about your snail? Why does it need to do that in order to survive?

Part Three

Preparation

Fill jar lids with butter or margarine. Smooth surfaces carefully. Cut enough apples to give each student a quarter.

You each have a piece of apple. Begin to bite into it, but stop! Remove the apple. Look carefully at what's there. What is it? Compare your teeth marks with your neighbors'. Are they just alike or is each a little different? Discuss with your partner how snails eat. Do you think a snail could make teeth marks? Take a good guess based on yesterday's findings. (Tabulate guesses.) Let's test these ideas. With your partner, find another guess you want to test out about snails. How can you test your idea?

1. Pass out butter and hand lenses, snails, trays, and newspaper.
2. Explain that snails like butter. Have students place one snail near the butter and be patient.
3. In the meantime, organize study topics. List pairs' topics on the board. Repeatedly ask, What do you suspect the answer is going to be? How do you propose to find out? Will you need a control to reinforce the notions of scientific method? Pairs unable to think of investigations can be helped by the additional questions in this activity's More section.
4. Pairs begin work on their investigations by writing their questions, their hypotheses, and their proposed tests in their Life Lab journals.
5. Examine butter dishes now with hand lenses. Discuss the results in cases where snails have eaten. Confirm the method of eating (as described in Teacher Background).

6. Assign students to make any further preparations and gather any materials or supplies needed for their pair investigations.
7. The work can proceed for the next two days.

At the final discussion, have each pair report their findings and discuss them with the class.

Additional questions to ask about snails:

Do they hear? What can they smell? What do the two sets of antennae do? How long is their digestion time? Do they prefer light or dark? Do they prefer lettuce, cabbage, or butter? How much food can a snail eat in a day? Do snails drink water? Will they eat meat? Can you teach a snail anything? Is the pod a suction cup? Can they walk on sand? Just how fast is a so-called snail's pace?

Following the Thread

Description

Students observe spiders in the garden.

Objective

To observe closely and record; to make distinctions between spiders and insects; to use observations for drawing conclusions and evaluating statements.

Teacher Background

In a single green acre, you could count as many as 50,000 spiders. There are 30,000 classified species, and perhaps double that number are still unnamed. All spiders have eight legs and a two-part body (cephalothorax and abdomen), whereas insects have six legs and a three-part body (head, thorax, and abdomen). In addition, spiders, unlike insects, have neither compound eyes nor antennae. Most spiders live only a year (although tarantulas may live 30 years).

Materials

Life Lab journals

Finding a spider to observe shouldn't be too difficult. They are everywhere—from high in the mountains to deserts to holes in the ground. They have even been found drifting above the water on strands of silk far out to sea. Since many spiders build webs, you can look for a web first and then find the spider who built it.

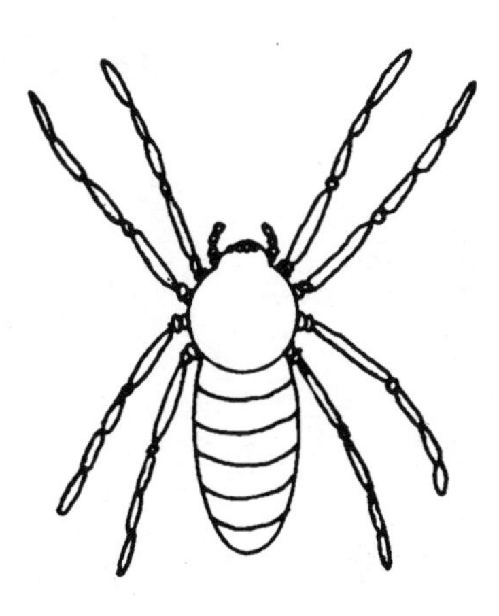

1. Have students find a garden spider and carefully draw and describe it and its web. Things to look for include
   - Number of legs
   - Number of eyes
   - Patterns on the abdomen
   - Where the silk for its web comes from
   - Kinds of things caught in the web
   - How the spider deals with insects caught in the web
   - Whether the spider hangs with its head up or down
   - Whether the web is a blob, a tangle, a hammock, a funnel, or a spiral

2. Have students look for egg cases and molted spider skins as well. Do spiders need to molt? How are they like or different from insects in this respect?
3. Have students try to find and observe a spider building a web—usually early in the morning. How long does it take for a spider to weave a complete web?

A friend describes a wonderful land where the spiders' webs come in all different colors, depending on the colors of the plants they eat. Given what you know about spiders and their lives, what two things are questionable about that statement? (First, spiders don't eat plants. Second, if they did eat plants, they wouldn't need webs to catch insects.)

1. Spiders are called *Arachnids*. Have students find and report on the Greek myth of Arachne who is the basis for this name. (Arachne was a weaver in ancient Greece and once challenged the goddess Athene to a contest. Athene was so jealous and enraged by the perfection of Arachne's weaving that she ripped it up. Arachne hanged herself in remorse. Athene brought her back to life as a spider, weaving webs.)
2. Have students read to find out why spiders don't get stuck in their own webs. (Ordinarily, spiders don't get stuck in their own webs. Spiders give off an oily substance at the tip of each leg that keeps them from getting stuck in the glue of the web.)

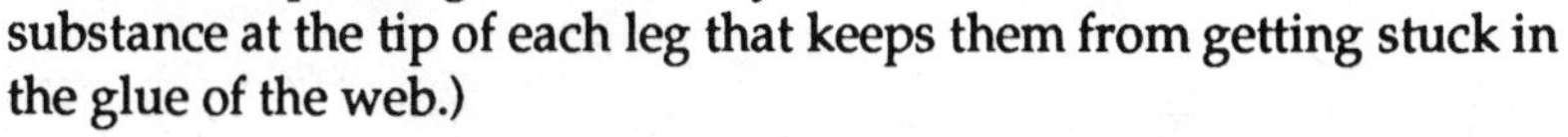

I 2-6 FS P

The Great and Powerful Earthworm

Description

Students conduct a simple earthworm experiment by preparing two boxes with organic materials and adding earthworms to one.

Objective

To illustrate the role of the earthworm as a soil tiller.

Teacher Background

Earthworms are true tillers of the soil. They digest organic matter and excrete it as castings, pellets that are excellent fertilizer high in nitrogen, phosphorous, and potassium. In this activity students will have the unique opportunity to see earthworms at work by setting up a home for them in a clear plastic or root view box. You will find information on constructing an earthworm box for your Life Lab garden in the Equipment Designs section of the Appendix. The box is easy to maintain and will give you an endless supply of castings, as well as serve as a lunch food recycling center.

Materials

Two clear plastic containers, terrariums, or root view boxes
One container of red worms from local fish and tackle shop
Compost, soil, leaves, and straw to layer in each box
Black cloth or paper to cover one box
Life Lab journals

Can you imagine eating your own weight in food every day? That's what the earthworm does! The earthworm improves the soil by eating it. When the food passes through the earthworm's digestive system it is changed into a material that plants especially like. The earthworm deposits this recycled organic matter in the form of soft pellets called *castings*. These castings make good fertilizer for the plants. As scientists, how can we test to find out if worms really are earthmovers and recyclers? (Discuss possible experiment designs.)

1. Have students set up two containers as shown in the drawing.

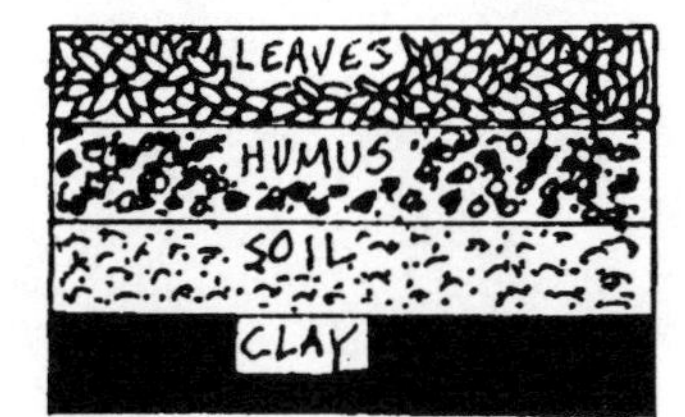

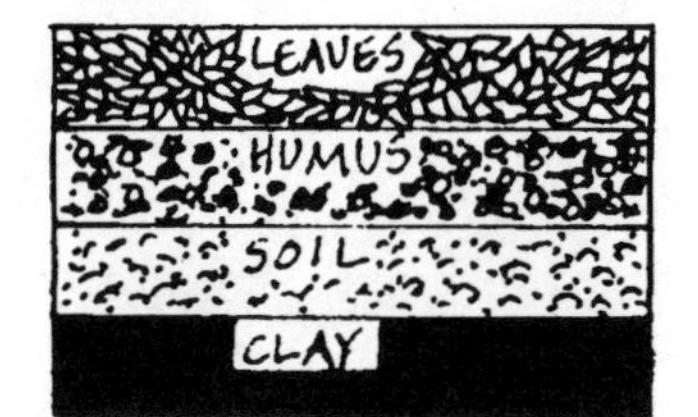

2. Have students place the worms in only one of the containers and keep the soil in this container slightly moist.
3. Have students describe and draw pictures of the soil in both containers. Ask them to predict what will happen to the soils.
4. Have students cover the worm container with black cloth or paper. Allow air holes for breathing.
5. Have students keep a log in their Life Lab journals, recording and drawing observations.
6. Have students remove the black cloth daily over the course of a few weeks to observe the effect of earthworms on the soil. Point out their tunnels and explain that these let in water and air needed by plants.
7. Have students compare the soil in the two containers. How is the earthworm like a tiller? How is it able to recycle materials in the environment?
8. Have students place the earthworms in the garden. Make sure the earthworms are covered with some soil.

How did you know the earthworms were changing the soil? What was the purpose of the control box? What did you learn about earthworms? How can earthworms help your garden? How are earthworms recyclers?

1. Have students construct an earthworm box for the garden. Have them feed the earthworms vegetable lunch wastes and keep the soil moist. In return, the earthworms will leave their castings on the surface of the soil for you to use as garden fertilizer. Extended activities include weighing the daily waste that gets recycled, comparing beds grown with earthworm castings to beds grown with compost or with no fertilizer, and writing stories about earthworms.
2. Have students hold an earthworm race. Put the earthworms in the center of a circular piece of paper. See which one crawls out of the circle first. What conditions make the earthworm move faster or slower? (light and darkness)

Climate

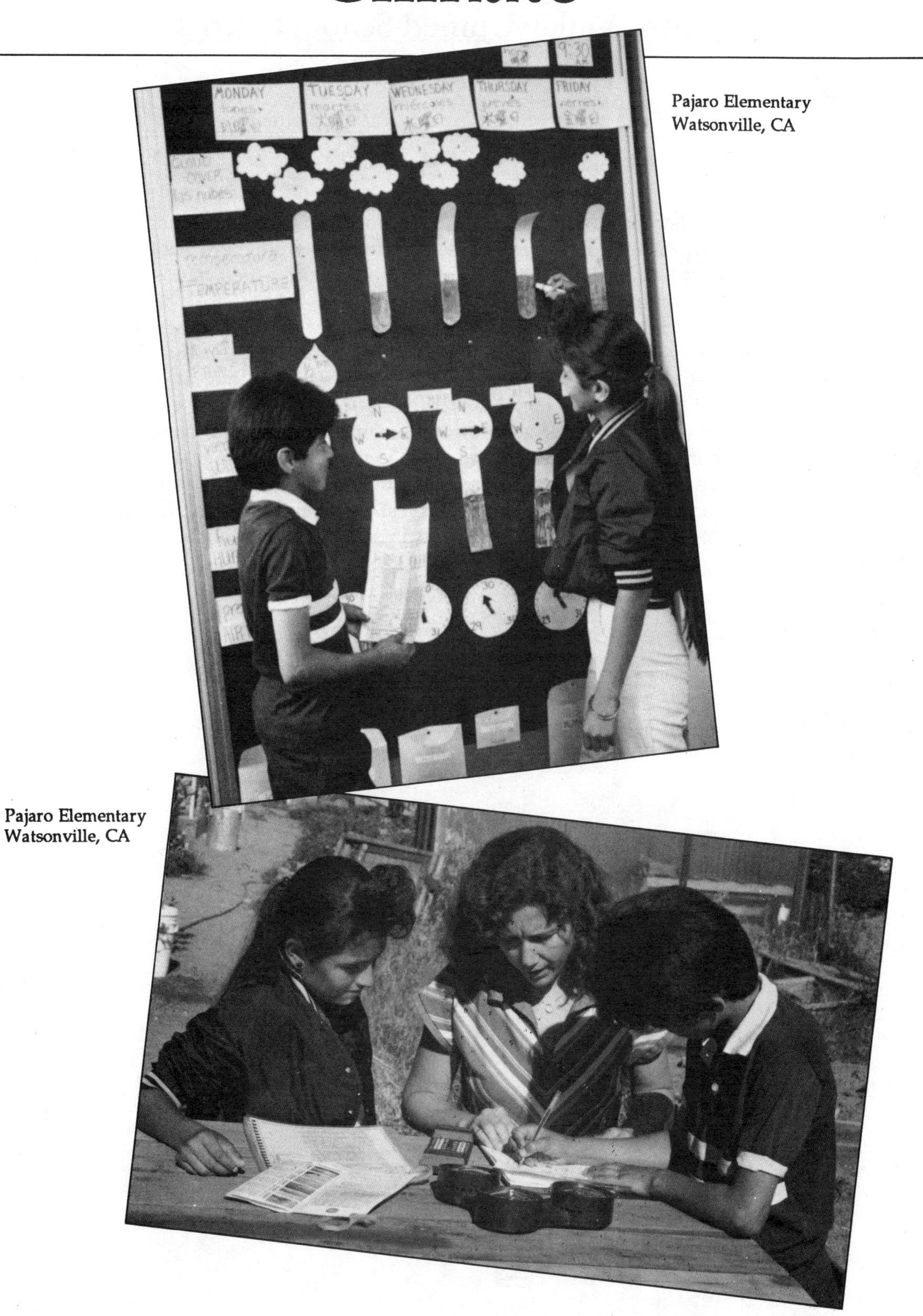

Pajaro Elementary
Watsonville, CA

Pajaro Elementary
Watsonville, CA

Pajaro School
Pajaro Valley Unified School District
Watsonville, California

Pajaro School is a bilingual school—99% of students are Hispanic—with the problems of many urban schools (overcrowding, low budget, and so on). The student population is motivated by their hands-on participation in Life Lab projects.

Pajaro School Life Lab is a cooperative effort on the part of nine elementary classes (K-5) and 4 middle school science classes. Through community support and the help of the older students, our site now includes 14 redwood garden boxes (each 10′ X 10′), a tool shed, a small fiberglass greenhouse, a coldframe, compost bins, an outdoor classroom work area with picnic tables, a weather station, and a lot of creative energy. Students set up experimental garden beds, predict and track the climate, grow and harvest fruits and vegetables, and work together to make Pajaro Life Lab a real living laboratory.

UNIT INTRODUCTION

Climate

Weather plays an integral role in our lives. It affects our daily activities and determines our food, clothes, and shelter. The science of meterology studies the physical components of the earth—land, water, and air—and analyzes weather patterns and predictions, as heat from the sun is absorbed and reflected by the earth and our atmosphere. The weather patterns of your region affect what you can or can't grow in your Living Laboratory.

In this unit, students will study their local weather patterns as well as global climate zones. The study starts with the construction of a weather station and learning to collect and chart data from various weather instruments. This data is used to analyze local patterns and to make predictions. The data is also exchanged and compared with that of another Life Lab school. Students experiment with how microclimates at their school affect plant growth. They model how the earth radiates heat. Understanding the effect of latitude on climate leads to research of major global climate zones to determine effects of climate on culture.

Some advance preparation is needed. "The Station Creation" (building a weather station) requires the purchase or prior construction of weather instruments. "Keeping Track" requires a class recording chart. "Wind, Rain, Sleet, or Hail" requires arrangements with an exchange school. "Keeping Track" and "Wind, Rain, Sleet, or Hail" involve extended collection and analysis of weather data. "A Ravishing Radish Party" is a garden experiment requiring data collection over time, "A Journey to Different Lands" can be an extended research project.

The study of climate allows integration of the phyical, biological, and earth sciences with the social sciences. The awareness of the influence of society on climate and climate on society is critical today as we face issues such as acid rain, drought, and the greenhouse effect. Students can grasp these concepts in the context of their Living Laboratory.

Activities	**Recommended Grade Level**
Degrees Count *(reading a thermometer)*	2,3
Temperature Hunt *(using a thermometer)*	3,4,5,6
The Station Creation *(making a Life Lab weather station)*	3,4,5,6
Keeping Track *(collecting and recording weather information)*	3,4,5,6
Wind, Rain, Sleet, or Hail *(school's exchange of weather information)*	5,6
I'm the Hottest *(heat retention and loss in different substances)*	3,4,5,6
A Ravishing Radish Party *(growing plants in different microclimates)*	5,6
We've Got Solar Power! *(construction of mini-solar collectors)*	4,5,6
A Shoebox of Sunshine *(construct and experiment with miniature greenhouses)*	5,6
What's the Angle? *(model of earth's revolution around sun)*	5,6
A Journey to Different Lands *(regional and world climates)*	5,6

This unit was developed by the University of California, Santa Cruz, under a project directed by Ronald W. Henderson, Professor of Education, and Ellen Moir, Supervisor of Student Teaching, as part of a National Science Foundation grant (MDR8550386).

Solar Energy Shout

Chorus

```
   G
Everybody's doing the solar energy shout...SOLAR ENERGY!
   C                                G
Everybody's doing the solar energy shout...SOLAR ENERGY!
       D7                         C                G
We can do it because we will never never run out...SOLAR ENERGY!
```

Oil is scarce and coal dirties the air...SOLAR ENERGY!
Nuclear waste is a chance we'd rather not dare...SOLAR ENERGY!
So, now is the hour solar power is everywhere...SOLAR ENERGY!

Chorus

The plants use the sun, just ask the birds in the trees...SOLAR ENERGY!
The sun moves the seas just by movin' the breeze...SOLAR ENERGY!
If it wasn't for the sun we'd all be in a deep freeze...SOLAR ENERGY!

Chorus

If it wasn't for the sun fire would never be found...SOLAR ENERGY!
There would be no oil pumped from out of the ground...SOLAR ENERGY!
The sky would be lonely without any clouds around...SOLAR ENERGY!

Chorus

The sun is the reason this whole world can live...SOLAR ENERGY!
The sun never takes, it's always willing to give...SOLAR ENERGY!
Let's meet the sun, learn to live and to give...SOLAR ENERGY!

Chorus

We can do it all day because we never never will run out...SOLAR ENERGY!
We can do it all day because we never never will run out...SOLAR ENERGY!

Degrees Count

Description

Students practice reading a thermometer on a large classroom thermometer model and taking outdoor temperatures.

Objective

To read a thermometer; to explain how a thermometer works.

Materials

Outdoor thermometers
A large piece of poster board
Black marker
70-cm white elastic strip, one inch wide
70-cm colored elastic strip, one inch wide

Preparation

Make a thermometer model:

- Draw a large thermometer on the poster board. Calibrate one side in degrees Fahrenheit, and the other side in degrees Celsius.
- Cut slits at the bottom and top of the scale of the thermometer not more than 70 cm apart. Make the slits slightly wider than the elastic.
- Attach the white and colored elastic strips so that they will hold together well. Thread the elastic through both slits. Then connect the ends of the elastic in a loop so that it easily moves but holds its place.

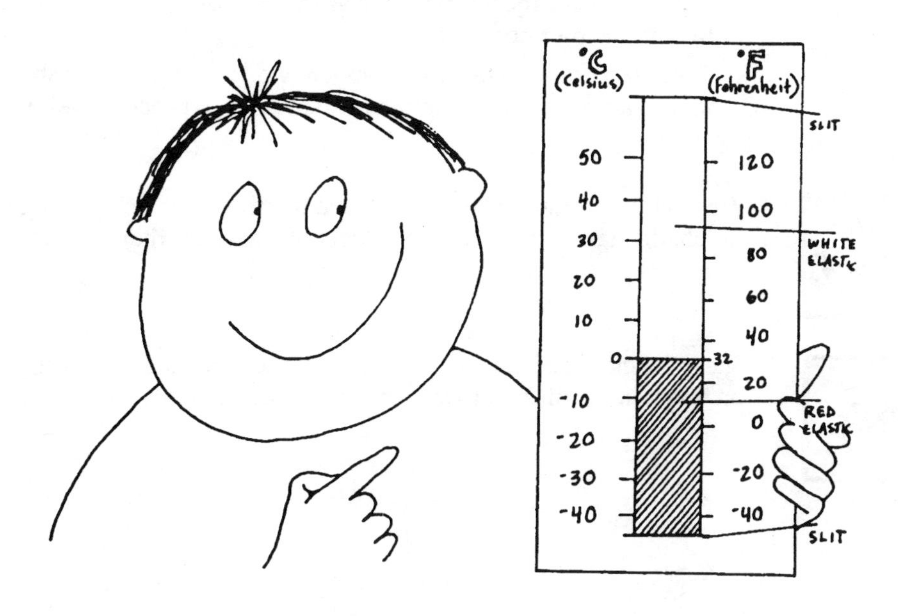

If you are not feeling well, does your body feel warm or cool? How can you tell if it is warmer or cooler than usual? Are there some places in the classroom or outdoors that seem warmer than others? (Ask students to stand in places they think may be warm or cold.) How could we tell if one place was warmer than another? Measuring temperature is a way of telling how hot or cold something is.

Air temperature is an important part of the weather. During the day, temperatures may be warm from the sun; at night the temperature is usually cooler. Temperature changes can cause other weather changes, such as wind, rain, or snow. We measure temperature with a thermometer. The Celsius thermometer, named after Swedish astronomer Anders Celsius, assigns the number 0 to the temperature at which water freezes and the number 100 to the temperature at which water boils. The space between is divided into 100 equal parts, called degrees. (Thus, it is also called a centigrade thermometer.) The Celsius scale is used by scientists worldwide, and is the standard scale in most countries. The United States still uses the Fahrenheit scale that designates 32 degrees as the freezing point for water and 212 degrees as water's boiling point. Most thermometers are made of a glass tube with red colored alcohol inside. When the air is hot, the liquid alcohol is heated and expands. The level of alcohol in the tube rises. When the air cools, so does the alcohol, and its level falls. This rising and falling can either be measured in degrees Fahrenheit or degrees Celsius.

1. Show the class the poster board thermometer. Show students how to read the thermometer by moving the colored elastic along the scale. Start at 0° C (32° F) and explain that this is the point at which water freezes.
2. Ask for student volunteers to change the thermometer to represent increasing temperatures. Point out the correspondence between degrees F and degrees C each time.
3. Take the students outside with the outdoor thermometers. Explain that to obtain an accurate reading, the thermometer must be left in a particular place for several minutes so that the alcohol will be heated or cooled to the same temperature as the outside air. Show students how to read the thermometer.
4. Take temperature readings from a few different places to show temperature variations, and allow students to practice taking readings.

Why is it important to be able to use a thermometer? Why is it generally cooler in the shade than in the sun? Why is it cooler at night than during the day?

Show the class different kinds of thermometers: dial, mercury, plastic strip, and so on. Explain the different ways in which these devices measure temperature.

Temperature Hunt

Description

Supervised groups of students search for the coldest and warmest locations in the garden or school yard.

Objective

To measure several different temperatures to identify some of the factors that cause temperature variations in the environment.

Materials

One thermometer for each group
Life Lab journals

If you were outside right now, do you think you could find some places that are warmer than others? What conditions make the temperature different in two places? (List responses on the chalkboard.) Do you think it would be easy to find the warmest place around the school? The coldest place? What conditions will you look for in each case?

1. Take the class outside into the garden or school yard. Ask them to name places where they would expect to be hot or cold.
2. Place thermometers in two or three places in turn. While waiting for the thermometers to register, ask the students to name factors they think may be influencing the temperature at each spot. Make sure that factors such as the following are considered: presence of shade or direct sunlight, surface color, wind-exposure. Show students the thermometers and explain how they will use them to discover the hottest and coldest locations.
3. Organize the students into small groups, each supervised by an intern or aide.
4. Tell students that the object of the temperature hunt is to try to find the hottest and coldest spots in the immediate area. Remind them to be creative! The most extreme temperatures may not necessarily be in the most obvious places.

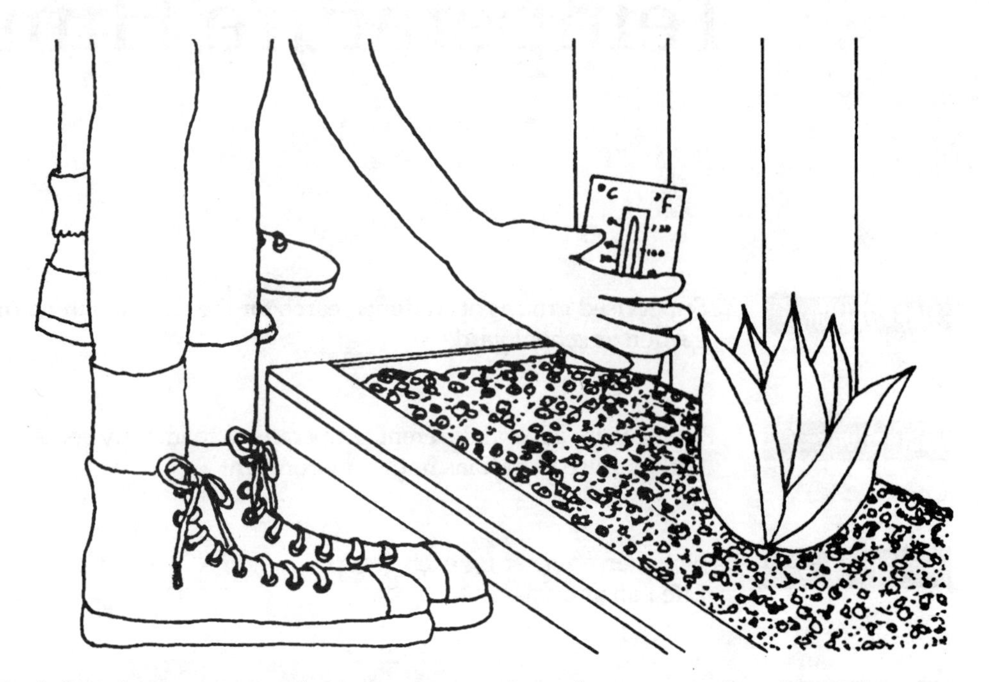

5. Ask each group to test the temperature at four places —two that they think will be hot and two that they think will be cold.
6. At each location, the adult supervisor will take the temperature reading. Students record location, temperature, and factors affecting temperature in their journals.
7. After all the readings are taken, gather the class together and compare findings.

Where were the coldest and hottest readings taken? What was the difference in temperature between the coldest place and the hottest? How do you think the temperatures will change during the day?

Have students come up with different temperature experiments in the garden. Be sure that each test is fair, with only one variable in each experiment.

The Station Creation

Description

Students build a Life Lab weather station.

Objective

To collect data from weather instruments; to describe the use of weather instruments.

Materials

Materials for weather station box (see Preparation)
Hammer and nails
Weather instruments (see Preparation)
Instrument Use Cards (see pp. 409-414)

Preparation

1. There are a number of options you can consider for this activity. You may want to purchase the instruments and have the class build the station box. You may want to have the class use science supply company catalogues to select the instruments. (See p. 480)
   Note: Instruments should be ordered at least three weeks in advance. You may want to have the weather box and instruments ready so that the class can simply assemble the station. Or you may want to have the class build the instruments.

2. A weather station is simple to make. Basically you need a box that can withstand rain and winds and a secure post on which to mount it. A heavy cardboard box spray-painted to make it rainproof works well. Both ends are open to allow air to circulate. Set the station at least one meter above the ground. Place instruments that need protection from rain and sun on the inside and mount the rain gauge and the wind vane on the outside. For a more secure station, build a plywood box with one side louvered. Hinge the louvered side for easy opening and closing.
3. Recommended instruments include:
   - Minimum-maximum thermometer to measure high and low temperatures
   - Barometer to measure rising and falling air pressure
   - Psychrometer to measure relative humidity
   - Rain gauge to measure precipitation
   - Wind meter to measure wind speed
   - Wind vane to determine wind direction
   - Pocket Weather Forecaster (Barron's Educational Series, 1974. Cloud cover and wind direction give clues to weather changes 12 to 36 hours in advance.)
4. Locate potential sites for the weather station.
5. If you are concerned about vandalism, you may want to purchase a padlock for the weather station and discuss ways to prevent vandalism with your class.
6. Pages 409-414 contain six Instrument Use Cards. Mount these cards onto poster board and use them for the instrument stations and for reference at the weather station. Adjust the directions on the cards to fit your own weather instruments.
7. Set up instrument stations in the classroom. At each station put the instrument and the card describing its use. You may also train one student per instrument so that each station can have a monitor.

(Divide the class into groups of six. Have each group silently pass around a piece of paper with each person adding one observation about today's weather. Set a time limit of two minutes.) How many different observations did your group come up with? List on the chalkboard the types of observations students made: temperature, wind speed, shape of clouds, and so on. What do you think the temperature is right now? How could we find out? How can we tell the wind direction? Scientists have developed instruments to help us accurately describe our weather. By accurately recording our weather we can compare it from day to day, average it to find out what it is usually like, and compare it with other places. We are going to build our own Life Lab Weather Station.

1. Show all of the weather instruments to the class and briefly explain their use. (Information is on Instrument Use Cards.)
2. Set up stations for each instrument and have students rotate to each station, examine each instrument, and study how and why it is used. A fan, a heat source, and a sprinkling can may be used to help demonstrate many of the instruments.
3. Take the class outside to find a good location for the weather station. A good location is free from obstructions, such as buildings and trees, that may affect wind patterns and rain measurement.
4. Secure the station box and mount the instruments. Be sure the instruments that need sun and rain protection are mounted inside the box and the rain gauge and wind vane are mounted outside. Some instruments, such as the wind meter, will be stored in the box and taken out to conduct measurements.
5. You may want to add a flag mounted on a pole near the station as a simple way of estimating wind speed and direction.

How can weather instruments help us? Which instrument did you find most interesting? How is it used? How will the weather station help the Life Lab garden? How will we take care of the station?

Now that you have a weather station, use the following activity, Keeping Track, to start recording information.

Keeping Track

Description

A group of four students collects and charts daily weather information for one week and presents a weather report to the class using the weather station.

Objective

To collect and chart data from instruments; to determine local factors affecting weather; to relate impact of weather on daily lives; to learn to make predictions based on current data.

Materials

Weather station, see p. 289
Large bulletin board space for graph
Graph
Markers, see p. 415
Velcro
Life Lab journals
Weather Symbols Key, pp. 416-417

Preparation

1. To make the graph, prepare a large bulletin board space. Students will use it to make a picture graph of their weather data. The graph will consist of six sections and be divided into five columns. Label the whole graph OUR DAILY WEATHER. The upper section will show cloud cover. It should be 20% of the board length and the background should be blue. Label this section CLOUD COVER. The next section will show daily temperature. This section should be 30% of the board length. You will need five thermometer outlines. (See p. 415.) Enlarge the outline if necessary. Label this section TEMPERATURE. Divide the remaining 50% into four sections. Label these sections WIND, RAIN, HUMIDITY, AIR PRESSURE. Divide the board into five columns. Label the columns MONDAY, TUESDAY, WEDNESDAY, THURSDAY, FRIDAY.
2. MARKERS. See patterns on p. 415 and Weather Symbols Key on p. 416. Clouds will be marked with the cloud outline. You will need 20. We suggest you use Velcro to attach them for easy removal. Temperature will be marked by red strips that will fit within the thermometer outline. You will need approximately one meter per week. Have students cut and color strips to fit. Rain will be marked with raindrops. Each raindrop outline equals 1.27 cm (1/2 inch). Estimate how many raindrops you will need for one week. Wind direction will be marked by the wind circle and rotating arrow. You will need five of these. Humidity will be recorded by the dry-to-wet circle with the rotating

OUR DAILY WEATHER	WEEK March 24 TIME 9 o'clock	MON	TUES	WED	THUR	FRI
clouds	CLOUD COVER					
bars	TEMPERATURE °C	45 25 0 -10 -25 HOTTEST COLDEST	45 25 0 -10 -25 HOTTEST COLDEST	45 25 0 -10 -25 HOTTEST COLDEST	45 25 0 -10 -25 HOTTEST COLDEST	45 25 0 -10 -25 HOTTEST COLDEST
raindrops	RAIN					
arrows	WIND	N W E S	N W E S	N W E S	N W E S	N W E S
arrows	HUMIDITY	DRY CRISP WET MUGGY	DRY CRISP WET MUGGY	DRY CRISP WET MUGGY	DRY CRISP WET MUGGY	DRY CRISP WET MUGGY
arrows	AIR PRESSURE	25 ⇨	27 ⇧			

arrow, plus scrap paper on which to record the actual humidity. You will need five dry-to-wet circles and cut-to-size scrap paper for each day of the project. Air pressure will be recorded by attaching scraps of paper with the barometer readings and arrows that will be placed in the direction the barometer is going (up, down, steady). You will need cut-to-size scrap paper for each day of the project and five arrows.

What is our local weather pattern like? What factors affect our local weather? (List these factors on the board. Responses can include global factors such as latitude and global wind directions, and local factors such as being near an ocean or near mountains.) To accurately determine our weather patterns, we will use our weather station. (Review how to read each of the weather instruments in the weather station.) Each day four students will go to the weather station in the morning. They will record the information on the picture graph for all of us to see. They will do this for one week. At the end of the week, they will give us a weather report. Each student will have a turn.

1. Choose four students to record data for the first week. Assign each student to a specific instrument. Students should rotate responsibilities during the week. Set up a specific time each morning for the students to collect and graph the data.
2. On the first morning, help the four students record and graph data on the picture graph. Post the Weather Symbols Key, pp. 416-417, by the graph.
3. Explain the picture graph to the class.
4. On Friday, have the group prepare a weather report for the week. Encourage them to be creative. Their report should include a summary of the week's weather and how it affected the class and different activities, explanations of environmental factors that affected their weather, a prediction for what the weather will be like on the weekend, and what types of activities would be appropriate considering the weather.
5. Repeat the activity with a new group of students each week until all have participated. To shorten project length, rotate students every two days.

How does cloud cover affect temperature? How does wind affect temperature? How does rain affect temperature? How do weather instruments help us? Why are graphs important to scientists? How does weather affect your life? What information did you use to make a weather prediction?

1. Ask students to listen to the weather report on the evening news. What information is given? What kinds of instruments provide this information?
2. Take a field trip to a local weather station. (Check with your local community college, TV stations, or newspaper to find its location.) Have a meteorologist explain the different weather instruments and average temperature and precipitation in your area.
3. Exchange weather information with another Life Lab school. See the next activity, Wind, Rain, Sleet, or Hail.
4. Educate the whole school. Change the weather information daily on a chalkboard in a corridor or in the library or lunchroom.

Wind, Rain, Sleet, or Hail

Description

Students exchange and compare weather data with a class at another Life Lab school. This activity is planned to be used in conjunction with the previous activity. The group of students responsible for collecting data from your school's weather station for the week will be responsible for sending the data to the exchange school and comparing it to the data received.

Objective

To compare and analyze data from two locations; to determine how climates vary with location.

Materials

Weather station, see p. 289
Life Lab journals
Exchange class, see Preparation
Bulletin board space
Newspaper weather map
National temperature chart from newspaper
Completed weekly weather graph
One Comparison Graph per week, blackline master p. 418
Envelopes and postage stamps

Preparation

1. Make an agreement with a Life Lab class at a different school to exchange weather data for a certain time period. Initial contact may be established through the Life Lab Science Program office or through The Growing Teacher newsletter (see p. 5).
2. Have the weather station set up and be sure students are familiar with its operation.
3. Prepare bulletin board space for an ongoing comparison graph and for posting letters.

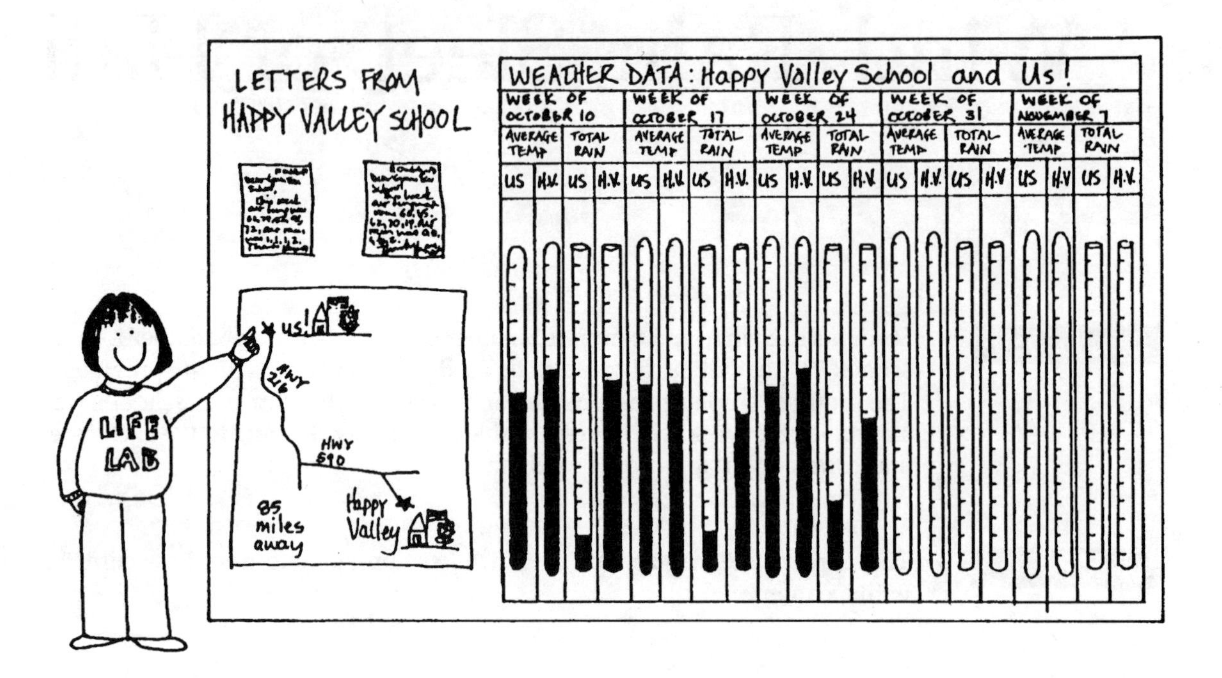

What are some examples of locations where weather is different from ours? How do you know? (Show the weather map and point out cities with different temperatures. Explain that students will be exchanging weather data with a class at another school to see how the two climates differ.)

1. Tell the students about the exchange class by showing them where it is located. Give a geographical description of the area. Ask the class to make predictions: over the next month which class will have warmer temperatures, more rainfall, more wind? Have students write their predictions in their journals.
2. Organize the exchange with your class. Once a week the group collecting weather data will send a copy of the data to the other class. When your class receives their data, it will be processed by the group that recorded data during that week.
3. Ask one student to write a letter introducing the class to the exchange class. Include the predictions your class made and a schedule for collecting and sending information. Note: it is important that the data be collected from the two schools at the same time of day.
4. At the end of each week, ask one person in the recording group to send a copy of the data to the other school. A letter can be included describing what the weather was like during the week and how it influenced students' activities.
5. When data is received from the exchange class, the appropriate group should average the temperature, rainfall, and wind speed for each area. These averages then should be recorded on the Comparison Graph, p. 418.

6. When the exchange period is over, each class can compare the actual data to their predictions and draw conclusions about the differences in climate. This information can also be shared through letters.

How may climates in different locations vary even though the season is the same? How can you record differences so that you can easily compare them? What geographical differences in the two schools contributed to the different climates?

1. Make a tape cassette to send to the exchange class. It can include your weekly weather report and weather sound effects.
2. Some Life Lab schools have received grants to purchase telecommunication modems so that their schools can communicate with each other by computer to exchange weather data. (Contact the Life Lab Science Program office for information about participating schools.)
3. Obtain computer software for weather recording and let interested students use it to record weather information.

I'm the Hottest

Description

Students measure the temperature of water, light-colored sand, and dark soil exposed to sunlight outdoors or to a heat lamp indoors to compare their heat-retaining capacities.

Objective

To observe, measure, graph, and compare the different heating and cooling of materials; to explain reasons for the uneven heating of the earth's surface within a region.

Teacher Background

The sun does not uniformly heat the whole planet. One major reason for this is the varied angles at which the sun strikes different regions of the earth. However, even within the same region, temperatures may vary. This is often due to the different heat absorption and loss rates of the substances making up the earth's surface. White surfaces reflect light and heat more slowly than dark surfaces. Dark surfaces absorb heat. Land surfaces heat and cool rapidly. Water heats slowly but also cools slowly. The different heating and cooling capacities of these surfaces will create breezes where they meet and affect the climate of the area.

Note: Another major reason for microclimates within a region is differences in altitude. Altitude will not be discussed in this activity.

Materials

Eight large paper cups
Scissors
Four thermometers
Water
White sand
Dark soil
Life Lab journals
40-cm square board to set the experiment on
Clip-on lamp (On a sunny day, the sun can be used as a light source; otherwise a lamp should be used.)

What heats the earth? How is heat measured? Do all places on the earth heat the same? (No, because the sun strikes different regions at different angles.) Even if you are in the same region, different places may have different temperatures. Any ideas why? Do you think color can affect how hot something will get? Materials of different colors absorb and reflect light differently. List examples of temperature differences in materials of different colors (black vs. white seats in a car, asphalt vs. concrete, and so on). Do you think lakes and oceans heat differently than land? These differences are important; they affect weather systems and microclimates within a region.

1. Cut the tops off of four paper cups, so that the remaining portion is about 4 cm tall.
2. Fill three cups with one of the following: sand, soil, water. Label them. Label the fourth cup AIR (Control).
3. Place the cups on the board in direct sunlight. Place a thermometer in each cup so that the bulb is covered by about 1 cm of material. Rest the top of each thermometer on an inverted cup, as shown. If a lamp is used, place it about 15 cm above the cups.

4. Use the graph to record temperatures. Put graph on chalkboard, and have students copy it into their journals.

HOLDING HEAT		BEGINNING TEMP.	5 minutes later	10 minutes later	15 minutes later	20 minutes later
SAND	SUN					
	SHADE					
SOIL	SUN					
	SHADE					
WATER	SUN					
	SHADE					
AIR	SUN					
	SHADE					

5. Assign pairs of students to take temperature readings at five-minute intervals. Record the results on the graph.
6. Move the board with cups and thermometers in place to a shady area. Take five temperature readings at five-minute intervals. Record the results.
7. When all of the readings have been taken, assist students as they work in small groups transferring the data to the graph in their journals. Help them use the graph to draw their conclusions.

Which material heated the most rapidly? Which heated the most slowly? Why? How can you account for the differences in heating between the sand and soil? How does this activity help explain the uneven heating of the earth's surface? What would happen to the temperature readings if all of the materials were left in the sun for several hours or outside all night? Name some locations nearby where you would expect to find different temperatures.

A Ravishing Radish Party

Description

Students place sown flats in the garden in five locations that have different microclimates and make biweekly observations.

Objective

To measure and observe the effects of different microclimates on soil temperature, soil moisture, and plant development.

Teacher Background

Microclimates occur when environmental factors affect the weather pattern within a given area. For example, a coastal area may have daily morning fog, while three miles inland there will be no fog. This creates two microclimates in the area due to temperature variation from one place to another. Even within a small area, such as your school yard, you can find microclimates caused by black asphalt, shade trees, building walls, and so on.

Materials

Five flats
Potting soil to fill the flats
Soil thermometers
Air thermometers
Radish seeds for all five flats
Five garden labels
Grease pencil
Five rulers
Life Lab journals

Even though we may not be aware of it, climate can vary in important ways within a relatively small area. Name factors that contribute to variations in a climate (shade, direct sunlight, exposure, reflection, barriers, exposure to wind, and so on). Each area that seems to have a different climate is called a *microclimate*. These differences in microclimate may affect plants. Different plants have different requirements. They will thrive in an environment that meets those requirements, and do less well where their needs are not met. We are going to test the effects of microclimates on radishes in three sections: Preparation, Maintenance, and Data Collection.

Part One

1. List the materials for this project on the chalkboard.
2. Divide the class into five groups. Ask one person in each group to be the recorder and another to be the reporter. Ask each group to design an experiment using the materials listed on the board that will test the hypothesis: Microclimates within a small area can affect a plant's growth.

3. Give the group a limited time to discuss their designs. The designs should include how the five flats will be arranged. Specify that all the flats should be prepared in the same way. The effect of the microclimate is the only factor being tested.
4. After the time is up, ask the reporter from each group to share the design. Then have the class agree on a class design to test the hypothesis. (A possible design would be for each group to plant one flat with the same number of radish seeds, using the same soil, and amount of water as the other groups. Each flat is placed in a different microclimate in the garden or school yard. Data is collected over several weeks.)

5. Have the class list the steps they will follow for the experiment. (For example: 1. Mix enough potting mix for the five flats. 2. Each group fill one flat and moisten the potting mix. 3. Plant eight rows of radish seeds in each flat. Put 3 cm between each row. Have 2 cm between each seed in a row. 4. Water with one liter of water. 5. Place each flat in a selected microclimate.)
6. Have the class determine five locations where they think different microclimates exist. (They can actually determine these by walking around the school or taking temperature readings.) They should also discuss the maintenance schedule for Part Two.
7. Have one recorder for each group copy the steps and then read them to the group as they carry out the preparations.
8. Each group prepares their flat and places it in the selected microclimate.

Part Two

1. Because it is important that all of the flats are maintained in exactly the same way, the class needs to agree on a maintenance schedule. Discuss why it is important to maintain the flats in exactly the same way. This is a good time to discuss variables.
2. The seeds will need more watering in the first week (approximately three times a day) than during the rest of the project. Decide how many times a day and at what times the flats will be watered. Decide upon the amount each flat should be watered (for example, one liter each time) and the watering technique. Have group members rotate monitoring their flats.
3. It is important—especially the first few days—that the flats are monitored carefully. If all of the flats are dry, then the number of times per day that they are watered may need to be increased. However, if only the flat in direct sunlight is dry, that is a result of the experiment.
4. To encourage similar maintenance of the flats, you can put a chart on the chalkboard with the maintenance schedule and a place for each group to check when they have completed a step.
5. After 12 days the seedlings may seem crowded in the flat. If so, the whole class may agree to thin each flat. Plants should be thinned so that they are the same distance apart in each flat.

Part Three

1. To compare the effect of microclimates on plant growth students will collect data throughout the project. Data will include measuring and recording specific data and writing observations. Students can collect data individually twice a week. It is important that all data be collected at the same time of day so that it can be compared.
2. Students can design their own table for collecting data or use the one illustrated. Twice a week each student should record the soil temperature, air temperature, number of plants, and average height of plants for the group flat. Students should also write down any

observations such as insect damage, health of the plants, soil moisture, and so on.

3. Continue the project until radishes are ready to harvest (approximately 25-40 days).
4. Have a radish party, comparing which microclimate grew the best radishes.
5. Compare data for each treatment using the comparison graph.

A RAVISHING RADISH PARTY

Flat Location

	WEEK 1		WEEK 2		WEEK 3		WEEK 4	
DATE								
TIME								
SOIL TEMP.								
AIR TEMP.								
# PLANTS								
AVG. HEIGHT								

6. Make radish salads and dips. Be creative!

What type of microclimate do radishes grow best in? Will all plants do better in this microclimate? How can weather affect the way plants grow?

1. Have each group average their soil and air temperatures, number of plants, and plant height so that they have one figure for each category. Then have the class make a bar graph comparing the data from the different microclimates.
2. Conduct a similar experiment using several different types of seeds—such as lettuce, corn, bean, tomato—in each flat rather than only one. This will demonstrate that different plants are affected differently by microclimates.

We've Got Solar Power!

Description

Students design and construct simple miniature solar collectors.

Objective

To apply concepts learned in previous activities; to design hot boxes; to collect as much heat as possible in two hours.

Materials

One shoe box for each group of four
White, black, and colored paint
Paintbrushes
Cellophane
Insulating material: puffed cereal, packing material, newspaper
Cardboard
Masking tape
Glue
Scissors
One thermometer capable of measuring at least 140°F per group
Aluminum foil

What are some of the factors that affect the temperature? (the relationship of the sun's angle to the region; the number of hours of daylight; the effect of surface color on reflection, absorption, and reradiation of light and heat; insulation) In this activity, you will construct boxes designed to get as hot as possible when placed in the sun for two hours.

1. Divide the class into groups of four and give each group a shoe box.
2. Make the rest of the materials available to the class.
3. Give students about an hour to build their miniature solar collectors. Design possibilities include batch collectors, in which the outside of the box is painted black and inside white; and glazed collectors, in which the box is topped with cellophane, painted black inside, and insulated. But don't give this away! Let students work out their own designs.
4. Place a thermometer in each box before closing it and have each group place their box in the sun. Some students may figure out they will get more heat if they angle their box toward the sun's path.

5. In two hours, check the thermometers and record the readings.
6. Ask the group that has the hottest box to describe their design to the class.
7. Students can graph the temperature variation.

Did any group place their box at an angle to maximize the surface exposed to direct sunlight? Which box got the hottest? Why? How could trapped solar energy like this be useful? In what ways is the earth similar to a hot box?

A Shoebox of Sunshine

Description

Students construct miniature greenhouses in shoeboxes and observe and measure their effects on plant growth. They relate these experiences to food production methods and to the earth's greenhouse effect, an important factor in the formation of our climate.

Objective

To describe how the earth radiates heat from the sun to our atmosphere.

Teacher Background

As the sun shines upon a greenhouse, radiant energy (mostly light and infrared rays) enters the structure through the glass. Absorbed by the inner surfaces, it is converted to heat energy and either stored by walls, soil, water barrels, and so on, or is radiated back into the interior air, which holds it and which is itself held in by the glass. Thus the greenhouse remains warm. If it becomes too warm, the operator can draw a blind across the glass or open a vent.

Similarly, the earth's atmosphere permits radiant energy to penetrate and be converted to heat energy by the earth itself but does not readily permit the heat to be radiated out to space. The principal atmospheric shield is carbon dioxide, and as CO_2 has built up in the atmosphere over the past few centuries of heavy burning of carbon-rich substances, the atmospheric blanket has become more dense. However, because CO_2 also blocks some of the incoming radiation, dispute has arisen over whether the earth is gradually warming or gradually cooling. At present, the "warmers" seem to be more numerous than the "coolers," but only time will tell, since the build-up is slow and worldwide averages are hard to obtain.

Materials

One shoe box per pair of students
Three baby food jars per pair
One 11" X 13" clear plastic food bag per pair
One thermometer per pair
Several pints black tempera paint
Several 1-1/2" paint brushes
Tape
Scissors
One potted pea or bean plant per pair
Energy Movement Diagram, blackline master, p. 419

Where do the earth and moon get their warmth? (sun) Compare the earth's climate to the moon's. If they receive the same sunlight, what makes the earth's climate so much more livable than the moon's? (the effects of atmosphere and water covering the surface) How can covering things help keep them warm? Give examples. Can coverings also keep things cool? Give examples. (ice chests, caftans, shade trees, and so on) Gardeners building greenhouses use these principles. We'll see how greenhouses work and how the greenhouse effect works in the earth's atmosphere.

1. Have students form groups of two. Appoint three groups as control groups. Their greenhouse assignment is described in step 4.
2. Have each pair make a greenhouse by painting the inside of a shoe box black, cutting windows in the lid and one of the long sides 1/2" smaller than the outside dimensions, and taping clear plastic into the windows as panes.
3. Have students place three covered jars of water in the box to act as heat sumps, along with one or two potted plants and a thermometer taped where it can be seen from the outside.
4. One control greenhouse will be just like the others but will not have plastic window coverings. The second will not be painted and will not have covered windows. The third will not be painted, will not have covered windows, and will not have water jars.

5. Place the greenhouses in position to receive sunlight for about three hours each morning.
6. Have each team prepare a record sheet for temperature readings taken three times daily.
7. Temperatures are taken three times: in full sun, 1/2 hour after removal from sun, and at least an hour later.

GREENHOUSE TEMPERATURES

NAMES:

DAY	DATE	IN FULL SUN	½ HOUR AFTER	1 HOUR LATER	NOTES
1					
2					
3					
4					
5					

8. Plants are watered every other day in uniform amounts.
9. At the end of the week, have students measure and average the growth of the plants in each greenhouse and draw conclusions about each greenhouse's effect.
10. Have students complete blackline master, p. 419, to diagram the movement of energy in the greenhouse system and the movement of energy through the earth's system to show similarities. Discuss their findings and how greenhouses may be applied to the cultivation of crops, especially in areas where greenhouses make wintertime crop growing in cold climates possible.

Which greenhouse stayed the warmest? Why? Which stayed the coolest? Why? What differences were there in plant growth? How did the sunlight become heat in the greenhouse? How was the heat held in the greenhouse? What was the purpose of the water in the greenhouse? How is the heating of a greenhouse similar to how the earth's air is heated?

1. Make two sets of wet and dry bulb thermometers. Place one of the thermometers on a greenhouse floor and the other in a control box. Compare the readings.
2. Investigate heat sumps. Prepare several identical cans with sand, water, dirt, paper shreds, and so on. Place a thermometer in each can, and put the cans in the greenhouse. Expose the greenhouse to the sun for at least a half hour. Remove cans but leave thermometers in them. Compare the rate of temperature drop in the various contents. The one that drops the slowest retains heat the best.

What's the Angle?

Description

Students observe a model of the sun and the earth to hypothesize effects of the light's angle and learn that latitude has a major influence on climate.

Objective

To determine potential differences in light intensity due to latitude.

Materials

Globe
Flashlight

We have a pretty good idea of what our climate is like throughout the year. Do you think different locations around the world have climates similar to ours? (Some do, some don't.) What are some reasons for these differences? (List all ideas.) We're going to use this globe and flashlight to help us develop our theory as to why there are different climates around the earth.

1. Work with the students in setting up a model of the earth in relationship to the sun. First examine the globe. Have them describe the earth's shape. Have them hypothesize as to how the shape may affect how much sunlight hits a specific location.
2. Have two students hold the globe and the flashlight (the sun) in whatever relationship they think is correct. Turn off the lights and ask students to make observations about how the light hits different locations on the earth.
3. Give the class more information. Explain that the sun is stationary and the earth revolves around it. The earth is always turning on its axis, and therefore is always at an angle to the sun. The globe will make the axis easy to observe. Point out some important locations on the globe: the equator, your region, how far you are from the equator, the Tropic of Cancer (an imaginary line 23.5 degrees north of the equator), and the Tropic of Capricorn (an imaginary line 23.5 degrees south of the equator).
4. Now have two more students hold the flashlight and globe in whatever relationship they think is correct. Turn off the lights and ask students to make observations about how the light hits different locations on the earth.

5. Give students this information about the relationship of the sun to the earth: The sun is always directly over the region between the Tropic of Cancer and the Tropic of Capricorn. On June 21 it shines directly over the Tropic of Cancer; on September 21 the earth's location has changed so that the sun is directly over the equator; on December 21 the southern hemisphere is facing the sun and the sun shines directly over the Tropic of Capricorn; after December 21 the northern hemisphere starts moving toward the sun, and by March 21 the sun is directly over the equator again.

6. Now have two more students hold the flashlight and globe in the correct relationship for March 21 and September 21, with the light perpendicular to the globe's surface at the equator and about 30 cm from the globe. Turn off the lights and have students make specific observations. Where is the light the most intense? How does it compare with the light in your region? How do you think this affects the climate? Where on the lit half of the globe is the light the least intense? How do you think the weather is there? Which direction is the sun from your region?
7. To verify what students observed in the model, orient them to north, south, east, and west; take them outside to observe which direction they have to look to see the sun's path.
8. NOTE: It is extremely important that students know *not* to look directly at the sun. Have students stand with their eyes closed and one side of their face aimed directly toward the sun. Ask them to imagine that their heads are like the earth. Which part of their heads feels the hottest? Why? Why aren't all places on their heads heated the same? What will happen when they turn their heads to a different angle?

Why does sunlight strike different parts of the earth differently? How do you think this affects climates around the world? Why would a solar collector in the Northern Hemisphere be set up at an angle facing south?

A Journey to Different Lands

Description

Students work in small groups researching the climates of specific regions. These climates are integrated on a climate zone map to illustrate the differences and common characteristics of worldwide climates. This project can be completed in a few sessions, depending on the extent of the research.

Objective

Students analyze the factors that affect global climate patterns and research major global climate zones.

Teacher Background

There are three factors that create differences in climates: (1) differences in latitude, (2) differences in land and water temperatures, and (3) differences in the surface of the land. Latitude is the major factor, since light intensity decreases as you move away from the equator; similarities in climate can be found at similar latitudes in the northern and southern hemispheres. The countries and regions selected for research will demonstrate the major types of climates as well as the similarities between northern and southern hemispheres. The latitudes and types of climates are provided for your information with the list of countries in Part One.

Materials

Large world map divided into climate zones, see p. 420
Encyclopedia or other reference materials
Blackline master of research questions, p. 421
Magazines to use to cut out pictures (optional)
Life Lab journals

Preparation

1. Enlarge the Climate Zone Map, p. 420, and hang it in the classroom.
2. Prepare access to references on climates of the following countries or regions: Philippines, Egypt, France, Finland, Zaire, Central Australia, Argentina, Antarctica.

What are some of the effects we have observed from the different angles at which the sun's light strikes the earth's surface? (Write answers on the chalkboard.)

- The area around the equator receives more direct sunlight each year than any other region.
- Areas above 30° north and below 30° south receive less direct sunlight per year than the equatorial region.

- Areas above 60° north and below 60° south receive even less direct sunlight each year. What are other factors that affect our local climate? (mountains, water, and so on) Let's remember these factors as we explore different lands.

Part One

1. Divide the class into eight groups. Tell the groups that they are to discover information about the location and climate of a certain region or country. Provide each group with the name of the country or region only. They will discover the other information in their research.

   Countries/Regions

   Philippines (10° - 20° N; tropical wet)

   Egypt (30° N; desert)

   Poland (40° N; temperate)

   Finland (60° N; subarctic)

   Zaire (0° - 15° S; tropical wet and dry)

   Central Australia (30° S; desert)

   Argentina (45° S; temperate)

   Antarctica (80° - 90° S; polar)
2. Set each group up as explorers about to visit a specific country or region. Before they go, they want to learn certain things about the country. Use the research questions as an outline for the research, p. 421. You may have the groups divide the questions among the members.
3. Familiarize the class with the research resources available to them. Encourage them to discuss the questions as a group and then to divide up responsibilities for finding the answers.
4. Give them a limited time to start their activity.

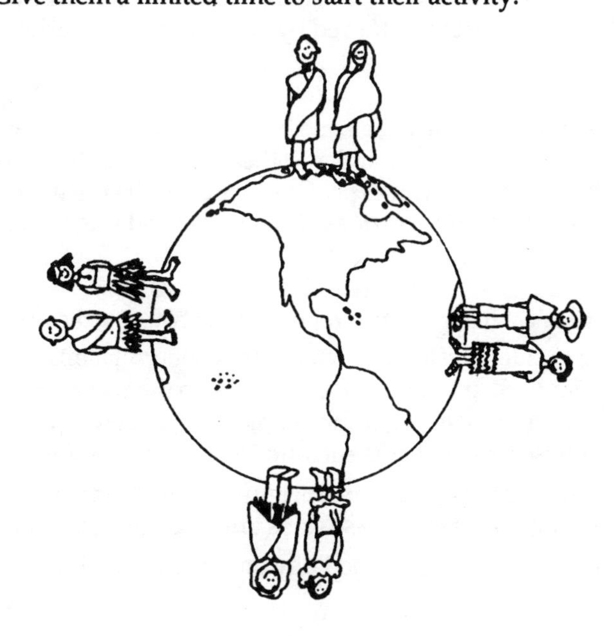

5. At the end of the time, have each group meet and share their answers among themselves. When they have answered all of the questions, the group should locate their country or region on the climate zone map and color it in according to the climate color key. In addition, they may cut out pictures of the country and its people from articles they find in magazines.
6. Provide additional time for research until this portion of the activity is completed by each group.

Part Two

1. Have each group give a short oral report describing the climate they found in their region. Start with the group closest to the equator in the Northern Hemisphere and move north to the Arctic region.
2. After these three presentations, challenge the class to find the pattern that causes these climate changes. Demonstrate to them how the hottest air is at the equator. This air rises with moisture and flows toward the North Pole, where it cools and sinks to the earth. This air then flows back to the equator again. This movement of air is our global wind pattern. It is divided into three large circular wind systems in the Northern Hemisphere because of the earth's rotation. The area between the equator and 30° N carries warm wind from the equator which cools and falls to earth and warms again as it flows back to the equator. The area between 30° N and 60° N is cooled by air flowing from the Arctic Circle. Ask students if they think there will be a similar pattern in the Southern Hemisphere.
3. Have the three groups representing the Southern Hemisphere give their reports. Start with the group closest to the equator.
4. Discuss with the whole class climate zones that are formed due to differential heating of the earth and global wind patterns. It is important to note that climate zones offer a general overview. Local climates are greatly affected by variations in the terrain and so on.

Have students identify the zone in the Southern Hemisphere where they would expect to find a climate similar to theirs. Why is light intensity important to wind patterns? Where do you expect the climate to be the warmest? Where do you expect it to be the coldest? Why does air move around the globe?

1. Have each group write a report on the country they researched.
2. Have students select a town in the Southern Hemisphere at a latitude similar to theirs. Have the class write to the mayor of this town, asking about the climate, what the people do, what crops are grown there. Try to choose a town that is surrounded by topography similar to your own.

   Note: The major difference between the Northern and Southern Hemispheres is that the seasons are at the opposite times of the year.
3. Have an international food day and prepare recipes from the different regions.

Food Choices

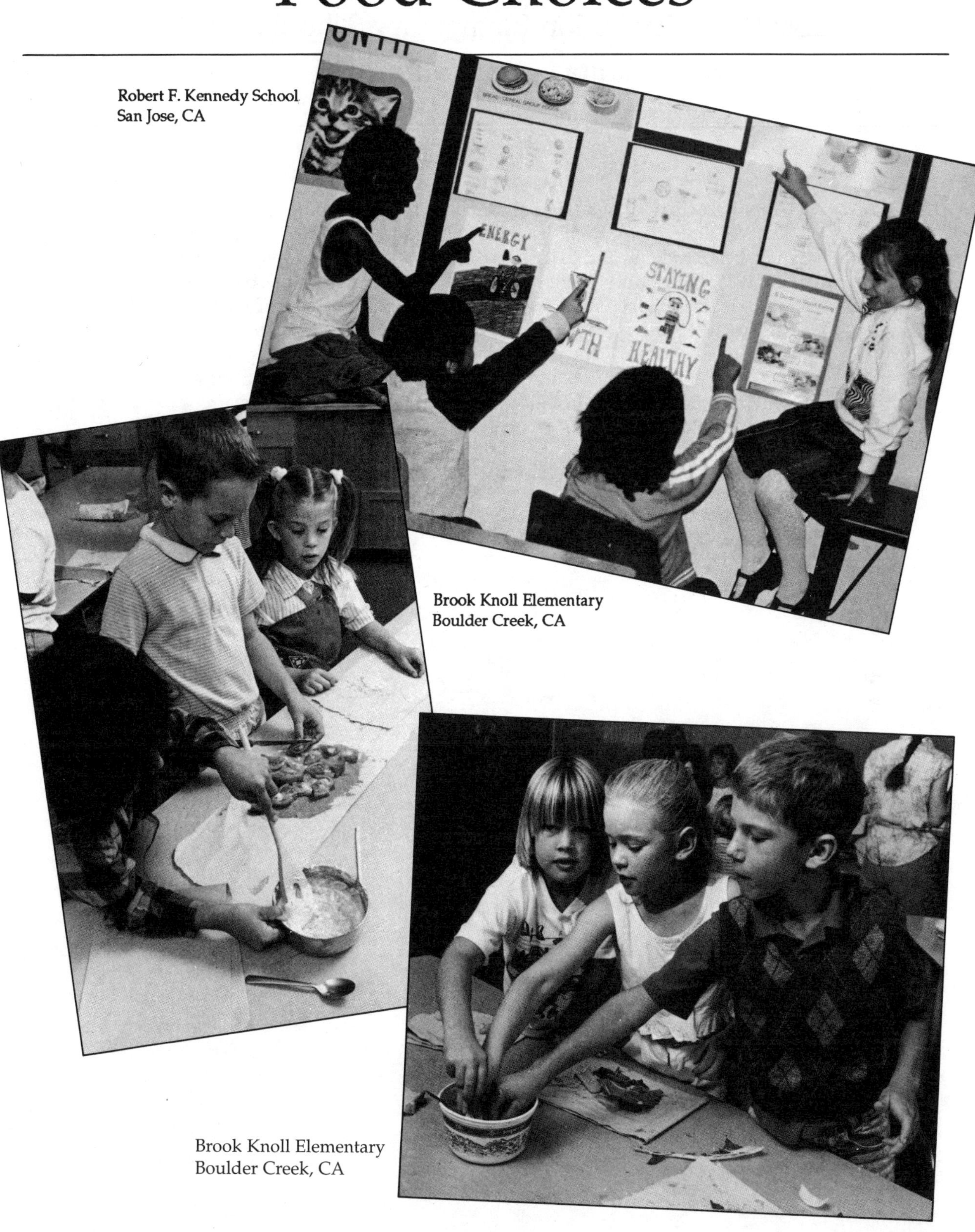

Robert F. Kennedy School
San Jose, CA

Brook Knoll Elementary
Boulder Creek, CA

Brook Knoll Elementary
Boulder Creek, CA

Brook Knoll Elementary School
Scotts Valley Union Elementary School District
Santa Cruz, California

After participating in a Life Lab Science Program two-day workshop in 1986, three Brook Knoll teachers decided to organize other faculty members, students, and parents to develop a Life Lab garden. On May 29, 1986, their efforts were realized when a formal groundbreaking and dedication ceremony was held. Brook Knoll's garden was dedicated to the memory of the seven crewmembers of the space shuttle "Challenger" who perished that year.

Since 1986, parent volunteers have built compost bins and raised garden beds, and have installed a drip irrigation watering system. A grant provided funds for a greenhouse.

UNIT INTRODUCTION

Food Choices

As part of the web of life, we need food to give us energy, keep us healthy, and help us grow. Learning about what we eat provides us with important knowledge for lifelong habits and can be lots of fun—especially when we get to devour our knowledge! The Living Laboratory lets us discover and interact with the processes of the world around us. It also connects us with the 10,000-year-old tradition of agriculture—the cultivation of food. The last three units of this book are oriented toward the harvest as we explore food choices, nutrients in different foods, and awareness.

This unit introduces our study of nutrition. Language arts, reference skills, graphing, art, and experiential activities are used to explore why we eat, what is available to eat, our food preferences, and what influences our choices. The importance of eating a good breakfast and planning a school lunch are included. A Basic Four food group is introduced as a guideline for determining whether meals contain necessary nutrients. Reducing intake of sugar, fat, and salt is emphasized.

All of these activities can be done in the classroom. Also, the activities can be reinforced at home by informing parents what students are learning about healthy eating choices. Be sure to obtain parents' written permission before allowing students to eat anything. Some students may have allergies or diet restrictions.

"You are what you eat" is a saying that develops more significance the more we learn about the science of nutrition. The more we share with our students about the bounty of a healthy harvest, the better we will prepare them for their own future.

Activities	Recommended Grade Level
Keep Me Running *(introduces how our bodies use food)*	2,3,4,5,6
I Eat My Peas with Honey *(individual food choice survey)*	2,3
How Do We Make a Horse into Jell-O? *(the sources of foods)*	2,3,4
Are You a Natural? *(nutritional value of foods due to degrees of processing)*	2,3,4,5,6
Mrs. Price's Bread *(bread recipe)*	
When I Was Little *(demonstrates eating habits through generations)*	2,3,4,5,6
The Roots of Food *(social influences on food choices)*	2,3,4,5,6
Hear Us, Hear Us *(graphing of food preferences)*	2,3,4,5
The Good, the Bad, and the Ugly *(categorizes foods by nutritional value)*	3,4
On Your Mark, Get Set, Breakfast! *(the need for a healthy breakfast)*	2,3,4,5
Button Up for Breakfast *(reinforces the importance of eating breakfast)*	2,3,4
The Fabulous Familiar Four *(the basic food groups)*	2,3,4,5
Searching for the Terrible Three *(effects of sugar, fat, and salt)*	3,4,5,6
Dear Diary *(analysis of food we eat)*	4,5,6
Food Planners *(reinforces the importance of a balanced diet)*	2,3,4,5,6

Bats Eat Bugs

Chorus:

```
G                         C
Bats eat bugs, they don't eat people
G                                  D7
Bats eat bugs they don't fly in your hair
G                         C
Bats eat bugs they eat insects for dinner
G                         D7         G
That's why they're flying up there.
```

Coyotes eat rabbits they don't eat people
Coyotes eat rabbits 'cause you're too big to bite
Coyotes eat rabbits they eat rabbits for dinner
That's why they're out in the night.

Snakes eat mice they don't eat people
Snakes eat mice that's why they're on the ground
Snakes eat mice cause we're too big to swallow
So they don't want you hanging around.

Bears eat berries they don't eat people
Bears eat berries they don't eat you or me
Bears ear berries and they'll steal your dinner
So you better hang it high in a tree.

Nothing out there wants to eat you
Nothing out there wants to make you its meal
Nothing out there eats people for dinner
'Cause they know how sick they would feel.

Chorus

Keep Me Running

Description

Students role play energy, growth, or maintenance of health.

Objective

To introduce how our bodies use food.

Materials

Energy, Growth, Keeping Healthy Posters, pp. 422–424
Drawing paper or heavy cardboard
Crayons
Scissors

In this activity, we will learn why we eat. Food is the fuel that keeps the human machine running. How does food keep us going? (It provides growth, energy, and maintenance.) (Discuss with students why they eat.)

1. What does food do for our bodies? Have students act out roles that show an object other than a person performing an activity that requires energy, growth, or maintenance of health.

   Examples:

   Energy: A car driving around. What does it use up as it moves? What do you do when it can't move?

   Growth: A seedling. What will the seedling need to grow larger? How does it get it?

   Maintenance: A machine that you use all the time. What would you do to it to keep it running smoothly?

2. Use posters to show that our bodies need fuel to help us grow, to keep us healthy, and to give us energy, just as the objects in the examples did.
3. To help students further define energy, growth, and maintenance, have them make their own food shield. Instruct them to draw as large an outline as possible of their favorite food on heavy cardboard and cut it out. Divide the shield into three sections: Energy, Growth, Maintenance. Ask students to draw pictures in each section, showing how the food fulfills their requirements, helping them grow, giving them energy, and keeping them healthy.

How does food help our bodies? Why do we need energy? What happens to a plant if it doesn't get what it needs to grow? Name one food that can provide everything your body needs to stay healthy, grow, and have energy. (No one food is perfect. A variety of foods is needed to provide all of the nutrients necessary to keep a body healthy. When our bodies do not have the necessary fuel going to all of their parts, they are under stress.) What would you like to learn this year about the food choices you make?

I Eat My Peas with Honey

Description

Students conduct a survey of food preferences.

Objective

To introduce the individual decision-making process in choosing foods to eat; to use writing and communication skills to demonstrate how and why students make certain food choices.

Materials

Life Lab journals
Pencil

Have you ever heard this rhyme?

I EAT MY PEAS WITH HONEY.
I'VE DONE IT ALL MY LIFE.
IT MAKES THE PEAS TASTE FUNNY.
BUT IT KEEPS THEM ON MY KNIFE!

People have different tastes in food. In this survey, we'll take a look at some of the foods we like and don't like.

Have each student conduct the following survey.

1. Name your favorite food. ______________________________
2. Find two people in your classroom who like the same food you do.
3. Find some classmates who do not like your favorite food. Ask them to try to explain why they don't like it. List two reasons:
   - ____
   - ____

4. Make a list of foods you don't like.

____

____

____

____

5. Look at the foods on your list. Can you think of reasons why you don't like each food? Write the reasons beside each one.
6. Now make a list of your favorite foods.

____

____

____

____

7. Write the reason you like each of these favorite foods beside each one.

Discuss with students how and why they made these choices. Do they consider some foods to be good or bad? Why?

Who chooses what you eat? What affects whether you like or don't like a food? Why can you like a food that isn't good for you? Do you like to try new foods? Why or why not?

Make a food sample tray of foods the students might not have tried: raw vegetables (broccoli, spinach) with dip; hummus (garbanzo bean dip).

How Do We Make a Horse into Jell-O?

Description

Students will identify the plants and animals from which a variety of foods are derived.

Objective

To understand that all food comes from plants and animals.

Materials

Pictures of foods; pictures of plants and animals that are the source of those foods
Life Lab journals

Today many foods are processed into forms so that their source is not recognizable. What do we mean by *processed*? Can you name any foods that are processed? Even though we eat processed foods, we are dependent upon the original plants and animals for all of our food.

1. Divide the class into small groups.
2. Hold up pictures of foods for the class to see. Give each group a few minutes to discuss the source among themselves.
3. Have each group draw the food source in their journals.
4. Share responses when done.

5. List all sources on the chalkboard. Have students categorize them into animals and plants. Is there any source that is not an animal or plant?

Suggested Foods

Food	*Source*
ketchup	tomato, sugar cane, spices
hamburger	cow
french fries	potato plant
peanut butter	peanuts
grape jelly	sugar cane, grapes
spaghetti	wheat
cornflakes	corn plant
crackers	wheat
cheese	cow
popcorn	popcorn plant
bologna	cow, pig, or turkey
gelatin	horse hooves
chili	beans, cow, tomato
pickles	cucumbers
ice cream	cow, sugar cane

Identify three foods you eat that are in the form of their natural sources. Identify three foods that do not resemble their sources. Why are plants and animals so important to us?

1. Have students classify plants into fruits, vegetables, nuts, legumes, and grains. Sort seed samples of each into five labeled cans. Prove they are plants by sprouting them.
2. Discuss meat-eaters and vegetarians. Find out why some people do not eat meat. Which of the foods listed above would a vegetarian not eat?

Are You a Natural?

Description

In this activity, foods are first defined according to the amount of processing. Then wheat is used as an example to demonstrate how processing affects nutritional value. Finally, students grind flour and bake bread.

Objective

To learn the nutritional value of natural foods compared to processed foods.

Materials

Samples of a wheat plant, wheat berries, whole grain flour, white flour
A hand flour grinder
Life Lab journal
Ingredients for breadmaking, Recipe p. 327

Introduce the following definitions:

NATURAL FOODS—fresh raw foods that do not undergo any changes from their plant or animal sources when eaten other than the outer covering being removed. The closer a food is to its natural state, the greater is its nutritional value. Examples: raw fruits, vegetables, nuts. Hold up the wheat plant and wheat berry as examples. Note: The term *natural* is often misused.

MINIMALLY PROCESSED FOODS—raw foods that are slightly changed from their original form into one that is more usable or available. Examples: whole wheat bread, peanut butter, baked potato. These foods retain most of their nutrients but can spoil quickly and may be harder to obtain at certain times of the year. Hold up the whole wheat flour as an example.

HIGHLY PROCESSED FOODS—foods that undergo considerable change from their original form. These foods are generally quickly prepared and easily available but lose nutrients in processing and often contain chemical additives. Examples: foods in packages, cured meats, pies, white bread. Hold up the white flour as an example.

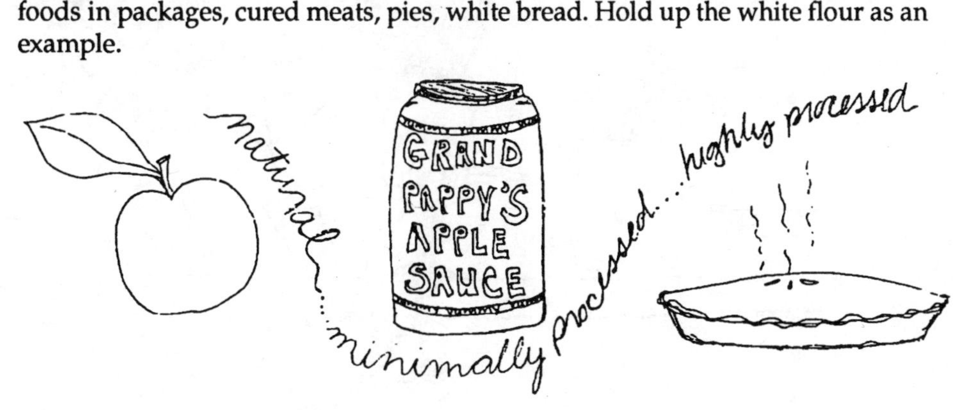

1. Explain that wheat is used around the world in basic foods such as breads, cereals, and spaghetti. Depending on how it is processed or milled into flour, it may or may not lose many of its nutrients.
2. Have students draw a whole wheat berry (the seed of the plant) and label its parts.

   1) The *germ* is the embryo or seedling plant within the grain. It contains the most nutrients in the berry.

   2) The *bran* is the seed coat and contains fewer nutrients than the germ.

   3) The *endosperm* is the large starchy interior and provides the food for the embryo, but it contains the least nutrients for humans.

   3 ENDOSPERM
   2 BRAN
   1 GERM

   Whole wheat flour has all parts of the wheat berry in it. White flour has only the endosperm in it. Even when the white flour is enriched by added nutrients, only four or five of the 20 nutrients removed in the processing are replaced.
3. Use a hand grinder to mill wheat berries into whole wheat flour. Seven cups of berries will produce ten cups of flour. Pass it through the grinder twice so that it is fine enough for baking. Use the whole wheat flour to bake Mrs. Price's bread on p. 327.

Why is whole wheat flour better for you than white flour? Give an example of a food in its natural form, in a minimally processed form, and in a highly processed form. List an advantage and disadvantage for the three types of food. Why are there so many highly processed foods used in our society?

Do a bread taste test with different types of bread. Have students compare nutritional value and taste preference.

MRS. PRICE'S BREAD

INGREDIENTS

3 CUPS HOT WATER
3 TABLESPOONS HONEY
3 TABLESPOONS OIL
3 TEASPOONS SALT
2 PACKETS YEAST
7-9 CUPS WHOLE WHEAT FLOUR

EQUIPMENT

MEASURING CUPS
LARGE BOWL OR POT
LID OR CLOTH
3 BREAD PANS
OVEN

DIRECTIONS

STIR TOGETHER HOT WATER, HONEY, SALT UNTIL DISSOLVED.

WHEN LUKEWARM, ADD YEAST.

GRADUALLY STIR IN FLOUR UNTIL YOU CAN NO LONGER ABSORB ANY MORE INTO MIXTURE (YOU MAY NEED MORE THAN 9 CUPS).

KNEAD DOUGH UNTIL IT HAS A SATIN GLOW AND IS NO LONGER STICKY (ABOUT 10 MINUTES).

COVER DOUGH IN BOWL WITH LID OR CLOTH.

PUT IN WARM PLACE (70°–75°) BUT NOT TOO HOT. BE SURE IT IS NOT IN A DRAFT.

AFTER A FEW HOURS, THE DOUGH WILL DOUBLE IN BULK. PUNCH IT DOWN AND LET IT RISE AGAIN FOR 30-45 MINUTES. PUNCH IT DOWN AGAIN AND KNEAD LIGHTLY.

FORM THREE LOAVES.

PUT THE LOAVES IN WELL-GREASED PANS.

LET THE LOAVES RISE AGAIN FOR 1 HOUR UNTIL THEY DOUBLE IN BULK.

BAKE AT 350° FOR 1 HOUR. RUB TOP WITH BUTTER OR MARGARINE. REMOVE FROM PAN IMMEDIATELY.

EDIBLE THOUGHTS....

WHAT MAKES BREAD RISE?

(THE YEAST ARE ALIVE! THEY ARE FUNGI AND THEY EAT THE HONEY IN THE DOUGH. AS A BY-PRODUCT THEY GIVE OFF CARBON DIOXIDE. THE CARBON DIOXIDE CAUSES THE BREAD TO RISE.)

DISCUSS THE DIFFERENCES IN MAKING HOME-MADE BREAD VS. STORE BOUGHT.

HOW LONG AGO WAS IT THAT PEOPLE ALWAYS BAKED THEIR OWN BREADS?

When I Was Little

Description

This activity uses writing and oral communication to research food changes over the past two generations. Students prepare a questionnaire about food and food habits. Using this questionnaire they will interview a grandparent or someone of that generation.

Objective

To demonstrate how eating habits have changed over the past two generations; to develop an idea of the roots of food choices.

Materials

Life Lab journals

Do you ever complain about having the same old thing for lunch? That's the short view. In the long view, the food we eat has changed quite a bit just in the last two generations. Many of the prepared foods in supermarkets of today were probably unknown to your great-grandparents or grandparents. Why not ask them and see? What questions would you ask them about the food they ate when they were your age? (List questions.)

1. Have students prepare a questionnaire about food and eating habits. Possible questions include: What foods did you eat when you were little that you seldom or never eat now? Where did you get your food? Were foods prepared differently than they are today? What foods were always made at home instead of purchased at a store? How often did the family eat together? Do you eat any foods today that seemed strange or unappetizing when first introduced to you? At what time of day did you eat your largest meal? Did you eat out? If so, where? How often? What was your favorite food as a child? Do you think foods are better or worse today?
2. Have students analyze the responses by comparing the changes and similarities in eating habits.

Summarize your interview. Compare how eating habits have changed over the years. Name an old eating habit that you would like to practice. Trace the development of an old eating habit into a modern eating habit.

1. Invite a panel of grandparents to school to be interviewed by the whole class.
2. Have some seniors work with a group of students to prepare an old-fashioned meal.
3. Have students research food preferences in different historical periods. Can modern habits be traced back to them?

The Roots of Food

Description

This activity introduces what influences students' own food choices and those of people in different cultures.

Objective

To discover sociological and environmental influences on food choice.

Materials

None

Maybe you've heard the saying, "You are what you eat." In another sense, however, "You eat what you are!" There are many factors other than availability that influence our food choices: culture, religion, agriculture, medicine, tradition, economics, history, geography, climate, and social status. For example, at the first Thanksgiving celebration, pilgrims and Native Americans shared certain foods. Americans continue to eat turkey because of that historical tradition. People who live by the ocean and eat fresh fish are probably influenced by geography. What influences your food choices? Try to think of these influences in terms of cultural and environmental factors.

1. Write the following lists on the chalkboard in random order.

I am a person who	*I choose to eat*
lives by the ocean	shrimp
has a garden	fresh salads
lives in Alaska	canned vegetables
lives without electricity	pickled vegetables, meat, fish
is always in a rush	processed foods, frozen dinners
lives in Hawaii	fresh fruits
has high blood pressure	food without salt
practices Hinduism	no meat from a cow

2. Have students match the appropriate person with the food that person might select.
3. Continue by having one student announce a cultural or environmental influence and another name a food that a person might eat because of that influence.

Name three possible differences between the food choices of a person who lives in Bangladesh or Ethiopia and you. What are the reasons for those differences? Discuss why you are able to eat your favorite food in your community. Describe a community where a person may not find your favorite food.

1. Research food preferences of different cultures within your community. Invite community members to the class to explain the development of those food choices.
2. Prepare a meal that is typical of a culture or historical time period the class is studying.
3. Create a community food root tree. Let the leaves of the tree represent different foods eaten in your community. The roots represent specific factors that influence that food choice. Connect the roots to the appropriate leaves.

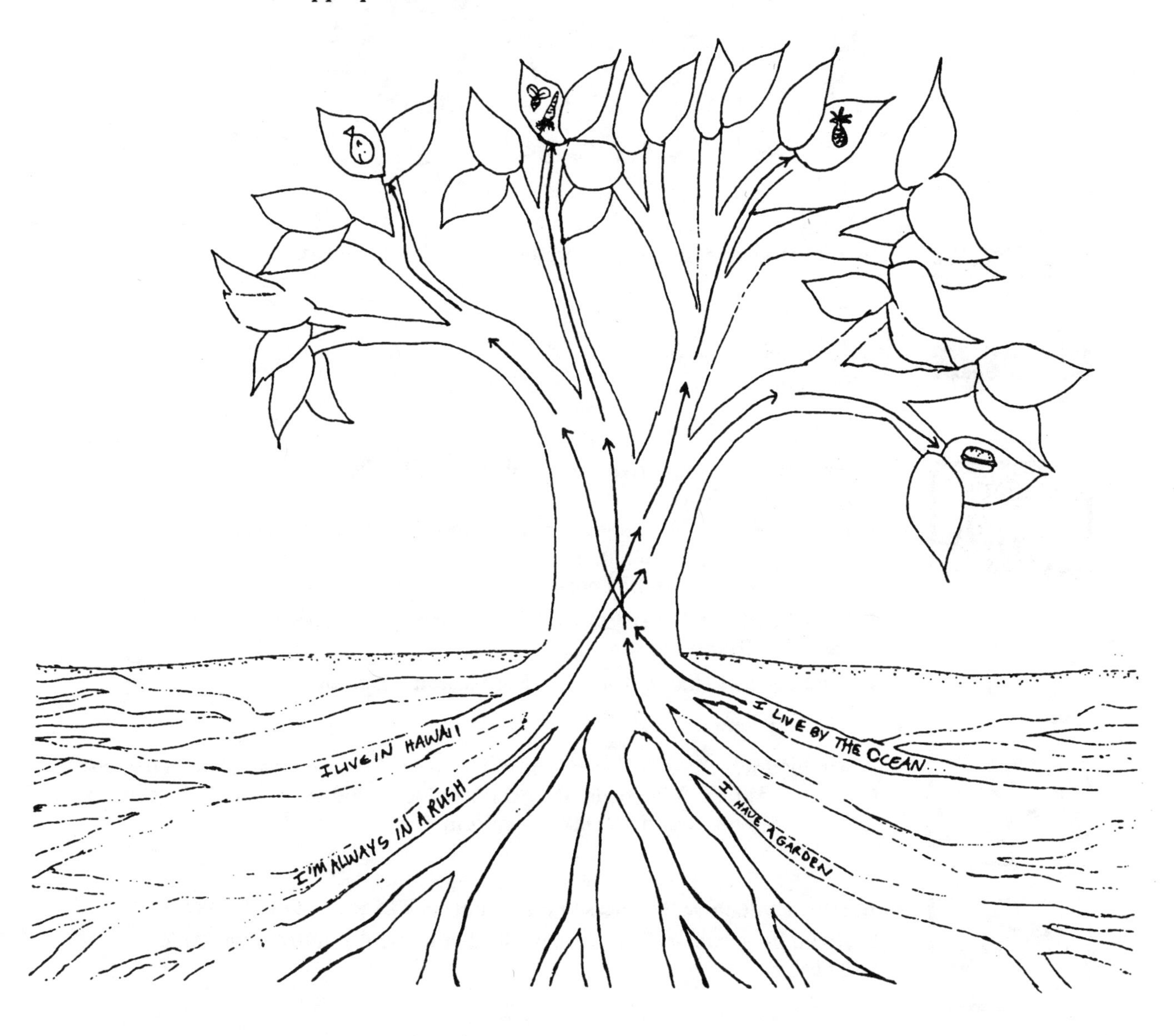

Hear Us, Hear Us

Description

Students graph their food preferences and present their graphs to the school cook (or appropriate representative).

Objective

To use graphing as a means of presenting food preferences.

Materials

Large sheets of newsprint
Felt-tip pen

1. Have students make five graphs labeled Fruits, Vegetables, Beverages, Grains, and Meats.
2. Have students take nominations for seven favorite foods in each category and list nominations on the bottom of each graph. Guide the students to make healthy choices.
3. Have them record on each graph the number of students who prefer each food listed.
4. Send results to the cook at the cafeteria, as appropriate.

Which food was the class favorite in each category? Name a grain that you like that is not on the graph. Which was the overall favorite food in the class? If you were the school cook, how would you use this information?

1. Have students write letters to the school cook to accompany the graph.
2. Invite the school cook to class to talk about how the information could be used.

The Good, the Bad, and the Ugly

Description

Students use reference skills to categorize breakfast foods according to their food value.

Objective

To examine the importance of breakfast; to determine which breakfast foods are good for you and which are not so good for you.

Materials

Breakfast Food Cards, available from Dairy Council, see p. 480
Boxes for cereals with sugar and without sugar
Blackline masters, pp. 425-426

How many of you ate breakfast this morning? Why or why not? Studies have shown that students who don't eat a good breakfast often have trouble doing as well as they could in school. If we start the day without any food energy, we will not get proteins to where they are needed in our body until 3 P.M. (if we eat lunch). That means from 6 A.M. to 3 P.M. stress is put on our bodies—stress that could have been prevented by eating a good breakfast.

In eating breakfast, it's important to choose foods that will get our bodies off to a good start. Some cereals are processed grains with many of the nutrients missing, and many cereals have sugar as a major ingredient, which is bad for the teeth and adds empty calories. If you compare the ingredients in a candy bar with some breakfast cereals advertised for children, you'll find that some of the cereals have a higher percentage of sugar (over 30 percent).

1. Using the food cards and the Breakfast Food list for ideas, have students categorize breakfast foods according to whether they are Good (good for us to eat), Bad (not good for us to eat but have some OK parts), or Ugly (very bad for us to eat).
2. Use cereal boxes to show examples of food labels and how students can tell if sugar or other sweetener is in the product. (See label reading lesson, p. 374.)

Why is it important to eat breakfast? Give two examples of a good breakfast. Compare three differences between a good cereal and an ugly cereal. Name three good breakfasts you like to eat.

Prepare a good breakfast with your class.

On Your Mark, Get Set, Breakfast!

Description

This activity uses a five-lane race to demonstrate how breakfast helps us through the day. It is a good follow-up to The Good, the Bad, and the Ugly.

Objective

To reinforce the idea that everyone needs to eat a healthy breakfast.

Teacher Background

Students will perform a five-lane running race to demonstrate how breakfast helps them through the day. Students who are in lanes with healthy breakfasts will have no trouble making it to lunch time. Students who are in lanes with no breakfast or a sugary breakfast will have to pick up buckets of water as they move toward lunch time. The water represents the added stress they carry by not eating a healthy breakfast. (You may want to substitute buckets of water with plastic gallon jugs filled with water or some other heavy material.)

Materials

Buckets or jugs of water

Preparation

On a blacktop area, mark off five racing lanes. Mark time lines across the lanes as shown.
Place a bucket of water at each lane.

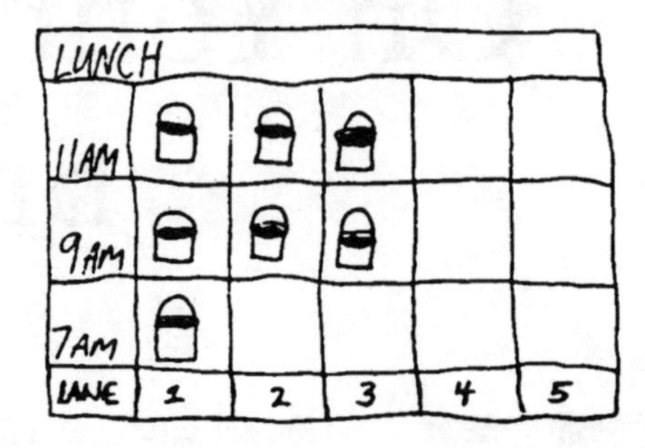

Who ate breakfast this morning? Tell me some examples of what you had. Would you say it was a Good, Bad, or Ugly breakfast? How will you feel by lunch time? How will people who didn't eat any breakfast feel? Do you want to run a race to find out?

1. Select five students to run the first race. Assign each one a lane.
2. Assign a breakfast to each racer. Let all students know the different breakfasts that were eaten.

   Lane 1: Ate no breakfast at all.

   Lane 2: Ate two doughnuts.

   Lane 3: Ate a sugary cereal product.

   Lane 4: Ate oatmeal.

   Lane 5: Ate whole grain cereal, fruit, and skim milk.
3. Have students begin the race at the same time, picking up any buckets that are in the lane. Each bucket is carried all the way to lunch time at the finish line. It will be very difficult for the person who ate no breakfast to reach lunch time carrying three buckets or jugs. The goal is to reach lunch time with plenty of energy and little stress.
4. You may repeat the race with another group of breakfasters.

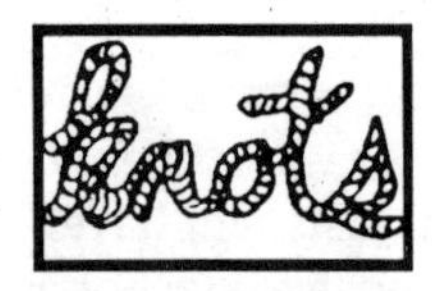

Why did Lane 1 have the most difficult time getting to lunch? Why was it easy for some people to get to lunch and more difficult for others? What kinds of foods are best to eat for breakfast?

Button Up for Breakfast

Description

Students make their own "Breakfast Buttons" with breakfast messages: "I 8 breakfast 2day" or "Breakfast is Good for Me." This activity uses art to reinforce the importance of eating breakfast. It is a good follow-up to The Good, the Bad, and the Ugly.

Objective

To reinforce the importance of eating breakfast.

Materials

Poster board or cardboard
Safety pins or button-making machine and appropriate materials
Colored markers

Let's start a campaign for better breakfasts. Like any campaign, our campaign for Better Beginnings with Breakfast needs campaign buttons. So we'll make buttons to show what we think about breakfast. What about "I break for breakfast"? Or maybe "Breakfast Comes First!"

1. Cut poster board or cardboard into campaign buttons.
2. Have students write a breakfast message on the buttons and decorate them.
3. Back the button with a safety pin or mount it and seal it with a button-making machine.
4. Start a Breakfast Button Box in the classroom. Have students choose a button in the morning when they come into class.

Describe what your button message means to you. Why is it important to eat breakfast?

The Fabulous Familiar Four

Description

Students learn the four basic food groups and identify and classify foods into these groups.

Objective

To introduce the basic food groups as guidelines for healthy eating.

Materials

Food pictures representing the basic food groups plus extra foods, available from Dairy Council, see p. 480
Velcro
Poster board labeled and divided into each food group and its function
Cardboard
Colored pens or crayons

Preparation

Place Velcro on poster board and on pictures of foods.

Why do we need food? How can we discover whether we are eating the variety of foods that will give our bodies what they need? Food gives us energy, helps us grow, and keeps us healthy. We need to eat a variety of foods so that our bodies will get all of the nutrients that they need. Food scientists have divided food into four basic food groups to help make sure that we are eating the variety of foods that we need. We will call them the Basic Four.

Milk Group—gives us strong bones and teeth and gives us energy. All foods that are made from milk are in this group. Ask students for examples such as cheese, yogurt, and other dairy products.

Vegetable and Fruit Group—helps to keep us healthy by helping our bodies function properly. Ask students for examples such as carrots, corn, oranges, and tomatoes.

Protein Group—helps us grow and builds strong muscles. Ask students for examples such as meat, fish, eggs, nuts and seeds, beans, and tofu.

Grain Group—gives us energy. Ask students for examples such as rice, tortillas, oatmeal, and bread.

Extra Group—This is the group for the foods that are not in the four main groups. These foods mainly consist of sugars and fats and are not good for us. Ask students for examples such as donuts, candy, cakes, and sweetened sodas.

1. Have students place the food pictures in the proper food categories on the poster board.
2. Have students create food pictures of their own using cardboard and crayons or colored pens and classify them according to food group.

How can we use food groups? What food groups give us energy? What food groups help us grow? What groups keep us healthy? Name your favorite food in each food group.

Make a class graph of the favorite foods in each of the four basic food groups. Send it to the cafeteria staff so that they will know the students' preferences.

Searching for the Terrible Three

Description

Students play a card game using critical thinking and math skills to plan a nutritious Basic Four meal.

Objective

To demonstrate that foods with added sugar, salt, or a high fat content detract from a good diet.

Materials

One set of Fortitude Cards for each group of three, pp. 427-438

Preparation

Cut sheets of Fortitude Cards into individual cards.

The four basic food groups provide a simple approach to checking whether or not we get the essential nutrients we need. However, this simplification allows foods to be included that are actually dangerous to our health. The three ingredients that can be dangerous when consumed in large amounts are sugar, salt, and fats.

SUGAR—Americans eat three times as much sugar as they did 100 years ago. Sugar is added to many canned and packaged foods. Eating a lot of sugar may cause tooth decay, obesity, heart disease, vitamin deficiencies, and diabetes. When we eat sugar, it is quickly used up by our bodies, giving a rush of energy that makes us feel very tired as soon as it is gone (sugar crash). Sugar gives empty calories. It does not provide any vitamins, minerals, or protein. The next time you're ready to bite into a cookie or drink some soda pop, think about it.

SALT—Salt is a mineral made up of two elements, sodium and chloride. Sodium is also found in many food additives. Check those labels! A high sodium diet can lead to high blood pressure, which is a main cause of heart disease. Salt is added to many packaged goods.

FATS— Some fat is necessary in the human diet. Fat contains needed nutrients, helps digest certain vitamins, is an excellent source of energy, and adds flavor to meals. However, too much fat can make us overweight and can be a major contributor to heart disease. We can reduce the amount of fats we eat by drinking skim milk instead of whole milk, not eating hot dogs or bologna, and cutting off excess fat from meat before and after cooking.

How can we tell if the Terrible Three are in foods? It is important to read labels on all packaged goods. Ingredients are listed in order of their percentage of weight in the product.

1. Divide students into groups of three. Give each group a set of Fortitude Cards.
2. Have them keep cards face up. Tell them not to look at the backs.
3. Have them use the cards to plan a meal that is balanced and that is low in sugar, salt, and fats.
4. To check their work, have them turn the cards over. Scoring on the cards gives nutritious foods 10 points and foods containing sugar, salt, or fats –10. Meals should have a food from each food group, and it should add up to 40 points.
5. Vary the activity by having students score foods they eat often. Or have students deal the cards to each member of the group. The first player starts with a nutritious food from any food group. The other players add to the meal until they have a food from each food group and a meal worth 40 points.

What are three ingredients in food that are dangerous to a person's health? How can you tell if any of these ingredients are in a food? Name two ways you could make your diet healthier. Predict what would happen if you ate a candy bar every time you had a snack.

1. Have students bring in labels from foods they eat at home. Circle any of the Terrible Three that are ingredients. Are those foods nutritious or dangerous?
2. Have students prepare a letter for the school newspaper or newsletter to educate others about the Terrible Three.

Dear Diary

Description

Students use reference and math skills to record what they eat for one day and analyze its nutritional value. Students should be familiar with the Basic Four and with sugar, fats, and salt before doing the activity.

Objective

To analyze the nutritional value of the food we eat.

Materials

Life Lab journals

Is it important for you to know if you are eating the food that is good for you? How can you tell if your diet is good for you? Pretend to be food scientists for one day and write down everything you eat. Then as a class, we can analyze our diets.

1. Have students write down everything they eat for one day. They should record all meals and snacks and include space for five columns: Basic Four, High fat, High sugar, High sodium, and Points.
2. Have students analyze the food they eat by giving points to each item:
   - 25 points for eating 100% of Basic Four (2 servings of protein; 4 servings of fruits and vegetables; 4 servings of grain; 3 servings of milk).
   - 5 points for each vegetable. Add 1 point if fresh (unprocessed) and another point if eaten raw.
   - 5 points for each protein food. Add 2 points for nonmeat protein foods.
   - Allow 5 points for each serving of grain or bread. Add 2 points for whole wheat grain products.
   - Subtract 5 points for each food high in fat and calories (processed meats, pastries, fried food).
   - Subtract 5 points for each food with added sugar.
   - Subtract 5 points for each food with added salt (either added in the processing or added by the student).

   A class average of 45-60 is excellent.

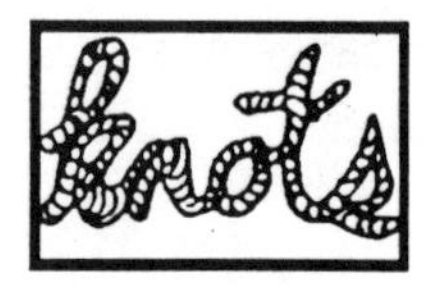

Why were points subtracted for foods with sugar, fats, or salt? Why is it important to have enough servings from the four food groups? Did your diet give you enough energy?

Food Planners

Description

Students use the guidelines for school lunches to plan a lunch menu that can be served in the cafeteria.

Objective

To reinforce the concept of balanced meals.

Materials

One Meal Planning Chart per group of 3, p. 439
Life Lab journals

The U.S. Department of Agriculture establishes specific guidelines that must be used in planning school lunches.

Protein

One serving of lean meat, poultry, fish, cheese, large eggs, cooked dry beans or peas, or peanut butter.

Vegetable and Fruit

Two or more servings of vegetables or fruit.

Bread or Bread Alternate

One slice of whole grain or enriched bread, whole-grain or enriched biscuit, roll, muffin; one cup of cooked whole-grain or enriched rice; macaroni, noodles, whole-grain or enriched pasta; other cereal grains such as bulgar or corn grits; or a combination of any of the above.

Milk

One serving of milk.

What factors should be considered in planning a menu? (nutrition, preparation, cost, taste, service, presentation)

1. Divide the class into groups of three. Give each group one Meal Planning Chart.
2. Have students consider the factors above and the school lunch guidelines, and plan a lunch menu that they would like to see served in the cafeteria.
3. Send the menus to the cafeteria staff as a suggestion for the school lunch. Remember to consider how you want the foods prepared. Fresh vegetables, for example, can be served raw or steamed. Chicken might be baked or fried.

Why do school lunches follow guidelines? Discuss the difficulties in planning school lunches. Discuss lunches students like and those that they do not like. How can the lunches be improved?

1. Prepare a written interview for the cafeteria manager. Invite the cafeteria manager to class for an oral interview. Use the answers as a language arts lesson for writing a newspaper article.
2. Using information obtained from the cafeteria manager, have the class create a budget for the whole school lunch program.

Nutrients

Pacific Elementary School
Davenport, CA

Pacific Elementary School
Davenport, CA

Pacific Elementary School
Davenport, CA

Pacific Elementary School
Davenport, CA

Pacific Elementary School
Pacific School District
Davenport, California

Pacific School began Life Lab in 1983. The project, which integrates nutrition and natural science lessons with gardening activities, augments the school's curriculum with its focus on experiential learning. The children love the garden and work very hard to make it successful.

In September, 1984, Life Lab took on new importance. Its concepts were extended to create a new program called "Food Lab." The purpose of Food Lab is to demonstrate through activity the cycle of food growth to include food use. The focus of the Food Lab curriculum is the preparation of nutritious meals and snacks on a daily basis. The children are actively involved in their food program: planning, planting, harvesting, cooking, canning, researching, and, of course, eating!

Each child has a weekly Food Lab and a weekly Life Lab group. The coordinators of each project work together to plan ways in which the children can learn to utilize the Life Lab garden to meet the daily nutritional needs of the school population. Both projects have been wondrous learning laboratories in which the children experience science.

UNIT INTRODUCTION
Nutrients

Science often gives us tools to discover and understand attributes and processes we weren't able to recognize previously. The scientific method offers a framework for analysis that enhances critical thinking. As we develop our critical thinking skills we become better observers of the obvious. Science also offers the opportunity to theorize and discover the not-so obvious. This is the case with nutrients. There are approximately 50 nutrients that our bodies depend on to give us energy, keep us healthy, and help us grow. These are categorized into six groups: carbohydrates, water, fats, minerals, vitamins, and proteins. It is important to balance these nutrients in our diet. We are able to do this by eating a variety of fresh foods from the different food groups.

In this unit, students perform simple experiments to reinforce their use of the scientific method, identify nutrients that are provided by the different food groups, and learn the functions these nutrients perform. They also explore how their bodies produce and burn energy and choose the necessary proteins for a vegetarian diet. The concept of nutrients is tied together at the end with a game of Nutrition Charades.

Some of the nutrient tests require the use of simple chemicals (iodine for complex carbohydrates and copper sulfate and sodium hydroxide for proteins). These are used safely when you follow the directions in the activities and exercise great care. It is a good opportunity to teach students the proper handling of chemicals. All of these activities can be done in the classroom. As in the previous unit, we recommend including nutrient-filled snacks with the activities and sending related information home to the parents.

"I like the nutrient lessons because I learn new things like testing starch. I never knew about starch until I did that test. Just because we can't see it doesn't mean we can't find it," said Tim DeSoto, Grade 4.

Activities	Recommended Grade Level
Invisible Gold *(introduction of the idea of food nutrients)*	4,5,6
The Cat's in the Carbohydrates *(test foods for complex carbohydrates)*	4,5,6
The Sugar Is in the Cat Who Is in the Carbohydrates *(test foods for simple carbohydrates: sugars)*	4,5,6
Fat Cats *(test food for fat)*	4,5,6
Burn Out *(to explore how our bodies produce energy)*	4,5,6
Water You Made Of? *(introduces water as a nutrient)*	4,5,6
The Body's Helping Hands *(introduces the body's need for vitamins and minerals)*	4,5,6
Patient Proteins to Grow On *(test food for proteins)*	4,5,6
We Complement Each Other *(to discuss amino acids as the building blocks of protein and how to combine foods for complete proteins)*	4,5,6
Nutrition Charades *(reinforce the need for nutrients)*	4,5,6

Invisible Gold

Description

Groups of students guess at and then compare the major nutrients in different foods, represented by grab bags with labels inside.

Objective

To introduce the concept of nutrients and the six nutrient categories. Prior to this activity students should be familiar with the purposes of food. (See "Keep Me Running," p. 319.)

Materials

Small brown bags
Pictures of food
Nutrient labels to be placed in bags
Nutrient chart, p. 440

Preparation

Prepare for the activity by making food grab bags. On the outside of the bag, attach a picture of a specific food. Inside the bag put individual labels of specific nutrients found in that food.

Examples

Meat patty	protein, fat, vitamins (B_1, B_2, niacin), minerals (iron), water
Corn tortilla	carbohydrates, vitamins (A), minerals (calcium), water
Swiss cheese	fat, protein, vitamins (A), minerals (calcium), water
Watermelon	carbohydrates, vitamins (A, C), water, minerals (calcium)
Carrots	carbohydrates, vitamins (A), water
Spaghetti with meat balls	carbohydrates, protein, vitamins (A, C, niacin), minerals (iron), fat, water
Baked Beans	protein, carbohydrates, water, minerals (iron)
Milk	fat, carbohydrates, water, minerals (calcium), protein, vitamins (B_2, D)
Egg	fat, protein, water
Peanut Butter	fat, protein, vitamins (niacin), water
Sugar	carbohydrates

What we really want from foods are the *nutrients*—the chemical substances that give us energy, help us grow, maintain our bodies, and regulate body processes. There are about 50 nutrients found in food. These nutrients are grouped into six categories: carbohydrates, water, fats, minerals, vitamins, and protein. The first letters of each word in the sentence "Cats wait for mice very patiently" might help you remember these groups.

Because each nutrient has a special job, we must take each into our bodies every day. No food contains them all. We must eat a variety of foods to provide our bodies with all the nutrients we need.

1. Give each group of three students a grab bag.
2. Start with a game of "If I looked inside my bag, what would I find?" Have each group guess at what is in their bag (use the Nutrient chart for reference).
3. Have groups compare nutrients. Are there some nutrients that all the foods have? Are there any that only one food has?
4. Use the grab bags to reinforce the concept that nutrients are an invisible part of food, that a food may have more than one nutrient, that different foods provide different nutrients and serve a variety of functions.

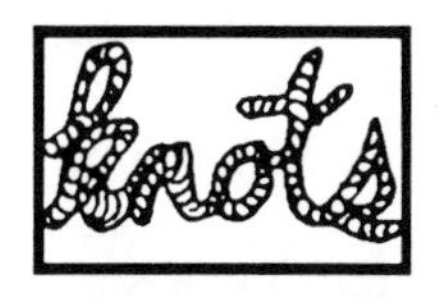

Where are nutrients found? Why are nutrients important? What are the six categories of nutrients? Do all foods contain the same nutrients?

Have students compare nutrients we get from food with nutrients plants get from soil. See Soil Doctors, p. 86.

The Cat's in the Carbohydrates

Description

Students review the use of carbohydrates in the human body and test for complex carbohydrates in foods. For this activity, students should be aware of why we eat and what nutrients are.

Objective

To use the scientific method to determine the types of food that contain complex carbohydrates (starches).

Teacher Background

The scientific method is a systematic way to analyze and develop information using what you already know to develop a hypothesis consisting of questions and guesses. These guesses can be checked by experiments, the results of which can be analyzed in order to reach a conclusion.

This activity illustrates the scientific method as follows (**I** **H**ave **E**instein's **R**ules **C**lear **A**lways).

Information: Starch is a form of carbohydrate. Iodine turns black when starch is present.

Hypothesis: Bread contains carbohydrates.

Experiment: Place drop of iodine on bread.

Results: The iodine turns black on the bread.

Conclusions: There is starch in the bread. Bread contains carbohydrates.

Application: I eat bread for energy.

Materials

Scientific Method poster, p. 391
Two clear cups
One teaspoon of cornstarch
Water
Five eyedroppers
Iodine
Food samples (bread, apple, potato, milk, tofu, and so on)
Nutrient chart, p. 440
Life Lab journals

Half the food we eat every day gives us nutrients called *carbohydrates*. Carbohydrates are the main energy source for our bodies. There are three types of carbohydrates: complex, simple, and fiber. They are known as "quick energy" (especially simple carbohydrates) because they are the first of the nutrients from the food we eat to be made available as energy for our bodies. In this activity we will test foods to see which ones have complex carbohydrates.

Complex carbohydrates are the starches that provide our bodies with steady energy. Usually foods that have starches in them also contain other nutrients.

1. Fill two cups 1/4 full with water.
2. Stir 1/2 teaspoon of cornstarch into one cup. The other cup is the control.
3. Add two drops of iodine to each cup. Have class observe what happens in each cup. (When iodine is added to the cornstarch, it turns blue-black.)
4. Divide the class into five groups. Give each group sample foods to test, iodine, and an eyedropper. (Handle iodine carefully; it can stain clothing.)
5. Have each student state a specific hypothesis about the food they will test. (Example: This bread has starch in it.)
6. Have each student put two or three drops of iodine directly onto their sample. The darker the iodine turns, the more starch the food contains.
7. Have each student state the results and draw a conclusion.
8. Have students place the tested foods into the four basic food groups: protein, milk, grains, fruits and vegetables (and the extra group). Ask them to draw conclusions as to which groups they think contain lots of carbohydrates.

What do carbohydrates provide for our bodies? What is starch? How can you tell if starch is present in food? Name three foods that will give you energy.

1. Have students study the chemical reaction between a complex carbohydrate and iodine to determine what causes the reaction.
2. Have students interview ten people about their attitudes toward bread or another carbohydrate. Do they think it is good for them? How often do they eat it? What nutrients do they think bread has? Compare your results with what you have learned about bread.

The Sugar Is in the Cat Who Is in the Carbohydrates

Description

Students will test for simple carbohydrates (sugars) and discuss the effects of sugar on their bodies. Prior to this activity students should have learned how their bodies use food, what nutrients are, and which foods have complex carbohydrates.

Objective

To determine types of food that contain simple carbohydrates (sugars).

Teacher Background

Sugar accounts for 1/4 (24%) of the calories consumed each day by the average American (1/3 of one pound!). Most of this sugar is refined (sugar that is added to foods like soda, cakes, cookies, candy, ketchup, ice cream, and so on). This sugar is often called "empty calories" because it provides no nutrients. Sugars also occur naturally in fruit, vegetables, and milk. But in these cases, sugars are accompanied by healthy supplies of vitamins, trace minerals, and fiber.

Materials

Scientific Method Chart, p. 391
Small cups
Tes-tape, available at any pharmacy
Food samples (flour in water, orange, sugar in water, milk, zucchini, soda, and so on)
Nutrient chart, p. 440
Life Lab journals

Carbohydrates are the nutrients that give us a steady source of energy. There are three kinds of carbohydrates: complex, simple, and fiber. In this activity we will learn about simple carbohydrates. Simple carbohydrates are the sugars that provide our bodies with energy rushes and energy let downs. Therefore, simple carbohydrates are not as healthy for us as complex carbohydrates. The purpose of the following experiment is to test for these simple carbohydrates.

How does our body react to sugar? Nutritionists have found that sugar promotes tooth decay and gives quick bursts of energy followed by low-energy let-downs known as sugar crashes. Sugar also gives our bodies lots of calories (about 24% of our diet) that turn to fat if we don't burn them up. This simple carbohydrate gives our bodies no nutrients, only quick energy.

We can tell when we are eating sugar by tasting sweetness or by reading the labels on packages.

Just as we used iodine to test for complex carbohydrates, we will use an indicator called Tes-tape to test for sugars. Tes-tape is a chemically treated paper that diabetics use to test for sugar. If the Tes-tape turns green, then sugar is present. The darker the green, the more sugar there is.

1. Give each group of students sample foods to test. Have each group state a specific hypothesis about their experiment, such as "Oranges contain sugar."
2. Have each group dip a 1-inch piece of Tes-tape in liquid or juice of the particular food.
3. Have each group state its results, draw a conclusion, and record results.
4. Have students place the tested foods into the four basic food groups or the extra group.
5. Ask them to draw conclusions as to which groups they think they would get the most simple carbohydrates from. Sugar is usually found in foods in the extra food group that includes white, brown, and raw sugar, honey, molasses, and so on, because simple carbohydrates are the sugary and less healthy form of carbohydrates. Note: Fruits contain complex and simple carbohydrates. However, fruits give us vitamins and minerals, too, and foods from the extra food group don't.

Why is sugar unhealthy? Name three foods that contain sugars. Why are fruits OK for us even though they have natural sugar in them? What can you do to cut down the amount of sugar-food you eat?

1. Give the class a healthy snack such as nuts and raisins or celery and cream cheese to take the place of other sugar snacks.
2. Have students research sugar substitutes, their uses, and their effects on our bodies.

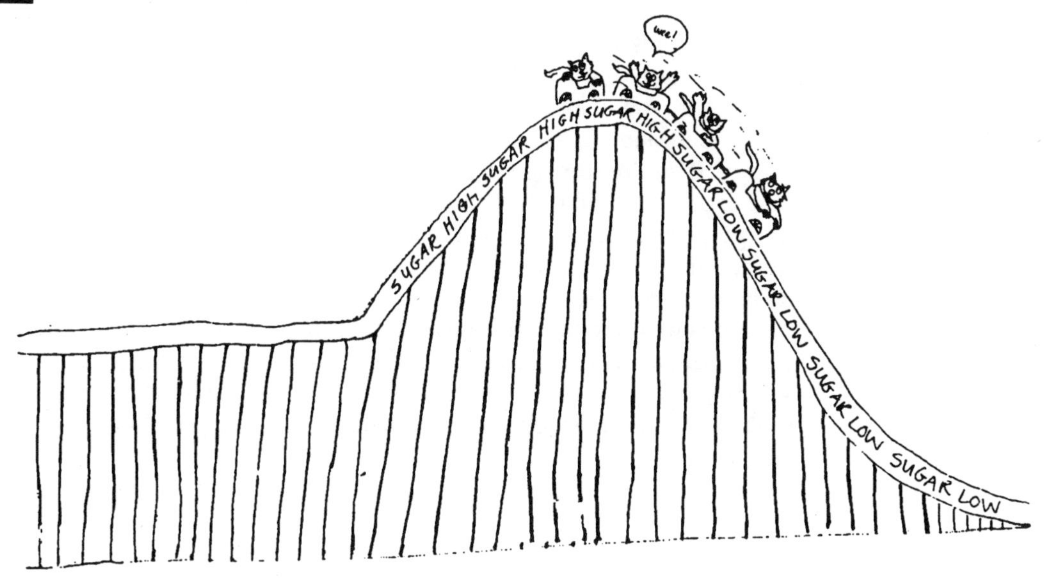

Fat Cats

Description

Students will learn the use of fat as a nutrient and test specific foods for fat content.

Objective

To introduce fat as a nutrient.

Materials

Scientific Method Chart, p. 391
Three-inch squares of brown paper
Food samples: oil, margarine, peanuts, milk, vegetables, and so on
Nutrient Chart, p. 440

Fats are another source of energy for our bodies. However, we need very little to meet our needs. Fats give a steady, slow-burning fuel to our cells. Carbohydrates, on the other hand, are quick burners and get used up more rapidly. We can store a limited amount of fat in our bodies. Where do we store fat? It is stored under every part of the skin, between muscles, around many of our organs, and even in the hollows of our bones. It is stored as an energy reserve of heat and fuel. When a body needs more fuel than it is getting , it converts the stored fat back into fuel for energy. If you're storing too much fat, how can you get rid of it? Cut down on energy-providing foods so that the stored fat can be used to provide your body's energy needs, or exercise so that your body burns more fuel.

Fat is found in food in two forms, visible and invisible. Visible fat can be seen with your eyes. Name some examples. (bacon, butter, margarine, fat on meats, and so on) Invisible fat cannot be seen. For example, some meats, poultry, eggs, whole milk, cheese, and desserts such as cakes, ice cream, and chocolate contain invisible fats.

1. Give each group of students foods to test for fats.
2. Have each group state their hypothesis about whether or not there is fat in a particular food.
3. On a brown paper square, have them rub a piece of food until it leaves a wet spot or put a drop on the paper if it is liquid.
4. Let the square dry.
5. Hold the square up to light. If there is a greasy spot—if it looks translucent—the food contains fat.
6. Have students draw conclusions about the fat content of their food.

How does your body use fats? What happens if you eat too many fatty foods? How can you get rid of excess fat? Name four foods that have fat in them.

1. Research the effects of too much fat in the body. (heart disease, strokes, obesity)
2. Collect menus from favorite restaurants and analyze them for their nutrient content. Are there a lot of fatty foods?

Burn Out

Description

Students learn how their bodies produce heat energy from food. They experiment with foods (via teacher demonstration) to determine if they are energy providers. They also discuss how they use that energy and the importance of a caloric-balanced diet.

Objective

To explore how our bodies produce energy.

Teacher Background

A calorie is a measure of energy stored in food. This energy is released in the form of heat. Scientists measure the calories in a food by burning the food and seeing how much heat is released. One calorie of energy will raise the temperature of one liter of water by one degree Celsius. Our bodies release this energy much more gradually.

Materials

Metal pie plate
Candle
Jar
One walnut
One sugar cube
One iron tablet (crushed)
One 2″ x 3″ piece of aluminum foil
Matches
Pencil with eraser
One straight pin
One teaspoon salt
One teaspoon oil
One cracker
One slice of bread
Tongs

(To demonstrate that heat is energy, hang the foil from the top of a pencil eraser with a pin. Light a candle and set it approximately four inches below the foil. The heat will rise, causing the aluminum to flutter.)

The energy produced in our bodies is similar to this candle and foil. Energy-providing nutrients (carbohydrates, fats) help to heat up each of the cells in our bodies. This heat gives us energy to function. If we eat foods that are high in carbohydrates and fats, we will have lots of energy to use.

Scientists measure the amount of heat food provides in *calories*. If a food has 100 calories, that means it will give us 100 calories of energy to burn up. Different types of food have different amounts of calories. On packaged foods, the calories will be written on the nutrition label.

We need to eat enough calories to provide us with enough energy. In a good diet the number of calories eaten equals the number of calories burned up in activities. Children aged 7-10 need about 2400 calories each day. The more active we are the more calories we burn. Running, for example, requires many more calories than sitting. Here's how we burn up calories:

Quiet things:	reading, watching TV, eating, school work	80-100 calories per hour
Light activities:	walking slowly, doing dishes	110-160 calories per hour
Medium things:	walking fast, household chores	170-240 calories per hour
Active things:	bowling, running, bike riding	250-350 calories per hour

If we don't burn up the calories we eat, we gain weight. A pound of body fat contains 3500 calories. If you eat 3500 more calories of food than you burn up, then you store a pound of fat. 3500 calories is a lot of food. Usually we gain weight a little at a time.

A lot of people today are overweight. This can be unhealthy. There are two ways to lose extra weight. One is to eat fewer calories in a day. If you cut out 500 calories each day, at the end of the week you would lose one pound. The other way would be to burn up more calories by exercising more. The two combined are very effective for losing extra weight and strengthening our bodies.

1. Have students form hypotheses as to whether each of the food samples will provide energy.
2. Test the foods by placing each in the pie plate. Use a candle to ignite the foods. Some items will "flame" more than others. The sugar will melt. The cracker and bread will burn since they contain carbohydrates. The walnut will burn for a particularly long time because it contains fat. The salt, water, and iron tablet will not burn since they are nonenergy-giving nutrients.
3. Have students state the results and draw conclusions.

How do our bodies use heat? Define *calories*. Which nutrients give us calories? Why is it important to watch how many calories we eat? Name three ways you like to burn up your calories. How did our great-grandparents burn up more calories than we do?

Have students keep a diary of what they eat and their activities for one day. Use a calorie book to determine the approximate number of calories eaten. Compare this to the number of calories burned up. Is it a balanced equation?

Water You Made Of?

Description

Students use experiments to demonstrate the significance of water in maintaining a healthy body.

Objective

To introduce water as a nutrient.

Materials

One teaspoon salt
Cotton
Rubbing alcohol

Water is the basic ingredient of the body's transportation system, the bloodstream. It carries nutrients to the cells and carries wastes from the cells. Water also helps to regulate body temperature. When we exercise strenuously or if it's hot outside, body temperature increases. When we sweat, the evaporating water cools our bodies and prevents overheating.

Two-thirds of body weight is water. If you weigh 90 pounds, 60 pounds is water. Water is in our cells, in our blood, and around our cells, so it is important for us to eat and drink foods that contain water.

Water is found in all the food we eat. Good sources of water are fruits and vegetables, juices, milk, and—of course—water.

1. To demonstrate how nutrients are dissolved in water, put a teaspoon of salt in the bottom of a glass. Have students observe what happens to the salt. (It stays on the bottom.) Add water to the salt and stir. Have students observe what happens to the salt. (It dissolves in the water.) Explain that in the same manner, nutrients dissolve in our blood and are carried throughout our bodies.
2. To demonstrate the cooling effect of water evaporating from our bodies, have students rub a little alcohol on their arms with a piece of cotton. How does it feel? (Cooler than the rest of the arm because alcohol evaporates more rapidly than perspiration—the same effect as water.)

Why is water so important to your body? What would happen if you did not sweat? Describe your bloodstream as if it were a river system: What does it transport? How does it get replenished? How does it cleanse itself (eliminate wastes)? How do animals that do not sweat cool themselves? (Dogs let water evaporate off their tongues. Snakes and lizards seek out shade in really hot weather. Birds spread their wings.) How do plants use water? Are plants' needs similar to our needs?

The Body's Helping Hands

Description

Students are introduced to several vitamins and minerals along with their food sources.

Objective

To introduce the body's need for vitamins and minerals.

Materials

Nutrient chart, p. 440
Vitamin sample plates, see Preparation

Preparation

Prepare a plate with the letter *A* in the center and food samples: carrot sticks, broccoli, yellow squash, Swiss chard, or spinach; a plate with the letter *C* in the center and food samples: orange slices, grapefruit slices, strawberries, tomatoes, green peppers; a plate with a piece of white chalk in the center and food samples: cheese, yogurt, broccoli. A plate with rusty nails (securely encased in a clear plastic bag) in the center and food samples: hard-boiled eggs, raisins, nuts.

Vitamins work in our bodies to keep us healthy. They lend a helping hand with food digestion, wound healing, blood clotting, good eyesight, and much more. There are 14 vitamins that are essential for good health. We can get them all by eating a variety of food. Plants can make their own vitamins, but animals must get their vitamins from their diet. Vitamins do not give us energy. They help our bodies function properly.

(Display each plate as you describe the vitamins and minerals.) Each vitamin has different functions. Vitamin A helps keep skin healthy, keeps eyesight good, helps us grow properly, and builds resistance to infection. Vitamin A is an example of a fat-soluble vitamin. Vitamins A, D, E, and K are all fat-soluble, which means they can be stored by the body for later use. Good sources of Vitamin A are dark green, yellow, or orange fruits and vegetables, beef liver, cheese, and milk. Sample the foods on the Vitamin A plate.

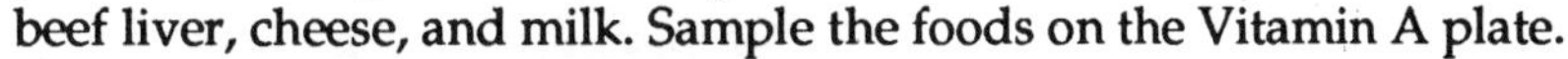

Vitamin C helps strengthen walls of blood vessels, keeps gums healthy, helps the body resist infection, and helps wounds and cuts heal. Vitamin C is an example of a water-soluble vitamin. The body does not store water-soluble vitamins, so we must eat food rich with them every day. Citrus fruits are good sources of Vitamin C. Cantaloupe, broccoli, strawberries, tomatoes, cabbage, green peppers, and potatoes also give us Vitamin C. Sample foods on Vitamin C plate.

Minerals, like vitamins, keep us healthy and our bodies functioning properly. Although there are minerals all around us, there are only 17 that our bodies need, and we get them from the food we eat.

Calcium is a very important mineral. It helps to build strong bones and teeth. A good source of calcium is milk and milk products. Broccoli is also calcium-rich. You can see what calcium looks like by examining the piece of chalk. Chalk is made from calcium. Sample the foods on the calcium plate.

Examine the rusty nails. Why are they red? (When iron comes into contact with oxygen, a chemical reaction takes place, turning the iron red.)

Iron is in our blood and helps to carry the oxygen from our lungs to all of the cells. That is why the blood that is carrying the oxygen looks red. The blood in our veins looks blue (look at your wrists) because this blood is returning to the heart and has no oxygen in it. "Iron-poor blood" can make you feel tired because the blood will not be able to pick up oxygen efficiently. The oxygen is needed to help burn food in the cells, which gives us energy. If the cells do not get enough oxygen, we don't have enough energy. Sample the iron-rich foods.

Vitamins and minerals keep us healthy by helping our bodies function properly. You can get all the vitamins and minerals you need by eating a variety of foods, the fresher the better.

1. Have students form groups to represent each plate.
2. Have each group write a story or make up a skit for the vitamin or mineral on their plate. For example, a cowboy named Rusty Nails might come to town to save it from the Tired Blood Gang.

Why are vitamins and minerals necessary for us? Name a food you like that gives you more than one vitamin. Name a good source of Vitamin A, Vitamin C, calcium, iron. Do we need to supplement our vitamin or mineral in-take? Why or why not?

Patient Proteins to Grow On

Description

Students use the scientific method to test foods for the presence of protein.

Objective

To introduce the use of proteins in the body.

Materials

Scientific Method chart, p. 391
Nutrient chart, p. 440
Sodium hydroxide solution, available from local high school or science supply catalogue
Copper sulfate solution, available from local high school or science supply catalogue
Eyedroppers
Five small jars
Food samples: milk, egg, zucchini, tuna fish, tofu, cookies, juice, and so on

Preparation

1. Teachers, be careful as you prepare solutions. Sodium hydroxide (lye) is poisonous and can burn. Put one layer of pellets on the bottom of a one-quart jar. Add 1 pint of water. Stir until all pellets are dissolved.
2. Copper sulfate is poisonous. Dilute 20 grams in 62 milliliters (1/4 cup) of water. (Be sure to return sodium hydroxide and copper sulfate to source, or store according to instructions.)

Proteins are absolutely necessary to our bodies. They are essential for proper growth. They help to form muscles, hair, bones, fingernails, brain, glands, teeth, and other solid matter in our bodies. Proteins also repair tissues.

Adults eat about 1,000 pounds of food in a year. Only about 100 pounds of that food is protein. How much protein we need depends upon our weight. If you're between 11 and 14 you need about 44 grams of protein each day. Look at the nutrition labels on the foods you eat to see how much protein you get. Good sources of protein are fish, poultry, nuts, beans, and milk products. To get enough protein, you should eat two servings a day from the protein group.

1. Divide students into groups of five. Have each group state hypotheses about the food samples and their protein content.
2. Pour a little sodium hydroxide solution into a jar so that it fills the jar about 3/4 of an inch.
3. Carefully add food to the jar. (Avoid splashing liquid.) Solid food should be crumbled.
4. Add a few drops of copper sulfate solution.
5. Stir. If protein is present, you will see a pinkish-blue color.
6. List the foods in which you found proteins. (Proteins are found in the milk group and protein group.)

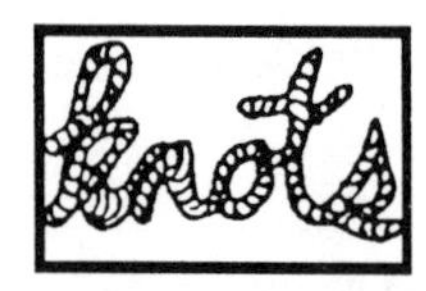

Why do we need proteins? Name four foods you eat that are good protein sources. What food groups contain proteins? Do you think all living things need proteins? Why or why not?

Have students make a high protein snack using nut butter.

We Complement Each Other

Description

Students learn about combining amino acids to create complete proteins. They simulate combining different foods to form a protein-rich meal, and they prepare a meal that provides all the necessary amino acids for a complete protein.

Objective

To discuss amino acids, the building blocks of protein, and to form complementary proteins.

Materials

Life Lab journals
One set of Protein Combination Cards for each group of six students (see Preparation)
International cookbooks
Recipe ingredients
Cooking materials

Preparation

Make one sheet of Protein Combination Cards in enlarged size (see next page for sample). Reproduce six sheets for group use.

People can get plenty of carbohydrates, vitamins, minerals, and enough fat from fruits, vegetables, and grains, but they must carefully plan their proteins. Proteins are made up of building blocks called *amino acids*. There are 22 amino acids. Our bodies can manufacture 14 of them. The other eight are called essential amino acids and we must eat these eight at the same time in order to be able to use the protein.

Animal proteins—meat, fish, poultry, eggs—have all eight essential amino acids. Plant foods have some of the amino acids, but usually all eight cannot be found in one plant food. However, by combining certain foods, we can get all the protein we need.

In order to get complete proteins, we can combine whole grains and legumes; milk products and whole grains; seeds and legumes. (Give examples of these combinations.)

For a multicultural lesson, consider ethnic variations in combining foods. Japanese: rice and soybeans (tofu); Mexican: corn and beans; Indian: rice and lentils.

1. Divide the class into groups of six students, with each group in a circle.
2. Give each group one set of Protein Combination Cards.
3. Have each group cut out the cards and place them face down in the center of their circle.

4. Each member draws one card and then must seek a partner to form a complete protein.

PROTEIN COMBINATION CARDS

whole grains	legumes
milk products	whole grains
seeds	legumes

5. Partners must brainstorm and list at least four possible meals that they could create together. Then they pick their favorite one to report back to the group. Each group can report back to the entire class.

Which nutrient might require menu planning? What are the building blocks of protein? Name a plant-food combination that forms a complete protein that you like to eat.

1. Have students prepare a meal that provides a complete protein. See recipes from international cookbooks.
2. Have students play the Fortitude Game (Searching for the Terrible Three, p. 340), earning extra points for forming complete proteins.
3. Have students research diets of different cultures and explore why certain foods are consumed. Look at international cookbooks to find recipes that combine different sources of protein.

Nutrition Charades

Description

Students use their knowledge of the purpose of nutrients and sources of specific nutrients to play a game of nutrition charades.

Objective

To synthesize the study of nutrients.

Materials

Nutrient chart, p. 440
Food cards from National Dairy Council, see Resources p. 480
Acting cards (see "Suggestions" list)
Answer cards (see "Suggestions" list)
3" by 5" blank cards

(Review the nutrient chart emphasizing the functions of specific nutrients and the food groups where they can be found. Let students refer to the chart during the game.)

1. Have team members guess what role the actor is playing and then hold up the card that tells the appropriate food that the actor should eat in his or her situation.
2. Participants: Director— teacher; actor—student who silently acts out situation on acting card; team—remaining students who guess what the actor is portraying and then hold up the proper food card.
3. Have students follow the procedure below:
   - Review the basic rules of charades: the actor cannot speak while the team tries to guess what he or she is doing.
   - The team will respond to the actor by guessing what role the actor is playing, naming what nutrient would be most helpful in the situation, holding up food cards that contain a high amount of that nutrient.
   - The director places a set of answer cards near the center of the team.
   - The director chooses the first actor and helps him or her to select an act from the acting cards.
   - While the actor prepares the charade, the team should study the answer cards.

- The actor silently acts out the charade.
- The first team member to hold up a correct food card is the next actor.

Which nutrients are good energy sources? Which nutrients help us grow? Which nutrients help maintain good health? Name the major nutrients found in each food group. What is the best way to make sure you get all of the nutrients you need?

SUGGESTIONS FOR NUTRITION CHARADES
(Role, types of nutrients, possible answers)

Acting Cards[1]	Major Nutrient Needed[2]	Appropriate Answer Cards[3]
Runner in a race, sweating	Carbohydrates	All fruits and vegetables, grains
Tired person	Carbohydrates, minerals	All fruits and vegetables, grains, meats
Bicycle rider	Carbohydrates	All fruits and vegetables, grains
Person loading large crates into truck	Carbohydrates	All fruits and vegetables, grains
Weight lifter	Protein	Protein group—meats, fish, poultry, legumes
Person with a cold	Vitamins, minerals	Fruits and vegetables
Pre-season football player	Protein	Protein group
Truck driver	Food without carbohydrates/fat	Vegetables and fruits, skim milk products, fish, chicken
Person reading	Food without carbohydrates/fat	Vegetables and fruits, skim milk products, fish, chicken
Person watching TV	Food without carbohydrates/fat	Vegetables and fruits, skim milk products, fish, chicken
Eskimo	Fat	Milk products, meats

[1]Make up acting cards by putting each action on an individual card. Add some of your own.

[2]All of the role-play people want a balanced diet (that is, all of the nutrients that are needed for them to function); however some will need more or less of a specific nutrient because of their activity.

[3]Answer cards should be individual foods within the listed groups. The Dairy Council Food cards are good to use because they provide pictures and specific nutrition content. However, just writing names of food on cards is sufficient for the activity.

Consumerism

Rock Creek School
Auburn, CA

Brook Knoll Elementary
Boulder Creek, CA

Rock Creek School
Auburn Union School District
Auburn, California

Rock Creek School, about thirty miles east of Sacramento, is located in the foothills of the Sierra Nevada. Our Life Lab project allows students the opportunity to learn many skills that provide an awareness of plants, soil, trees, and the survival of plants in the great outdoors.

This Life Lab has created an excitement not only in the students but also in our teachers who are willing to try new and different projects. We started in 1985 with seven teachers participating, and have grown to 14 teachers with about 400 students involved.

Two State grants were instrumental in implementing our Life Lab project, along with the Parent Club's financial and labor support. Various local civic clubs have donated funds, and our local nurseries have been very instrumental in providing seeds and tools.

In the summer, we have a local Grower's Market and have had students sell their own school-grown produce. This enables students to participate in the full cycle of planting to harvesting to consumerism.

UNIT INTRODUCTION

Consumerism

We live in a fast-moving society that has grown dependent upon convenience foods. Over half of the food products on supermarket shelves have been invented in the last 25 years. Although we may know which foods are good for us and which are not, it is difficult to avoid the influences of advertising. We must virtually become detectives to research information about the many foods available to us.

This unit culminates our efforts to have students become aware of why and what they eat. It connects the nutrition units with students' everyday world of food shopping with their parents, opening cans of food, and watching TV. Skills that range from reading and graphing to group cooperation are incorporated in activities: taking a supermarket field trip, learning how to interpret ingredient labels, making processed foods, influencing another class with advertising, planning and developing a market garden.

As we make students aware of our tremendous options as consumers, it is also important to create awareness of those people who do not have enough to eat. For this reason we have included a lesson on hunger in the world. We encourage you to explore this with your students. Currently, there is enough food grown in the world to feed all of the inhabitants satisfactorily. It is up to us and to the generation of students we now teach to make wise decisions about the distribution of this food so that all around the world children and their families can share in a bountiful harvest.

Activities	**Recommended Grade Level**
Supermarket Snoop *(supermarket field trip survey)*	4,5,6
Yes, It's Got No Bananas *(learning to read and understand package labels)*	2,3,4,5,6
The $1,000,000 Orange *(demonstrates why processed food costs more)*	4,5,6
This Little Lettuce Went to Market *(demonstrates steps from farmer to market)*	4,5,6
The New *Improved* Madison Avenue Diet *(effects of advertising)*	2,3,4,5,6
Try It, You'll Like It! *(experiment with the influence of advertising)*	4,5,6
Buy Me! Buy Me! *(influences on personal selection of food choices)*	4,5,6
Feast or Famine *(develop awareness of hunger in the world)*	4,5,6
Market Garden: Grow and Sell Time *(set up market of Life Lab vegetables)*	2,3,4,5,6

Supermarket Snoop

Description

The class takes a field trip to a supermarket, and students use research skills to gather information on products available to American consumers. The information gathered can be analyzed in activities throughout this unit, and adapted to other regions around the world.

Objective

To introduce responsible consumerism.

Materials

Life Lab journals

A *consumer* is someone who buys and uses things offered for sale, such as food, services, or clothing. We are all consumers. We choose what we consume, whether it is a certain TV program, a hamburger, a banana, or a new shirt. What are some things that influence our choices? (advertising, parents, friends, pretty packaging, health, and availability) To be good consumers, we must learn as much as we can about the products we are consuming, realize what influences our choices, and then make a responsible decision as to what we will and will not consume.

1. Plan a field trip to a local supermarket. If possible, arrange a time for the students to interview the manager. Be sure to plan time for students to do their own investigations in the store.
2. Divide the class into groups of three and have each group prepare a list of investigations to do at the store. Decide which questions they will answer themselves by investigating, which they will ask the manager, and which they will both investigate and ask the manager about. Suggested investigations:
   - How many products are sold in the store?

- How many products are in their original form (not processed)?
- How does the price of a nonprocessed food compare with that of the same food in a processed form, such as 1 lb potatoes and 1 lb potato chips?
- How does the store try to draw your attention to certain products?
- Which is the least expensive jar of peanut butter?
- Which size and brand of peanut butter is the least expensive per ounce?
- How many different breakfast cereals are sold?
- How many of the breakfast cereals list sugar as the first or second ingredient?
- What different materials are used to package foods?
- What products from different countries are available?
- What ten products do you think are shelved so that little kids can reach them?
- What kinds of things are sold in the bins nearest the cash register? Why?

3. Discuss the information students have gathered.
4. Make a class list of what students need to know to be good shoppers.

Why are you a consumer? Name one item you consume. What are three things that influence your choice of that item? Why do you think there are so many products in a store? Make up a checklist that would help you to be a good shopper.

1. Invite a consumer advocate to class. Many county and city governments have a consumer division. Find out what is investigated by consumer advocates and what consumers can do if they have a complaint.
2. Have students visit other types of food stores, such as a cooperative or farmers' market. Compare available products, prices, packaging, and atmosphere.
3. Have students research the history of supermarkets. They started in the 1930s. Where did people buy food before supermarkets existed?
4. Have students choose three different countries and research what people in those countries eat and how they buy their food.
5. Discuss being a producer vs. being a consumer. If you eat from your own garden, you're a producer. Make a list of producers and consumers.

Yes, It's Got No Bananas

Description

Students investigate the information found on packaged foods and start to analyze the contents.

Objective

To introduce package label reading.

Materials

Variety of packages with ingredient labels
One can banana pudding
Life Lab journals
Chemical Cuisine Chart from Let's Get Growing (optional)

The government requires all package labels to include the name of the product; the variety, style, and packing medium; weight of what's in the package; the name, city, state, and ZIP code of the manufacturer, packer, or distributor.

Most foods must list all of the ingredients used in the package. These ingredients will be listed in order of quantity: what was used most must be listed first, and so on.

Many foods also have nutritional information that tells how much of specific nutrients you get from one serving of that food. If we learn how to use all of this information, it can help us make informed decisions about the food we eat.

1. Begin by holding up the can of banana pudding. Have students list the ingredients they think are in the pudding and number them in order from most to least quantity. Discuss the lists.
2. Slowly read the banana pudding ingredients to the class, allowing them to discover that there are no bananas in the can. Are they surprised? How can something be called banana pudding if there are no bananas in it? Are there ingredients in the pudding they have never heard of? How do they feel about eating food that is made up of ingredients they know nothing about?

3. Divide students into groups of three. Give each group one package. Have each group find the ingredient label on the package. Have students name the primary ingredients. Does the food contain sugar, dextrose, glucose, corn syrup, sucrose? These are all sugars that have been added to the product. Have students compare the ingredients with the name of the food product. Is it what they thought it would be? Can they tell from the ingredients whether the food is good for them or not? Is there more they want to investigate?
4. Have students use the Chemical Cuisine Chart to learn more about some of the ingredients in the packages.

Why is it important to know the ingredients of packaged foods? If you read the label of everything you ate, do you think your eating habits would change? If you were not allowed to have salt in your diet, how would labels help you?

1. Set up a display of labels students bring from home. Have them circle in red any ingredient that means sugar.
2. Make a smoothie, or other multiple ingredient snack, with the class. Have them design a package label for their snack and prepare an advertising campaign.

The $1,000,000 Orange

Description

Students make orange "juicicles" to learn the steps and costs involved in food processing from original form to final form.

Objective

To demonstrate the increased cost of food as it is processed.

Materials

One orange with price tag
1/2 fresh orange per student
Hand juicer
Bowl
Pitcher
One small paper cup per student
Masking tape
One coffee stirrer or juice bar stick per student
Freezer space
Knife
Cutting board
Measuring cup
Water

Preparation

Draw the following chart on the chalkboard. Have a recorder fill in the chart as the activity progresses.

JOB	PROCESS	MATERIALS	LABOR	ENERGY	COST

What's the difference between an orange and an orange juicicle? What steps do you think are taken to make juicicles from oranges? (list on chalkboard) What do these steps involve? (energy, money, labor) If we turned our class into an orange juicicle factory, how much do you think we would need to charge for a juicicle in order to cover the cost of production? (Record predictions.)

1. Divide the class so that each group of students has a job from item 4.
2. Take the class through the process of making orange juicicles. With each step, discuss the materials, labor, and energy that would be involved.
3. Have the processors at each step determine how much they would charge for their part of the process. The recorder should record this information and then add up the final cost of the orange juicicles.
4. Jobs and steps:

Farmers	Grow and harvest oranges, truck them to processor.
Slicers	Cut oranges in half until there is one half per person.
Juicers	Squeeze orange halves into a bowl.
Blenders	Stir in 1/3 cup of water to the juice of each orange half.
Packagers	Pour mixture into a small paper cup.
Labelers	Put tape on each cup with product name and ingredients label.
Truckers	Carry orange juicicles to freezer.

5. Have students insert sticks into juicicles after 20 minutes.
6. Discuss what price the class would sell their product for in order to pay their costs and make a profit. Why would it be cheaper to buy the ingredients and materials and make it at home?
7. Enjoy the orange juicicles as an afternoon snack.

Why is it more expensive to buy processed foods than unprocessed foods? Name three advantages of buying processed foods. Name three disadvantages. Describe how having so many processed foods affects our society in terms of jobs, costs, energy use, health. Is the cost of food related to nutritional value?

1. Have students research food preparation in other cultures such as that of Native Americans in precolonial times. Compare it with food preparation today.
2. Have students investigate prices of foods in original form and the same product in processed forms. How many processed products can they find for one original food?
3. Have students research the actual breakdown in cost of a food product from farmer to store. Which is the most expensive step? The least? (This will vary for each product, because there are different processing and transporting requirements. Your County Agricultural Extension should be able to provide you with information.)

This Little Lettuce Went to Market

Description

Students investigate and compare the roads to market for local produce and produce grown far away. The lesson can be enhanced by inviting a farmer and a supermarket produce manager to class.

Objective

To investigate the steps from farmer to supermarket in marketing produce.

Materials

Seasonal list of locally grown fruits and vegetables, available from County Agricultural Extension or Agriculture Commissioner
One grocery store newspaper ad per group of four

Review with students a list of locally grown fruits and vegetables and their seasons.

1. Divide class into groups of four. Give each group a grocery store ad from the local newspaper.
2. Have students list fresh produce advertised and where they think it was grown.
3. Have students take one item grown locally and one transported from far away and list the different steps each had to go through to get from harvest to the supermarket. What are the costs and energy uses with each step?

4. Invite a farmer to class to explain how local farmers sell their produce. Have the farmer trace the steps from the farm to the market, and the costs along the way. How much of the produce is sold locally?

5. Invite a supermarket chain's produce manager to class. Interview the manager to find out how stores purchase produce. Can they buy direct from local farmers? How does out-of-season produce get to the store from where it is grown?

Why don't stores carry only local produce? How many people handle the food between the farmer and the store?

Have students harvest some produce from the garden and determine its price at a farmers' market or produce stand.

Make a seasonal stew with fresh fruits or vegetables that are in season locally.

The New *Improved* Madison Avenue Diet

Description

Students analyze newspaper and magazine ads to discover the techniques companies use to sell their products.

Objective

To use deductive reasoning to determine how advertising influences buying habits.

Materials

Whole food ads from newspapers and magazines
Parts of ads
Construction paper
Glue
Crayons

Let's talk about the purpose of advertising. Where do we find food advertisements? (TV, newspapers, magazines, radios, stores)

Advertising affects all of us in the products we choose. The food industry spends $6 billion annually in advertisements. Action for Children's Television, a consumer group, counted 7,000 TV ads for sugar products in one year. In analyzing all of the food ads directed toward children, two thirds are for high-sugar products. It's important to learn that advertising does not necessarily reflect nutritional quality.

Look at the magazine and newspaper ads. What techniques are used to sell the food? (Make a list on the chalkboard. Possible choices include low prices, healthy for you, tastes good, will make you be like someone else, catchy slogan.)

1. Help students think of an imaginary type of food they would like to sell.
2. Divide the students into groups of four. From the parts of ads, have each group make their own ad to sell one of their imagined products.
3. Share ads with the rest of the class. What selling techniques are used?

Why do companies spend so much money advertising? Name three techniques that are used to sell products. How do you think advertising influences you? Name four things you might watch for in TV ads. What does advertising have to do with nutrition?

Have students make up some class slogans about good nutrition. Contact a local radio station to find out how the class can tape the slogan and have them play it as a Public Service Announcement.

Try It, You'll Like It!

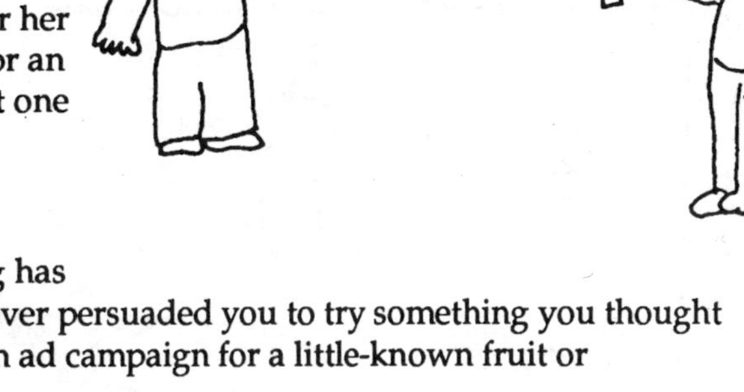

Description

Students design an advertising campaign by selecting a little-known fruit or vegetable and trying to influence another class to eat more of it.

Objective

To experiment with the influence of advertising.

Materials

Samples of food to be advertised
Art supplies

Preparation

Arrange for a test class by asking another teacher to allow the ad campaign to be presented to his or her class. The campaign should last for an extended time period. We suggest one month.

Name some ways that advertising has influenced you. Has advertising ever persuaded you to try something you thought you wouldn't like? Let's design an ad campaign for a little-known fruit or vegetable.

1. Have your class design the ad campaign:
   - Select a fruit or vegetable that isn't well-known or popular but is available, such as kohlrabi, Swiss chard, kiwis, turnips.
   - Select techniques students want to use to convince more people in the test class to eat the fruit or vegetable. Suggestions include free samples, posters in their classroom, brochures, slogans, skits.
   - Establish a time line of when and how the techniques will be used.
2. Have your class poll the other students on their previous experience with the fruit or vegetable, i.e., have they ever, never, or frequently tried the food?
3. Supply the fruit or vegetable, and advertise the food, using suggestions from Step 1.
4. Then conduct another poll as in Step 2.

5. Try different techniques with different classes or students.
6. Discuss the influence of advertising on consumers.

What techniques worked best with the class? Do companies use those same techniques a lot? If you were selling the food you advertised, how much would you have increased or decreased your sales? Do you think it is fair to influence a consumer through advertisements?

Have students make a checklist of techniques they can look for when watching ads.

Buy Me! Buy Me!

Description

Students analyze the influences on their food choices by filling in a chart.

Objective

To use charting and graphing skills to determine what influences individual food selection.

Materials

Life Lab journals
One blackline master per group of three, p. 441

1. Have groups of three students imagine that they will be in charge of all the food for their household for three days. Have them plan the meals and snacks and then list all of the foods they will need.
2. Have them list each food and check the reason for buying it.
3. Add up the checks under each category and discuss why they chose certain foods and how advertising influenced them.

WHAT YOU BOUGHT...	INEXPENSIVE	EASY TO FIX	ADVERTISED ON T.V.	HAS HIGH NUTRITIONAL VALUE	HAS NOTHING ARTIFICIAL IN IT	MY PARENTS BUY IT	WANTED TO TRY SOMETHING NEW	CATCHY PACKAGE (PRIZE INSIDE)	FRIEND SAID TO TRY IT	ON SALE	MY FAMILY LIKES IT	OTHER
1. WHOLE WHEAT BREAD				X	X							
2. CEREAL			X					X	X			
3. CARROTS	X	X		X	X	X					X	
4. TUNA		X		X		X					X	
5. RICE	X		X	X	X	X					X	
6. JUICE MIX		X	X				X				X	
7. PICKLES		X								X	X	

What reason influenced your food choice most often? Name two reasons that did not influence you much but that you may think of next time. What do you think are the three most important reasons for choosing food items?

Feast or Famine

Description

Students work together to develop solutions to the hunger problem that exists around the world.

Objective

To develop awareness of hunger in the world.

Teacher Background

The reasons for hunger are complicated. They involve politics, economics, ecology, and technology. People need to take action to create a solution. We all should be aware that there are people the world over who do not have enough food. Although many people face bountiful supermarket shelves and learn to be critical consumers, the World Bank estimates more than one fourth of the world does not have enough food to eat. The world's population is 4,760,000,000. Nearly one billion people go hungry each day. Twenty million people die from starvation or hunger-related diseases each year. Three fourths of those who die are children.

Enough food is produced right now to feed all of the people in the world. The world produces enough in grain alone to feed all of the people more than 3300 calories a day. So why are people going hungry?

Materials

Enough peanuts with shells for every student in the class

1. Distribute peanuts according to the percentage of hungry people in the world. One fourth of the students do not receive any peanuts. The peanuts are given to the remaining three fourths of the class, with some students receiving more than others.
2. Tell students the peanuts represent how food is distributed around the world. Some people have an abundance while others have nothing.
3. Have the class work together to solve their food distribution problem. Have students discuss a number of solutions and means of redistribution. How can the students with an abundance help the hungry students without giving their food away? (Increase farming production at the local level.)

How would it feel to be starving? How does it feel to know there are people going hungry right now? Name two solutions your class developed to the hunger problem. How could you help initiate those solutions?

1. Have students choose four countries with hunger problems: Mexico, Guatemala, Peru, Nigeria, Kenya, Somalia, Turkey, India, Bangladesh, Laos. Research what the people eat and what is grown in each country and compare the diets in these countries to the diets in other more fortunate countries.
2. Have students find out if there are hungry people in your own town. What is being done to help them? Can your class help?
3. Have students write for "Who's Involved with Hunger: An Organization Guide" from the American Freedom from Hunger Foundation and World Hunger Education Service, 1625 Eye Street NW, Washington, D.C., 20006. Investigate what other organizations are doing to help.
4. Make a list of what you can do to help others to improve their diets.
5. For more related activities, see the *Food First Curriculum*, p. 479.

6. Conclude the activity with the following story.

   A child asked his grandmother about hunger.

   "I will show you hunger," said the grandmother. And they went into a room that had a large pot of stew in the middle. The smell was delicious and around the pot sat people who were famished and desperate. All were holding spoons with very long handles that reached to the pot, but because the handles of the spoons were longer than their arms, it was impossible to get the stew back into their mouths. Their suffering was terrible.

   "Now I will show you plenty," said the grandmother. They went into an identical room. There was the same pot of stew and the people had the same identical spoons, but they were well nourished, talking, and happy. At first the child did not understand. "It is simple," said the grandmother. "You see, they have learned to feed each other."

   Adapted from *Earth Wisdom*, by Dolores La Chapelle

Market Garden: Grow and Sell Time

Description

Students raise their own vegetables especially for the market, either their own or a local outlet.

Objective

To develop awareness of some of the concerns of farming.

Materials

Garden plot
Garden tools: shovel, rake, and trowels
Seeds for chosen food crops such as beans, peas, squash, and tomatoes
Water hose

Teacher Background

Most of the world's plant foods come from only 20 kinds of plants. In fact, although there are 300,000 plant species in the world today, only about 150 of them are grown commercially. People have been actually growing rather than simply gathering plant foods for about 10,000 years. Until fairly recently, most people grew their own food. Today only 3 percent of the people in the U.S. are involved in agriculture. This project will give students an opportunity to use math and science skills in a real-life situation.

What do you think the Top Twenty food crops around the world are? (potatoes, sweet potatoes, tomatoes, beans, peas, peanuts, soy beans, corn, wheat, rice, oats, barley, buckwheat, millet, sorghum, sugar cane, sugar beets, coconuts, bananas, and tapioca) Are there any foods in the list that we could grow in this climate? Are there other vegetables or herbs we could grow to sell? Why would crops such as wheat, oats, or barley be difficult for us? What about coconuts? Finally, where could we sell or distribute our crops? (Set up your own stand, or sell at local stores or a Farmers' Market.)

1. Develop a garden plan for market vegetables such as potatoes, squash, corn, onions, radishes, lettuce, peppers, carrots, cucumber, pole beans, tomatoes, and specialty herbs.
2. Prepare soil in a group and then divide the class into individual crop groups or teams. They might want to pick a name for their group—named after their chosen crop or other garden topic: The Has-Beans, the Squash Squad, and so on.
3. Have each group plant their particular seeds and label the bed.

4. Have each group care for their plot by pulling weeds, keeping soil loosened, fertilizing, watering, controlling pests, and mulching.
5. Have students decide on a marketing or distribution idea. Make arrangements for marketing their produce before harvest time. Markets might include a local farmers' market, which would probably involve getting up very early; a small vegetable stand of their own; a nonprofit distribution to food banks or senior citizen groups. Local restaurants or the school cafeteria might even be interested in buying good, fresh vegetables.
6. Have students harvest the crops as the vegetables ripen and take them to market.

How did the climate in our area affect your choice of crops? How much of your crop would you have to grow in order to make a living? How do farmers market their crops? What expenses do farmers have that a small garden doesn't?

Have students consider more ways to make a small garden's produce marketable and maybe even profitable. Fresh herbs such as dill, basil, and oregano, or fancy foods such as rocket salad (arugula) and radicchio might be grown and packaged.

Appendix

The Scientific Method

Guess, Test, and Tell

1. **Guess**: What will happen?
2. **Test**: Try out your idea.
3. **Tell**: What happened?

Tell It Like It Is

Pattern Sheet. Do Not Duplicate For Students.

"Who Am I" Riddles

Directions for teacher:

1. Prepare six riddle cards (2" X 3") that give six different bits of information about an object. To begin, you may use the clues listed on this sheet. There are enough clues to prepare 4 different sets of riddle cards.
2. Label each card in the first set as 1-1, 1-2, and so on, for ease of identification. Begin the second set of cards with 2-1, 2-2, and so on.
3. Use these sets of cards with student groups as described in the "Who Am I?" activity.

Clues for riddle cards:

Set 1

1-1 Who am I waiting for?
1-2 This time of year I have a very sweet smell.
1-3 I have leaves.
1-4 People say I am pretty.
1-5 Some people sneeze when they're near to me at this time of year.
1-6 I am very heavy now with tiny little grains called *pollen*.

Set 2

2-1 I am easily camouflaged in my home.
2-2 I have six legs.
2-3 I have long and powerful hind legs, larger and stronger than my other legs.
2-4 I can jump over one meter when I need to get away from something. That's a big distance considering my size!
2-5 I like to live in the garden and munch on leaves.
2-6 I have four wings.

Set 3

3-1 I live in a colony.
3-2 My colony is made of underground tunnels.
3-3 I have six legs.
3-4 There is a queen in my colony.
3-5 I came from an egg.
3-6 My friends and I work very hard, and sometimes we get stepped on.

Set 4

4-1 Plants need me for food.
4-2 You can see me only during the daytime.
4-3 It looks like I travel a lot, but I'm actually very still.
4-4 I'm very far away.
4-5 I'm hot stuff!
4-6 I am an important part of the water cycle.

Clay, Silt, Sand Chart

The Nitty-Gritty

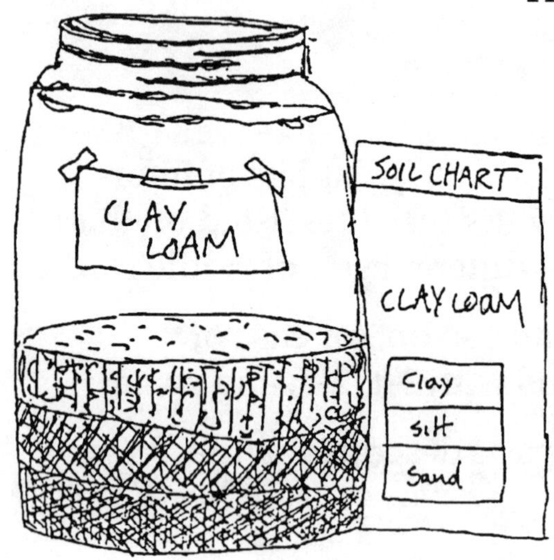

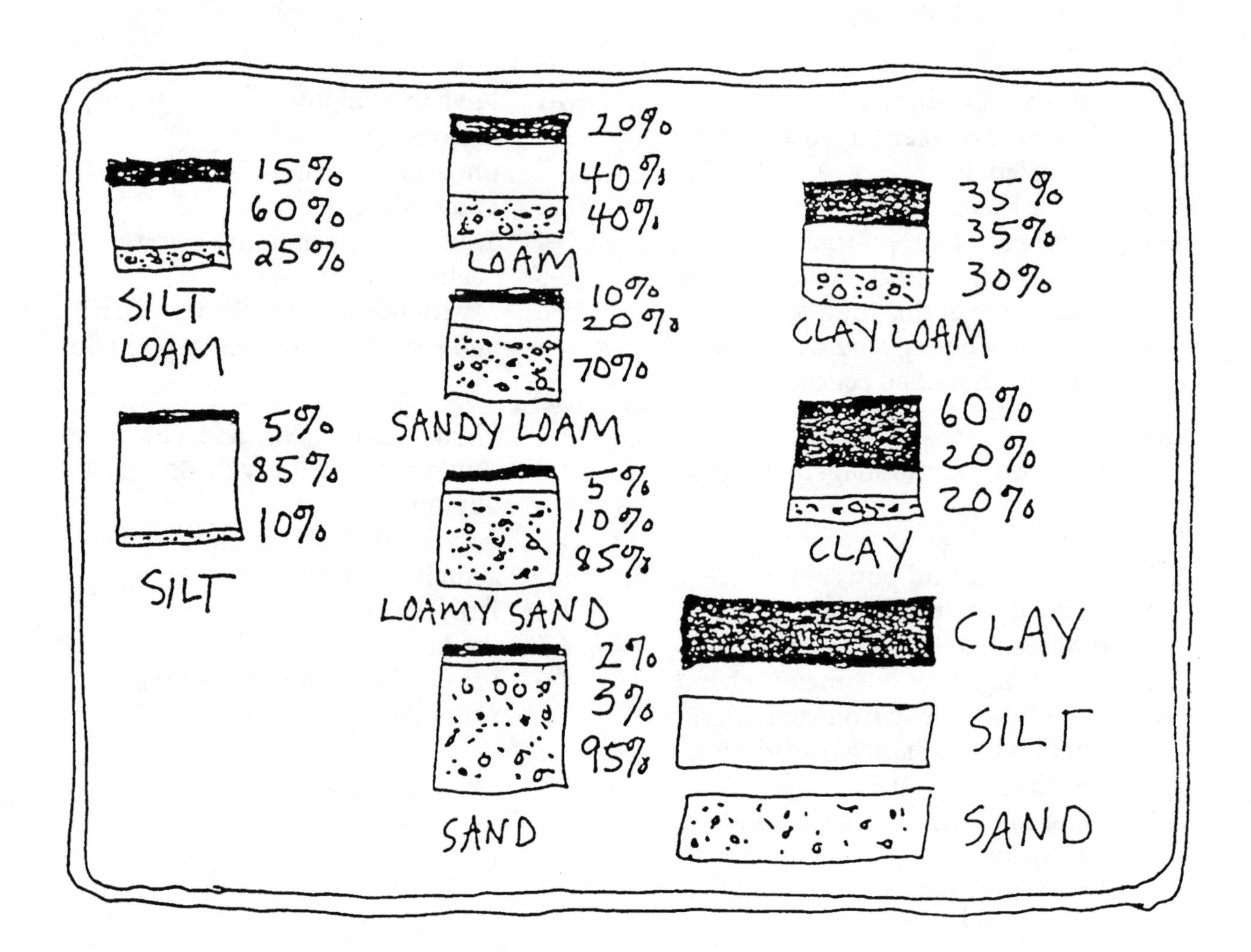

NAMES DATE

Soil Prescription

The basic formula for compost is approximately **50 percent dry organic material + 35 percent wet organic material + 15 percent soil.** Just like the foods we eat, some compost materials are high in one nutrient while others are high in other nutrients. Some decay very quickly while others take a very long time to decay, just as some foods stay in our bodies a much longer time than others. Your task is to make a recipe for compost that includes dry and wet organic materials and soil. All materials must be available for free at your school or in your town. Be sure to think about what you learned in your soil test! Possible ingredients that you might use are listed below. (Key: C=carbon, N=nitrogen, K=potassium)

Dry material:	**Good for:**	**Where found:**
Fallen leaves	C, N, P, K	under trees
Grass hay	C, N, P, K	lawns and fields
Legume hay	C, N, P, K	lawns and fields
Egg shells	Calcium	home, restaurant
Sawdust	C	woodshop, lumber yard
Straw	C, N, K	fields
Sugarcane fiber	C	sugar factory
Corn stalks	C	farm, garden
Nut shells	N	nut processing company
Cocoa husks	N	chocolate factory
Wood ash	K, Calcium	fireplace
Pine Needles	C, N, P, K, Iron	forest

NAMES DATE

Wet material:	Good for:	Where found:
Food scraps	N	home, restaurant, school cafeteria (no meat, grease, or dairy products)
Green plant wastes	N	garden, yard
Animal manure	N, P, K, Sulfur, Calcium, Iron, Magnesium	farm, stables, zoo
Fish Meal	N	cannery, fishmarket
Brewer's grains	N, P	brewery
Seaweed	K	ocean
Hoof and horn meal	N	slaughterhouse
Blood meal	N	slaughterhouse
Bone meal	P	slaughterhouse
Cottonseed meal	N	cotton mill
Soybean meal	N	soybean mill
Feather meal	N	chicken, duck, goose slaughterhouse
Soil	**Good for microorganisms**	**Found at school**

Recipe for Compost at ______________________ School

Dry/Wet Organic Materials	Found at: (Where in your town?)
____________	____________
____________	____________
____________	____________
____________	____________
____________	____________
+ Soil	

Seed Ordering Chart

ZIP Code Seeds

Catalogue			
Seasons we are ordering for:			
Hybrid or open-pollinated seeds:			
	Name of Seed	**Amount**	**Cost**
Grows Best Here			
I Like to Eat			
Best Variety for Our Soil and Climate			
Companion Plants			
Easiest to Grow			
Fun to Grow			
Challenging to Grow			
Edible Root Plant			
Edible Leaf Plant			
Edible Stem Plant			
Edible Flower Plant			
Edible Seed Plant			

The Nutrient Cycle

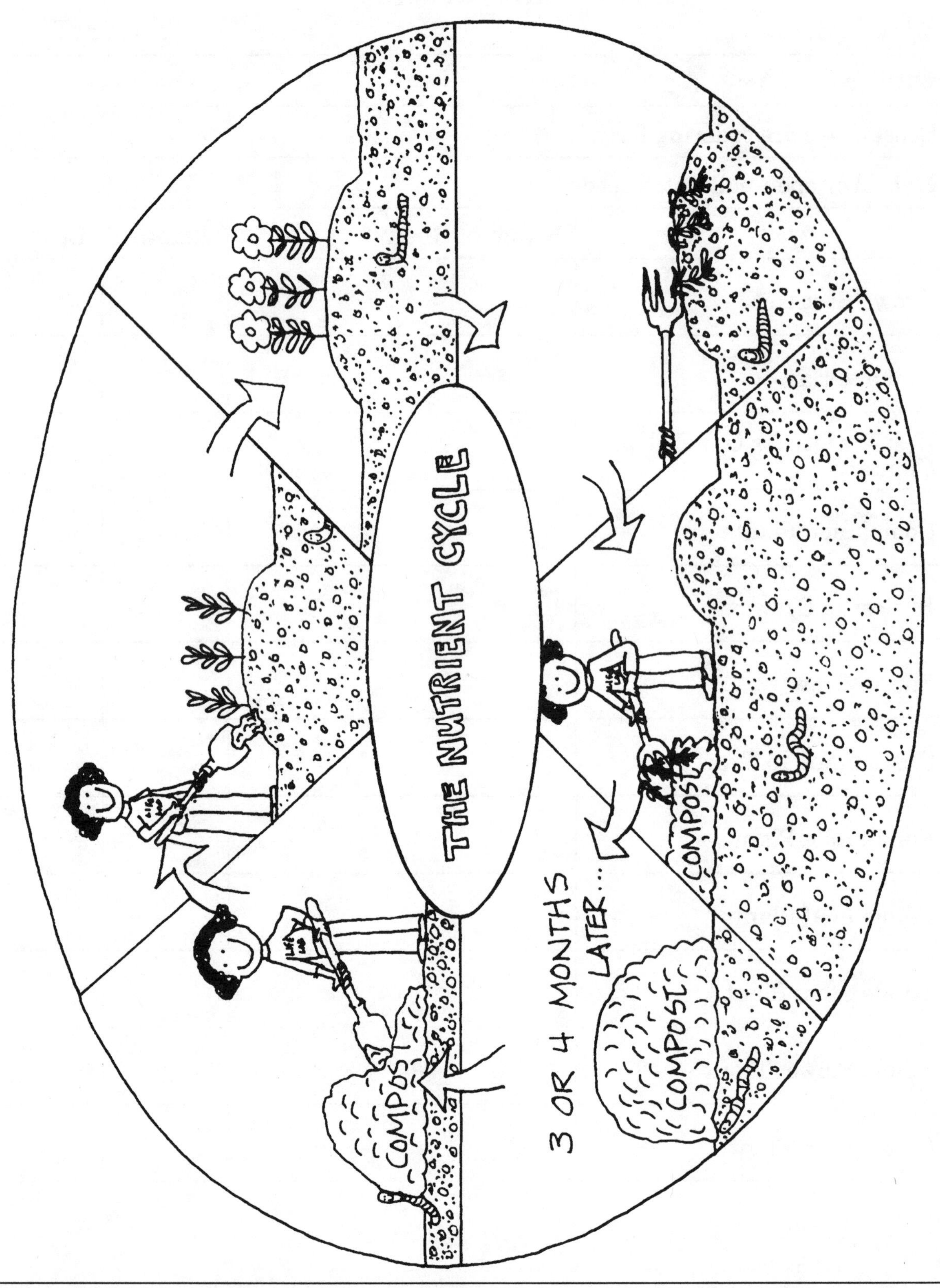

The Oxygen Cycle

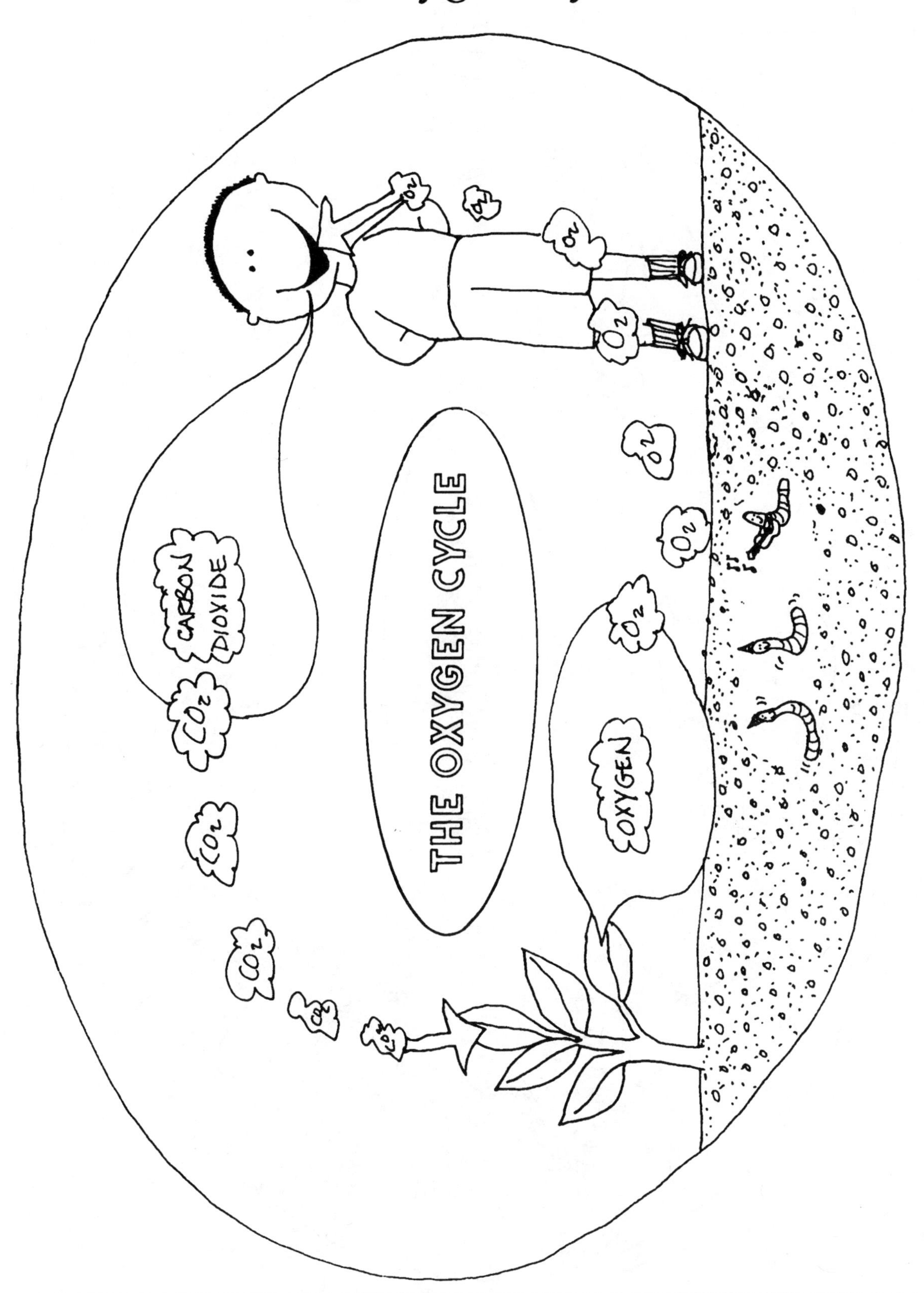

The Water Cycle

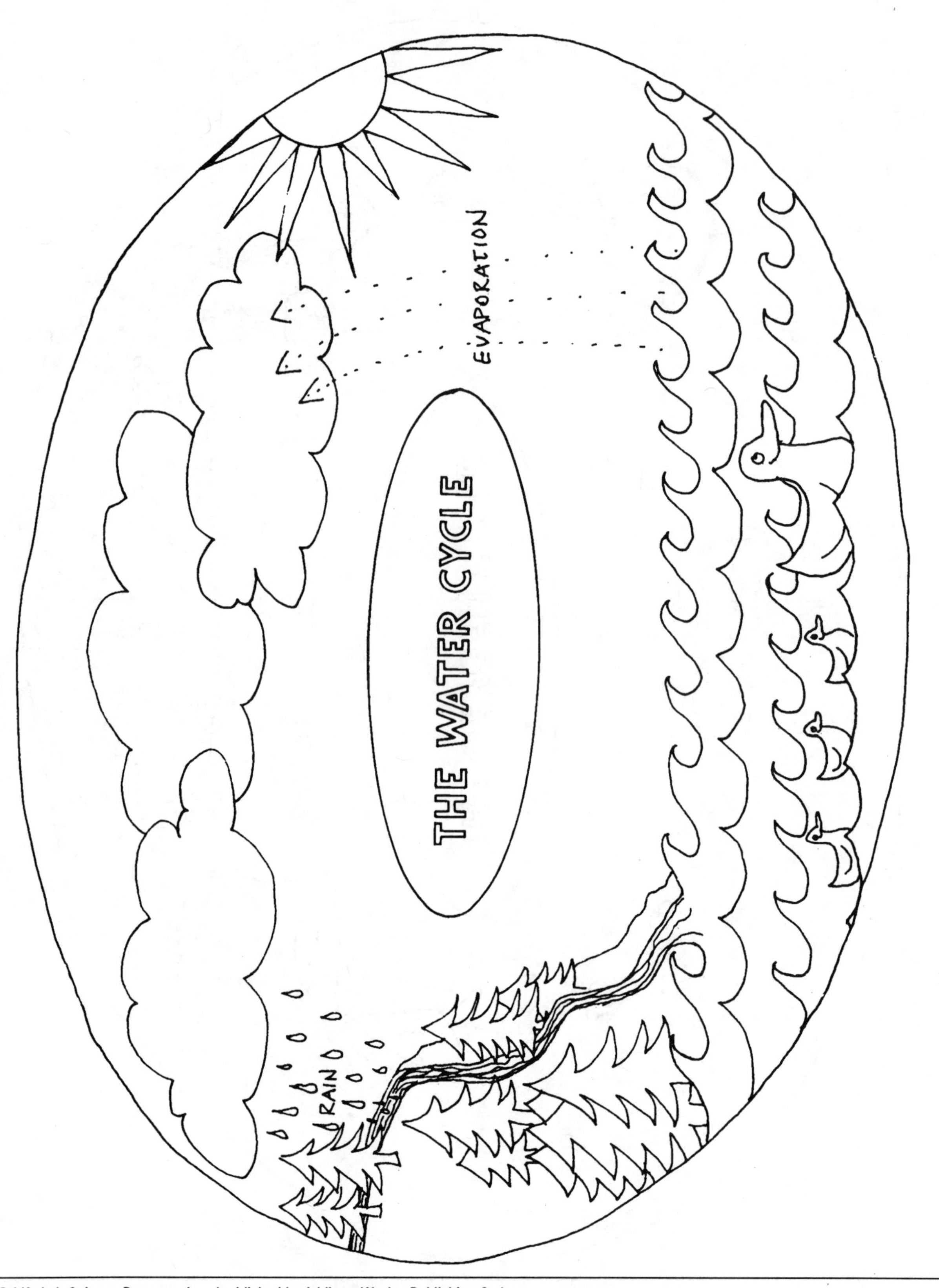

Me and the Seasons

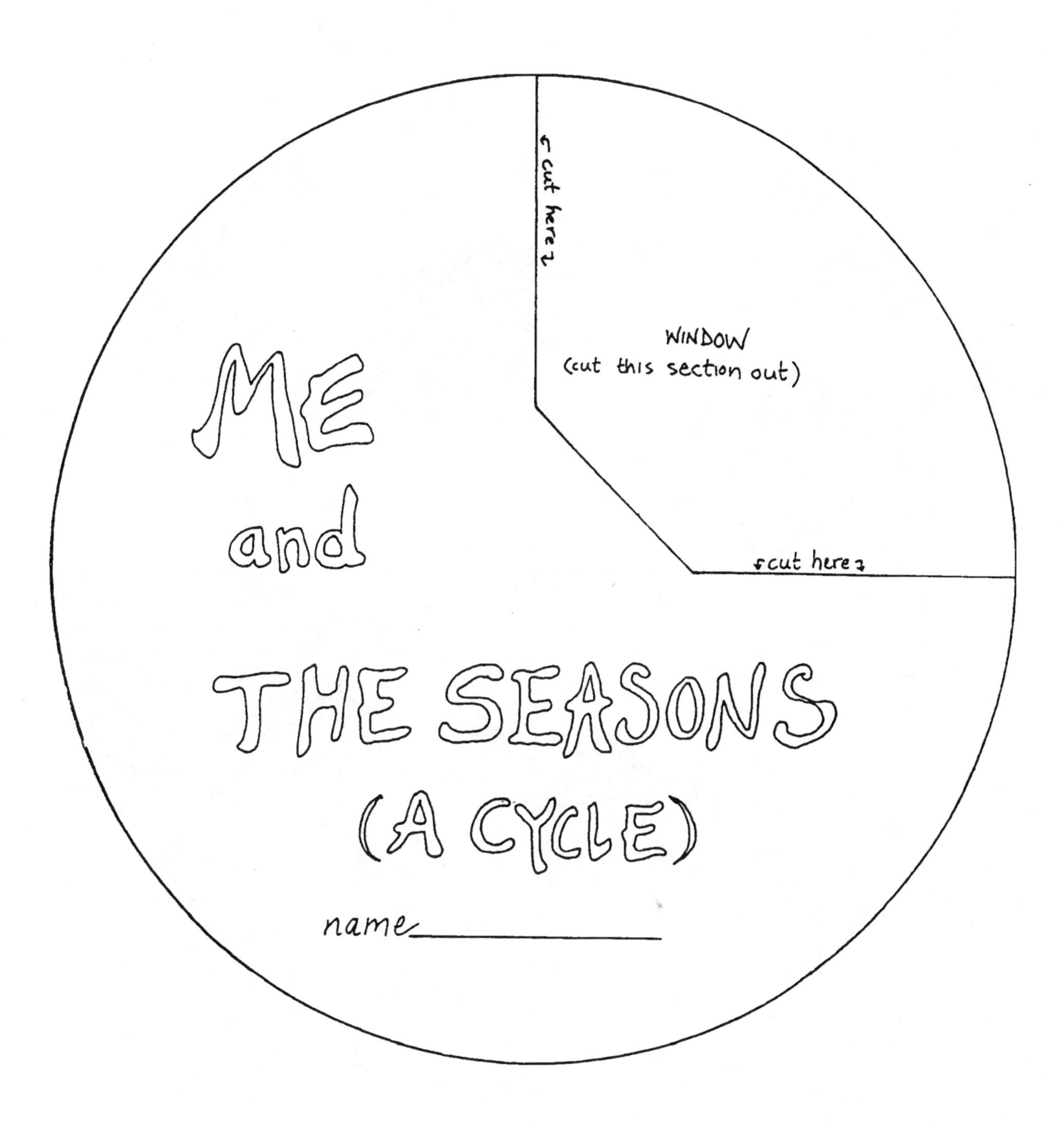

Me and the Seasons

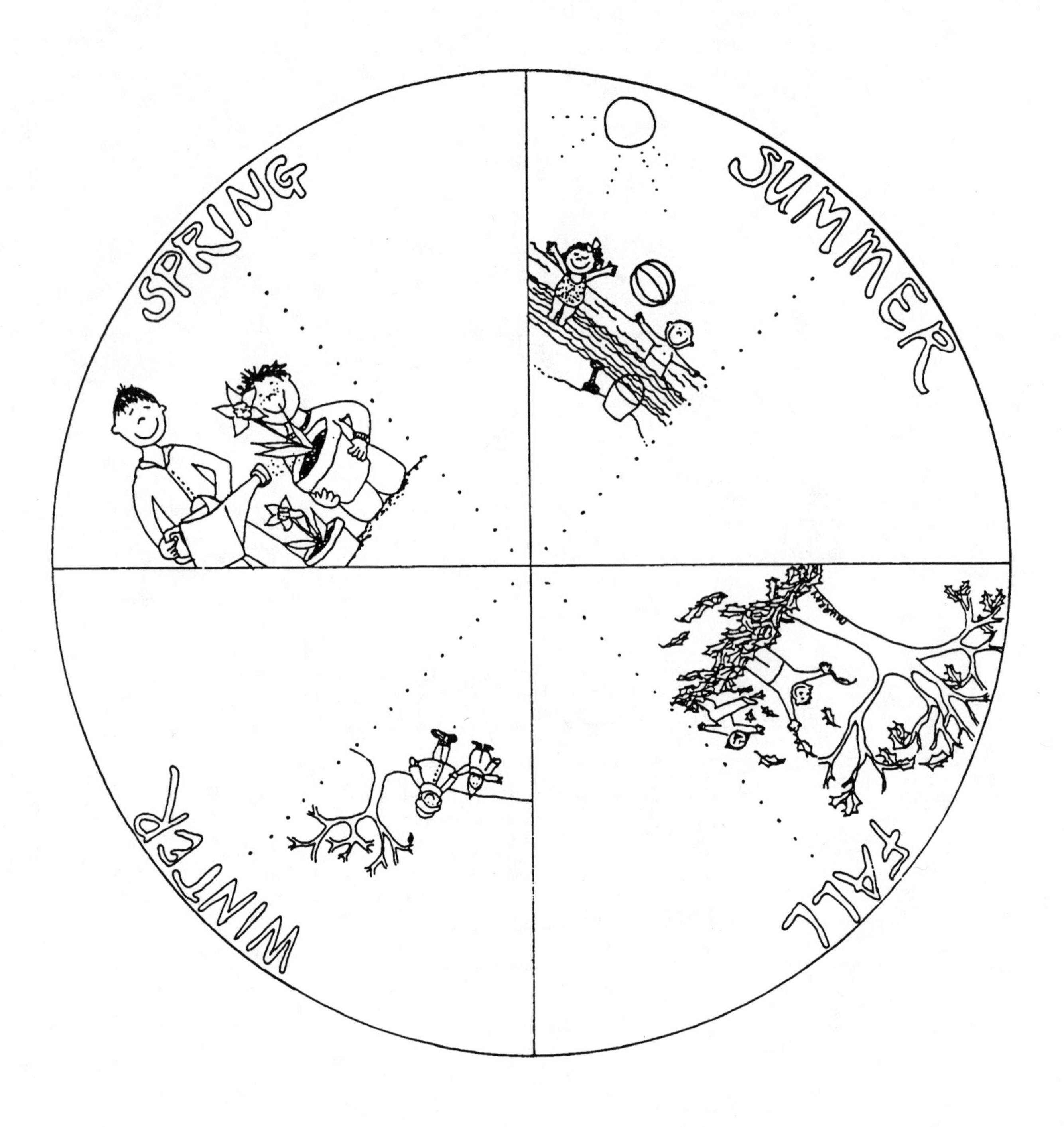

NAME DATE

Self-Survey

Every day you make decisions about how much garbage you throw away and how much you recycle. Most of the time these decisions are just habits. What are your trash habits? Find out by taking this self-survey. When done, add up your score and read the Self-Survey Results for more information.

DO YOU . . .	Never	Sometimes	Often
1. Use both sides of a paper before getting a new piece?	3	2	1
2. Avoid eating take-out food that is wrapped in many different packages?	3	2	1
3. Use dishes instead of paper plates?	3	2	1
4. Reuse plastic and brown paper bags?	3	2	1
5. Save newspapers to recycle?	3	2	1
6. Keep a compost pile at home for kitchen scraps?	3	2	1
7. Shop at garage sales and thrift stores?	3	2	1
8. Use cloth napkins instead of paper napkins?	3	2	1
9. Recycle soda cans instead of throwing them away?	3	2	1
10. Recycle scrap papers in your classroom?	3	2	1
11. Try not to buy plastic products?	3	2	1
12. Fix things instead of throwing them out?	3	2	1
13. Give outgrown clothes to someone smaller?	3	2	1
14. Remind your friends to recycle?	3	2	1
15. Reuse glass jars in your kitchen?	3	2	1
16. Give magazines to other people after reading?	3	2	1
17. Think of new ways to reuse old things?	3	2	1
18. Buy products that will last a long time?	3	2	1
19. Think about what happens to things sent to the dump?	3	2	1
20. RECYCLE AS MUCH AS YOU CAN?	3	2	1
Add up the columns for your Grand Total	+	+	=

Circle the number

NAME DATE

Self-Survey Results

My Grand Total is __________.
The *lower* your score, the better your habits are.

Less is better when it comes to **trash.**

If your score is

50 or more

Your trash habits could use some improvement. You are adding more than you need to the city dump and sending little to the recycling center. Try reusing items before trashing them. Buy fewer things. Every little bit counts.

30-50

Your habits are pretty good, but you could be doing more to reduce, reuse, recycle. Try composting your food scraps at home or recycling at school to cut down your trash.

30 or less

Your trash habits are GREAT. You must be very involved in recycling. Keep up the good work. Talk to your parents and friends about how to recycle. Think up some creative ways to make even LESS trash.

NOW go back and do the self-survey again. Use a different colored pencil or pen to circle the numbers. This time ask yourself "Could I . . . " before each question. Which of your trash habits could you change? Which ones do you want to change?

Remember: Every soda can you collect on the street, every plastic container you decide not to buy, every paper you recycle at school adds up to money saved, trees saved, resources saved!

A Very Short History of Trash

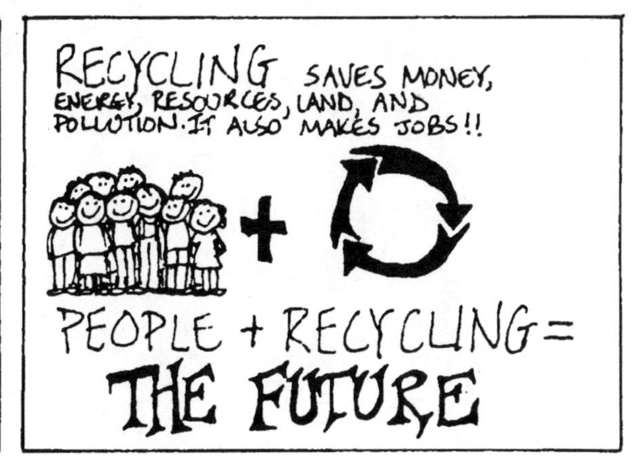

Concept adapted from MacDonald's Corp.

Insect Anatomy

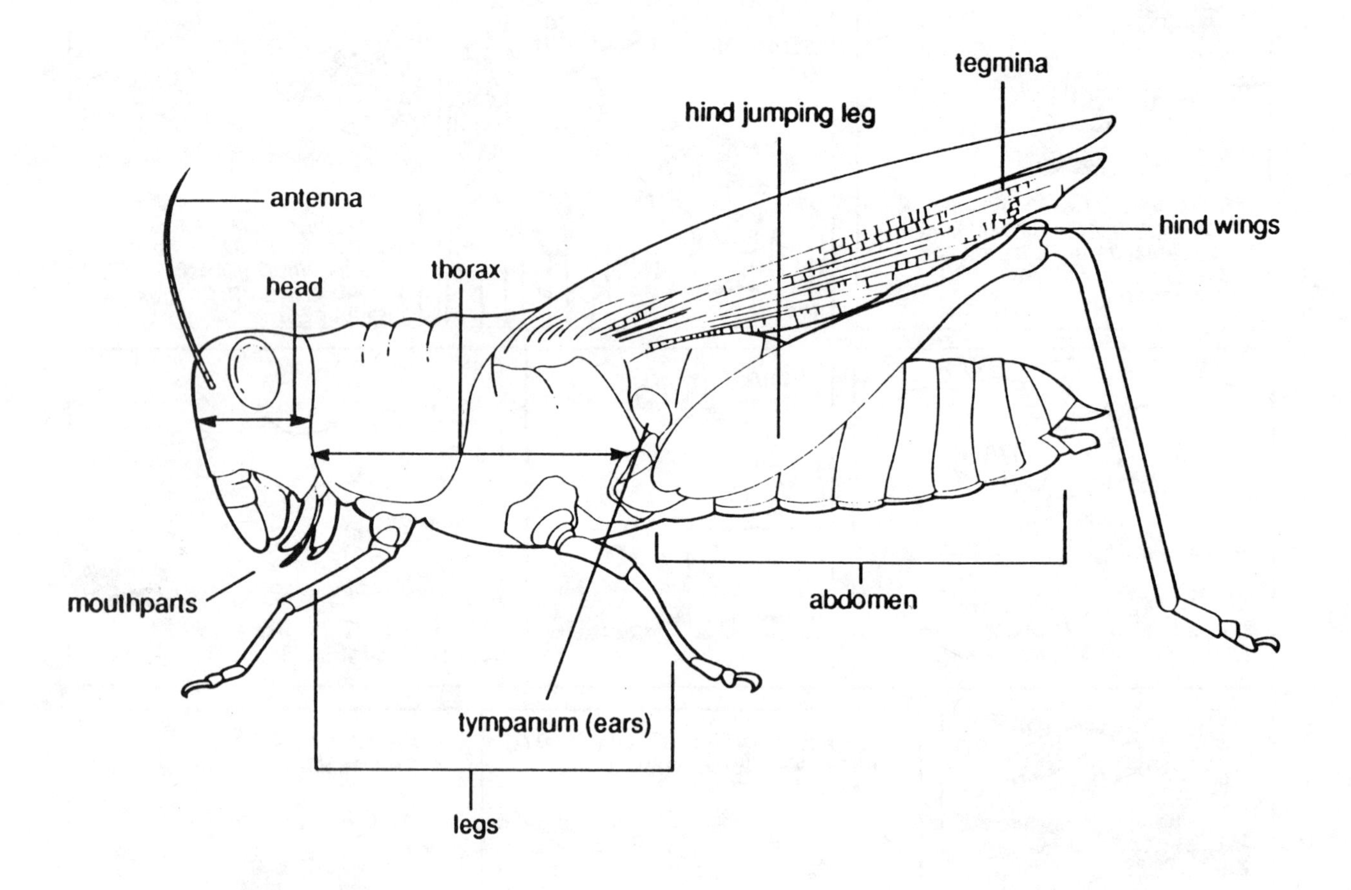

Beneficial Insects

ASSASIN BUG

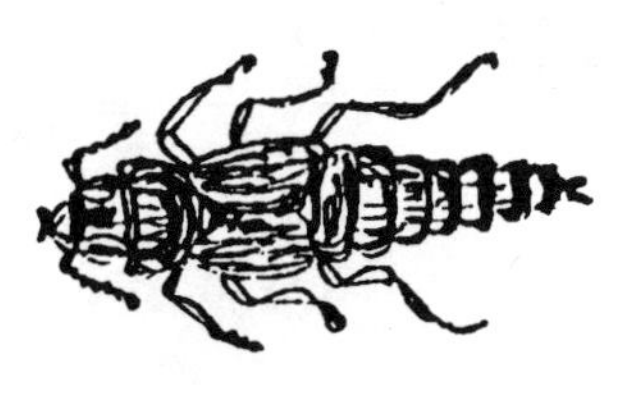

ROVE BEETLE

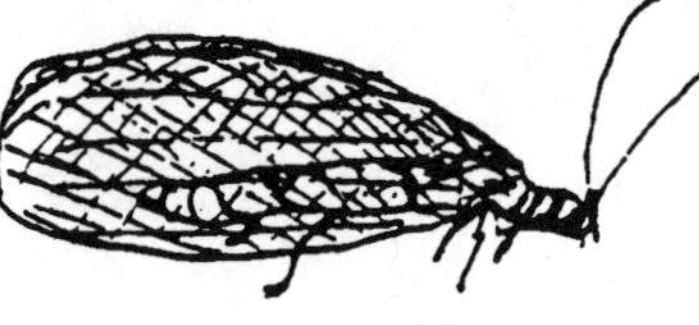

ADULT LACEWING

SYRPHID FLY

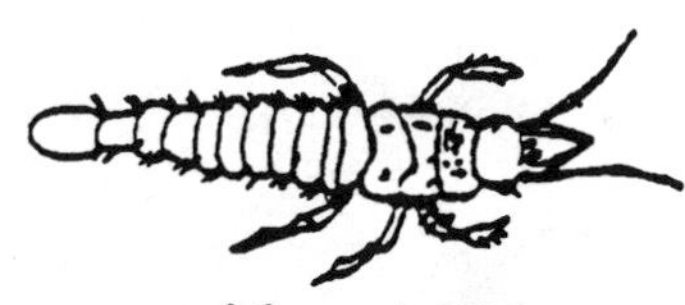

LACEWING LARVA

BEETLE

SYRPHID FLY LARVA

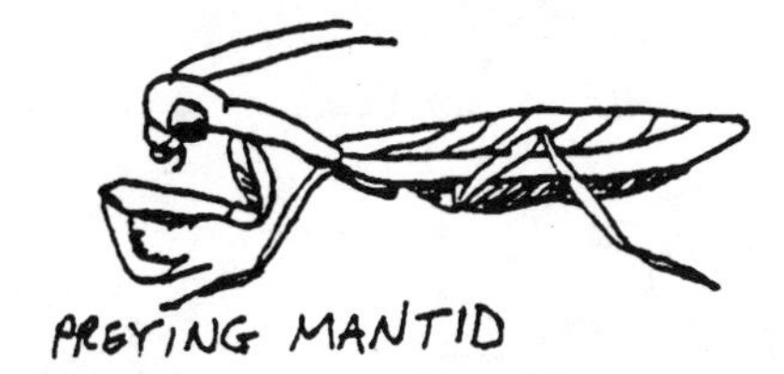

PREYING MANTID

ADULT WASP STINGING APHID

ADULT MINUTE PIRATE BUG

LADYBIRD BEETLE LARVA

ADULT LADYBIRD BEETLE

Common Pests

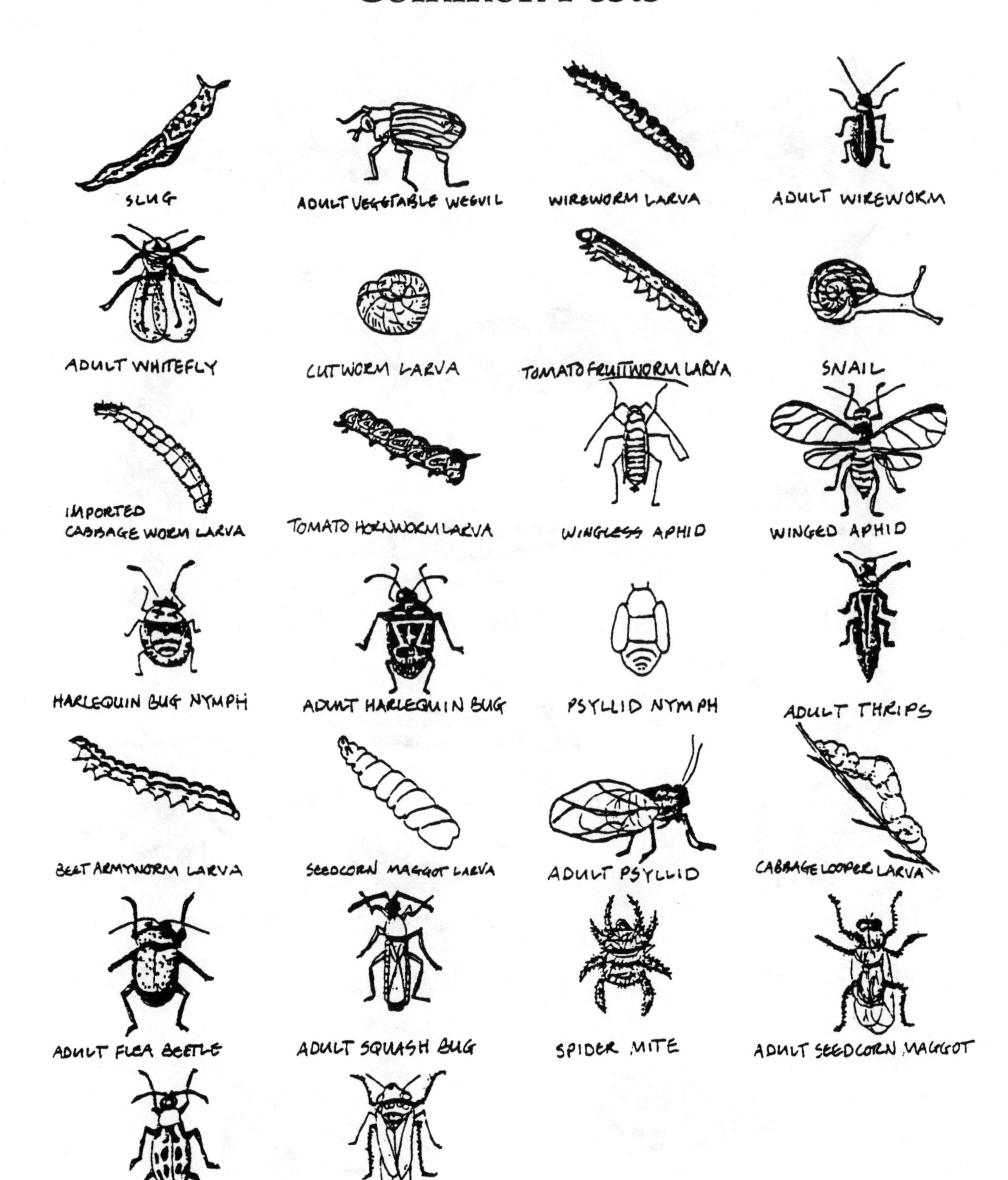

Minimum-Maximum Thermometer

Station 1

When you read the thermometer, do not touch the bulb. Keep the thermometer in the shade. This thermometer will give you three types of information.
One: The hottest temperature
Two: The coldest temperature
Three: The temperature right now!

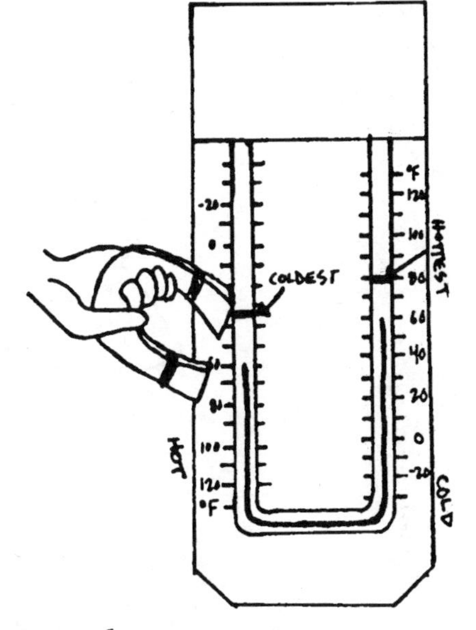

The minimum-maximum thermometer is really two thermometers joined at the bottom. The right thermometer starts at the bottom with the lowest temperature and goes up to the highest. As the temperature gets warmer, the liquid in this thermometer is pushed higher. As it gets colder, the liquid moves lower in the thermometer, but the metal stays at the hottest spot. This side of the thermometer tells you the hottest it has been since the last reading.

The left scale starts at the bottom with the hottest temperature and goes up to the coldest temperature. The liquid in the thermometer is pushed higher as the temperature gets colder. The little piece of metal indicates the coldest temperature since the last reading.

The thermometer also gives us a third piece of information. Look at the liquid level in both thermometers. Each scale should show the same reading. That is the temperature right now!

After you record the information from this thermometer, use the magnet to push both pieces of metal down to the liquid. This resets the thermometer so that it can mark the hottest and coldest temperatures until it is reset again. Be sure to store the thermometer in a cool, secure place.

Wind Vane

Station 2

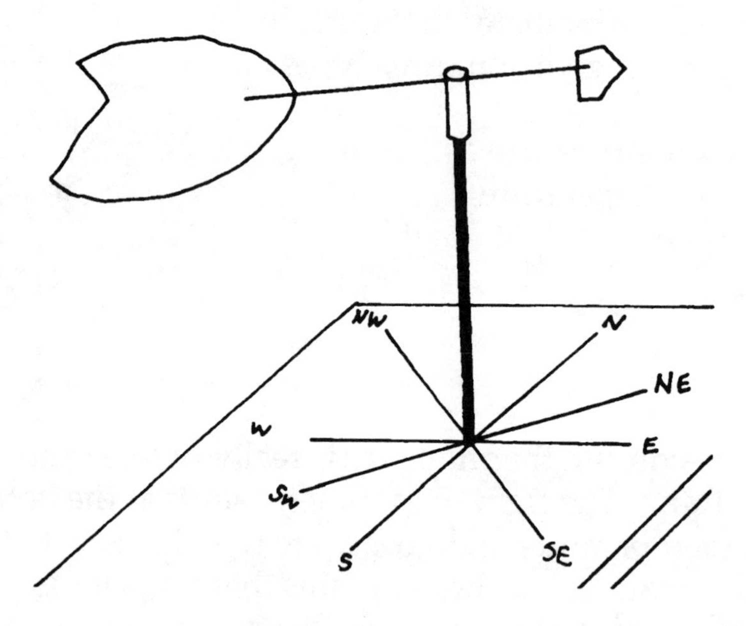

A wind vane measures the direction the wind is blowing from. For example, a wind blowing from the northeast toward the southwest is called a northeast wind. The *arrow* points *toward* the northeast.

Watch the wind vane for about one minute to determine which of the 8 compass directions it points to most often.

Record 999 if the wind direction cannot be determined.

Wind Speed: Sensing the Wind

Station 3

Card A

Different instruments can be used for measuring wind speed. You can also use your own senses and skills of observation to determine the wind speed. First determine the wind direction. Stand so the wind strikes you directly in the face. Estimate the wind speed. Observe how the wind moves objects that you can see. Compare the effects that you see with the effects described in the table on Card B. Look at the table on Card B. Wind speed can be recorded in five ways. You can use a descriptive word, miles per hour, meters per second, the weather service symbol, or the Beaufort number. The Beaufort number is a rating of wind force ranging from 1 to 12 or higher and is often used in meteorology.

Card B

WIND EFFECTS OBSERVED ON LAND	TERMS USED IN FORECASTS	BEAUFORT NUMBER
CALM; SMOKE RISES VERTICALLY. DIRECTION OF WIND SHOWN BY SMOKE DRIFT BUT NOT BY WIND VANES.	light	1
WIND FELT ON FACE. LEAVES RUSTLE. ORDINARY VANE MOVED BY WIND.		2
LEAVES AND SMALL TWIGS IN CONSTANT MOTION. WIND EXTENDS FLAG.	gentle	3
RAISES DUST AND LOOSE PAPER. SMALL BRANCHES MOVED.	moderate	4
SMALL TREES WITH LEAVES BEGIN TO SWAY. WHITECAPS FORM ON WATERS.	fresh	5
LARGE BRANCHES IN MOTION. WHISTLING HEARD IN WIRES BETWEEN UTILITY POLES. UMBRELLAS USED WITH DIFFICULTY.	strong	6
WHOLE TREES IN MOTION. INCONVENIENCE FELT IN WALKING AGAINST WIND.	gale	7

METERS PER SECOND	MILES PER HOUR	WEATHER SERVICE SYMBOL
0	0	
1	2	
2	4	
3	6	
4	8	
5	10	
6	12	
7	14	
8	16	
9	18	
10	20	
11	22	
12	24	
13	26	
14	28	
15	30	
16	32	
17	34	
18	36	
	38	
	40	

Measuring Humidity

Station 4

Card A

Humidity is defined as the percent of moisture that is in the air. When it is raining, there is 100 percent moisture. A wet and dry bulb hygrometer/ psychrometer is one of the few instruments that measures humidity in the air. It consists of two thermometers mounted near each other. One thermometer has a damp cloth around its bulb. The other is left exposed to the air. To determine the humidity from these thermometers, you will use the relative humidity table on Card B. First read the temperature on both thermometers. Then subtract the difference between the two thermometers and write down your answer. Look at the table on Card B. Find your dry-bulb temperature on the vertical axis. Now find your answer when you subtracted on the horizontal axis. Use your fingers to follow the vertical axis down and the horizontal axis across. Where these two lines cross tells you the relative humidity for your air!

Card B

Even though we cannot see water vapor with our eyes, our other senses can feel it. On days of high humidity, we say the air feels "muggy." When it is cold and damp we shiver, not so much from the cold as from the moisture in the air. If it is dry and cold, we enjoy the crisp weather. In very dry places the human body can withstand much higher temperatures without feeling uncomfortable. You can train your senses to notice moisture in the air. Compare your estimate with the readings from the instrument. Store and read your wet and dry bulb hygrometer or psychrometer in the weather station as long as it is well ventilated.

Relative Humidity Table

Dry-Bulb Temperature (C°)	Degrees Difference Between Wet- and Dry-Bulb Thermometers									
	1	2	3	4	5	6	7	8	9	10
10	88	77	66	55	44	34	24	15	6	0
12	89	78	68	58	48	39	29	21	12	0
14	90	79	70	60	51	42	34	26	18	0
16	90	81	71	63	54	46	38	30	23	15
18	91	82	73	61	57	49	41	34	27	20
20	91	83	74	66	59	51	44	37	31	24
22	92	83	76	68	61	54	47	40	34	28
24	92	84	77	69	62	56	48	43	36	31
26	92	85	78	71	64	58	51	46	39	34
28	93	85	78	72	65	58	53	48	42	37
30	93	86	79	73	67	61	55	50	44	39
32	93	86	80	74	68	62	57	51	46	41
34	93	87	81	75	69	63	58	53	48	43
36	94	87	81	75	70	64	59	54	50	48

Barometer

Station 5

Barometers measure the pressure of the air. Air is pulled to the earth's surface by gravity. The force of this pull is the air pressure. Air pressure is less as you move away from the earth's surface. However, even in one place the air pressure is always changing. When it changes, it is a sign that the weather is changing. When the pressure is high, the air keeps other types of weather from moving in. But when the air pressure drops, changes occur.

Learn how to read your barometer. Record your barometer readings and see if the pressure drops, rises, or stays the same. Compare changes in your weather with changes in the barometer reading. The barometer can help you be a good weather predictor.

Rain Gauge

Station 6

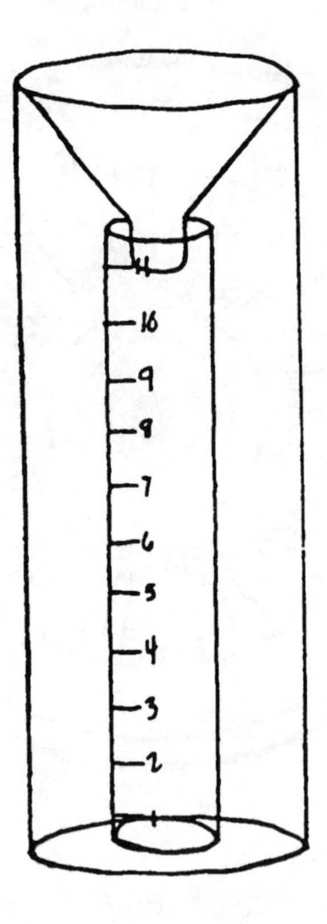

The rain gauge can tell you how much rain has fallen.

Examine your rain gauge. Learn how to read the scale on it.

After you read your gauge, record your reading. Empty the water so your gauge will be ready for the next rainfall. Try to read your rain gauge as soon as the rain has stopped, because the water will start to evaporate.

Attach the gauge to the outside of the weather station. Be sure that there is nothing in its way so that it will receive the exact amount of rainfall. Trees and buildings might change the amount of water in the gauge. Attach the gauge so that it is easy to remove for emptying water.

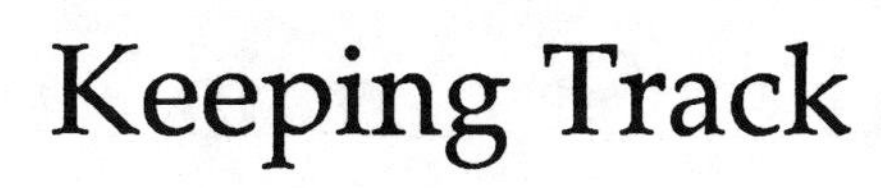

Keeping Track

Weather Symbols Key #1

CLOUDS

NO CLOUDS = NO CLOUDS IN SKY

= SOME CLOUDS IN SKY

= HALF THE SKY HAS CLOUDS

= MOST OF THE SKY HAS CLOUDS BUT STILL SOME BLUE IS SHOWING

= THE WHOLE SKY IS COVERED WITH CLOUDS

TEMPERATURE

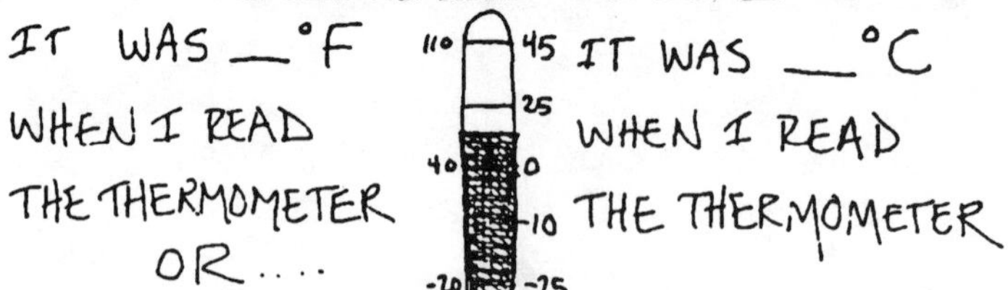

RAIN

= 1.27 cm or ½ inch of rain fell

= 2.54 cm or 1 inch of rain fell

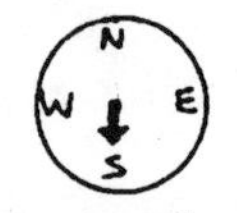

THE WIND IS COMING FROM THE NORTH

THE WIND IS COMING FROM THE SOUTH

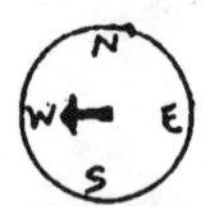

THE WIND IS COMING FROM THE EAST

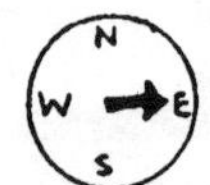

THE WIND IS COMING FROM THE WEST

Weather Symbols Key #2

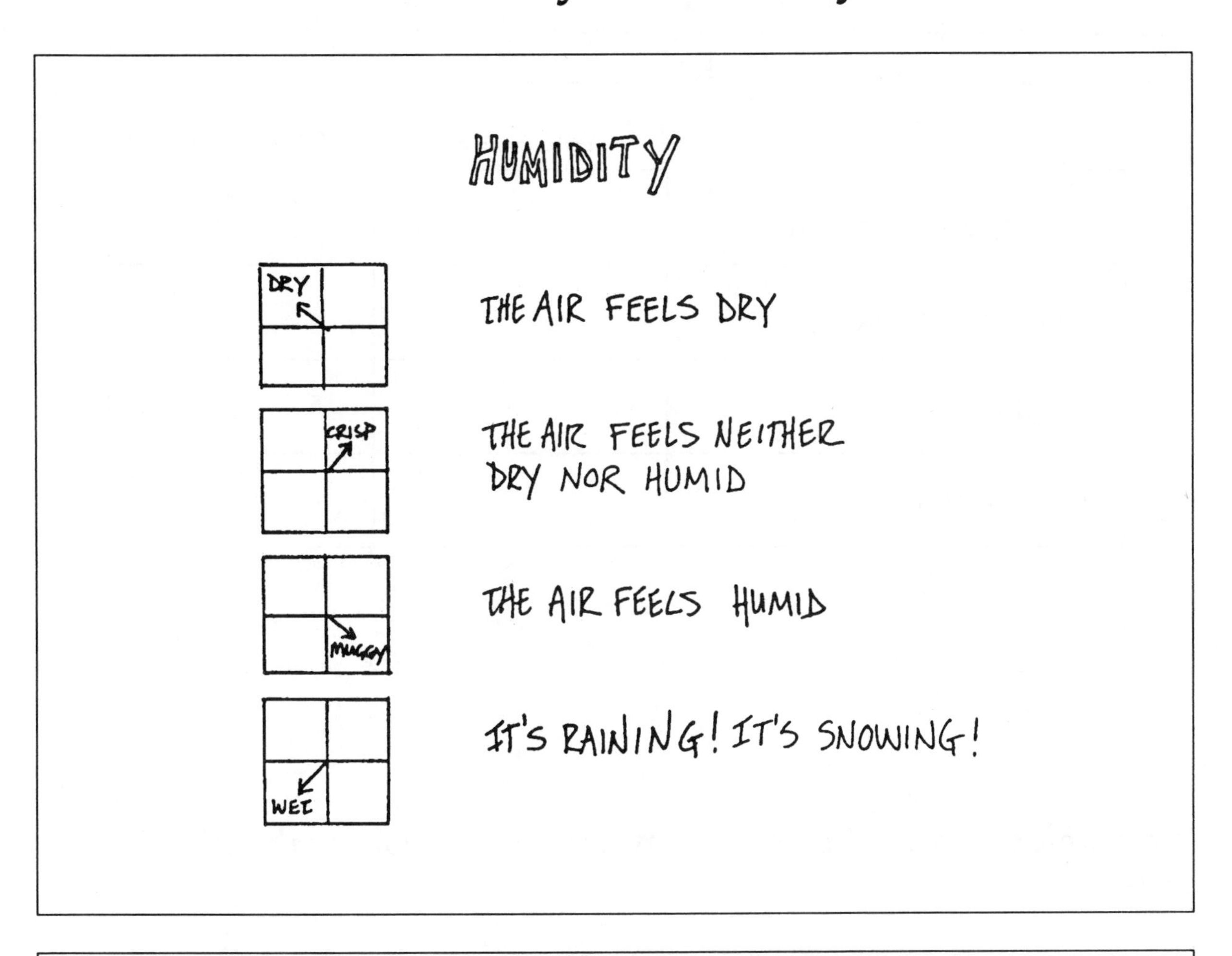

AIR PRESSURE

25 ↑	PRESSURE IS RISING
25 →	PRESSURE IS STAYING THE SAME
25 ↓	PRESSURE IS DROPPING

Wind, Rain, Sleet, or Hail

Week of __________

	Your School	Other School
Average Temperature		
Average Humidity		
Total Rainfall		
Fastest Wind Speed		

What were the major differences between your school and the other school?

____

____

____

____

Why do you think these differences occurred?

____

____

____

____

Energy Movement

Shoebox of Sunshine

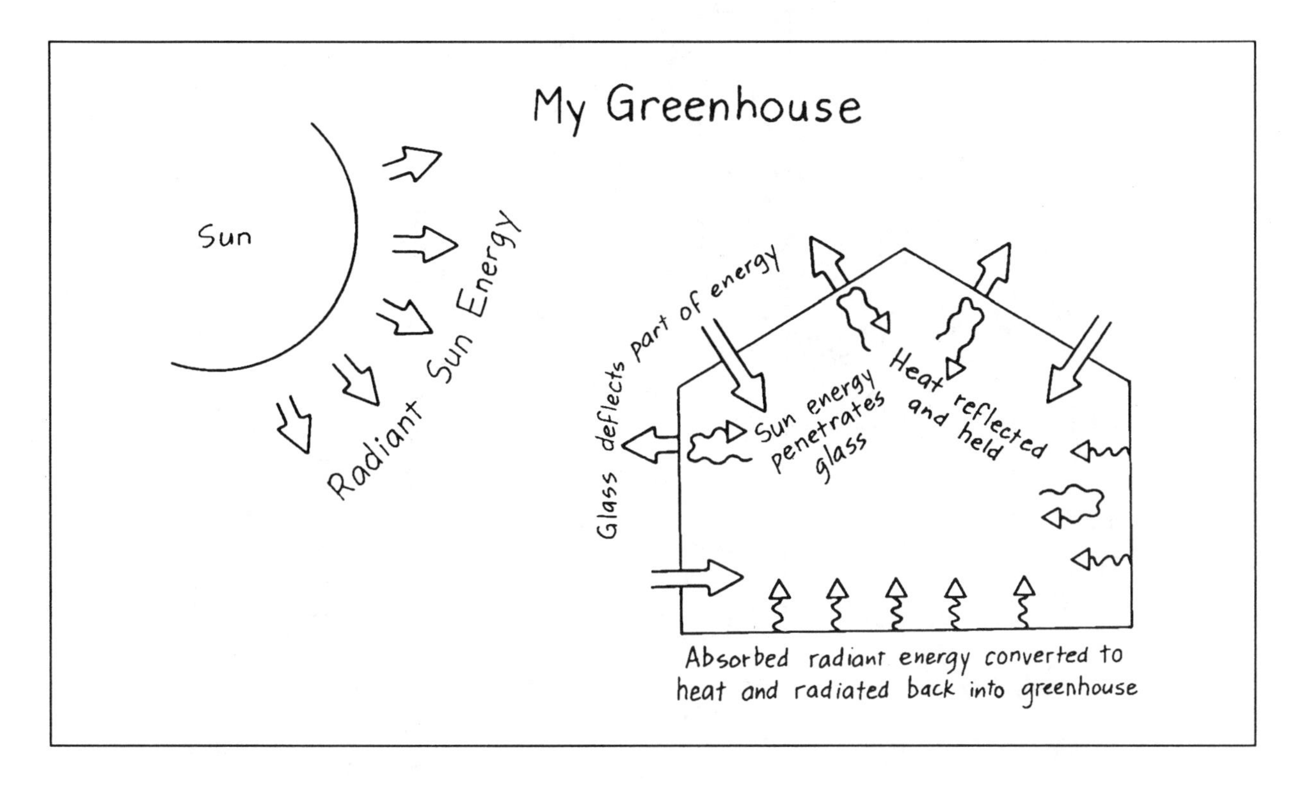

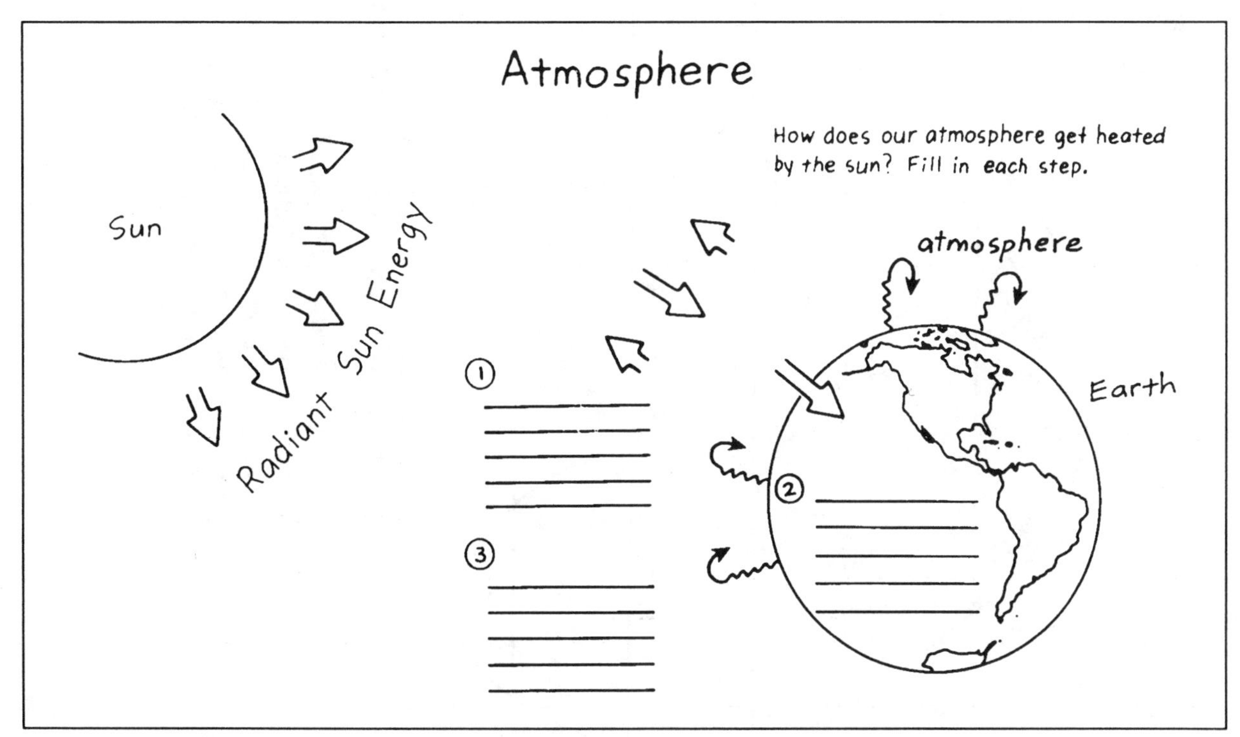

Climate Zone Map

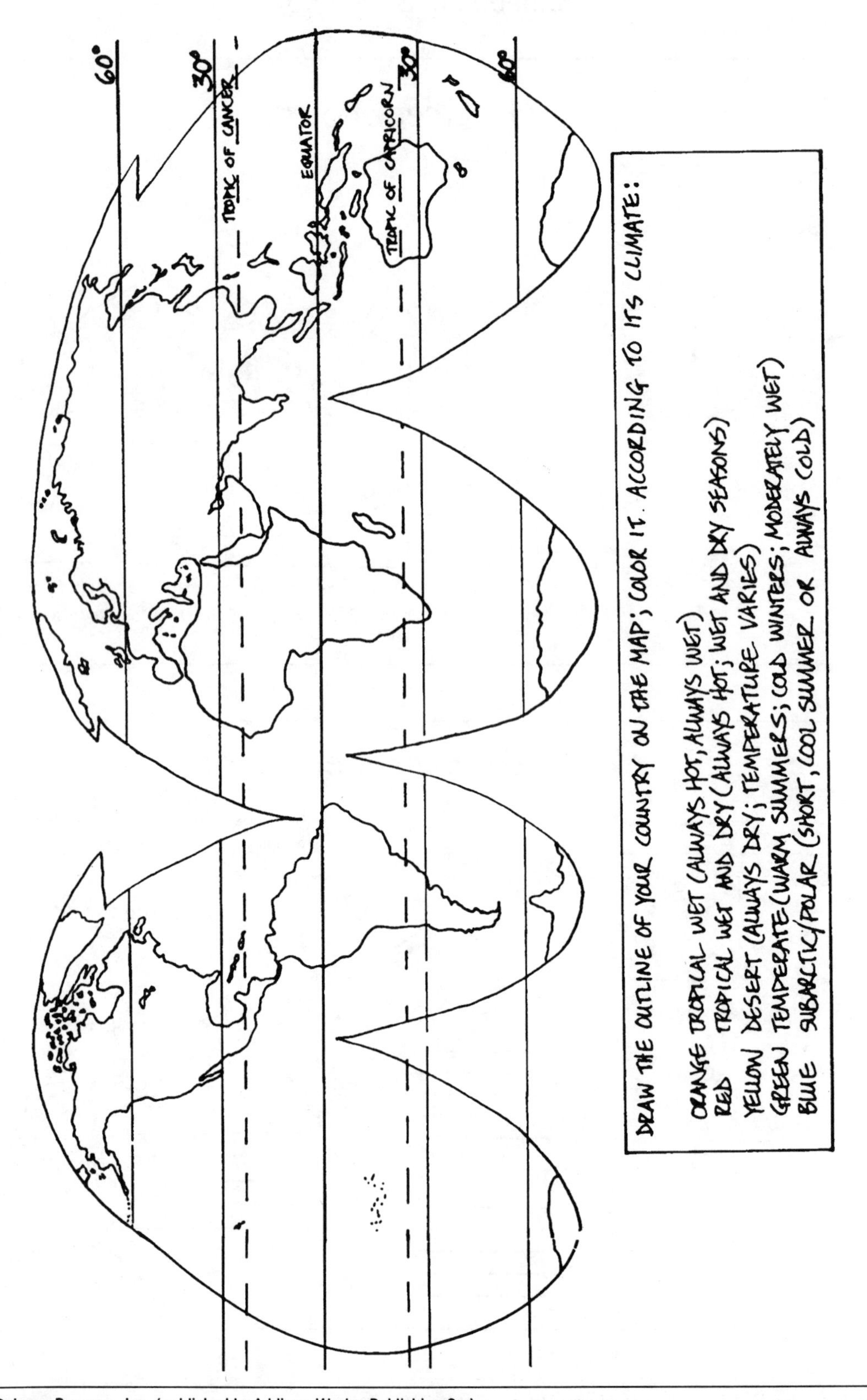

NAME DATE

Journey to Different Lands

Research questions for your region:

1. What is the latitude of this region?
2. How many miles is it from the equator?
3. What is the warmest time of the year in this region?
4. What are the high and average temperatures at this time of year?
5. What is the coldest time of the year in this region?
6. What are the lowest and average temperatures at this time of year?
7. How much precipitation does this region get in one year?
8. Give a general description of the climate.
9. Are there large bodies of water near this region?
10. If so, how does the body of water affect the climate?
11. How many people live in this region?
12. What styles of clothes do the men wear?
13. What styles of clothes do the women wear?
14. What types of food do they grow?
15. What types of houses do they live in?
16. How does this climate affect the life style of rural people in this region?
17. Other things we want to know:

____

____

____

Poster #1

Poster #2

Poster #3

Good (good for us to eat)	Bad (not good for us but have some OK parts)	Ugly (very bad for us to eat)
Cereals Whole grain cereals. These cereals have all the nutrients of the grains (not processed) and little or no sugar. Shredded Wheat Puffed Wheat Granola (some) Grape Nuts Flakes All Bran Team Wheat Chex Grape Nuts 40% Bran Flakes 100% Bran Raisin Bran Oatmeal—good variety of nutrients. Hominy Grits—good variety of nutrients.	Refined cereals with low sugar content—Many of the grain nutrients have been processed out. Puffed Rice Post Toasties Rice Chex Corn Flakes Cheerios Corn Chex Rice Krispies	Presweetened cereals—These cereals are high in sugar and use processed grains. Super Sugar Chex Sugar Frosted Flakes Frosted Mini Wheats Super Sugar Crisp Froot Loops Honeycomb Apple Jacks King Vitamin Sugar Frosted Cornflakes Sugar Pops Cap'n Crunch Crunch Berries Cocoa Puffs Cocoa Krispies Frosted Flakes Trix Cocoa Pebbles Fruity Pebbles Sugar Smacks
Main Dish Eggs—good protein source. Peanut butter—high in nutrients, especially protein. Cottage cheese—high in nutrients, especially protein. Yogurt—high in nutri-ents. (Note: fruit flavor yogurts have sugar in them. Use plain yogurt and add fresh fruit.)	White flour pancakes—some nutrients, but not many, and the extras (butter, syrup) are not good for us. White flour waffles—some nutrients, but not many, plus the extras (butter, syrup) are not good for us. Salted peanuts—good in nutrients but salt is bad for our hearts.	Bacon—very high in fats. Sausage—very high in fats.

Good (good for us to eat)	Bad (not good for us but have some OK parts)	Ugly (very bad for us to eat)
Main Dish (cont.) Whole wheat pancakes—many nutrients, but the extras (butter, syrup) are not good for us. Whole wheat waffles—many nutrients, but the extras (butter, syrup) are not good for us.		
Breads Whole wheat bread—low in calories, variety of nutrients. Corn tortilla—low in calories, variety of nutrients.	Cornbread—high in calories, but variety of nutrients. Biscuits—variety of nutrients.	Doughnuts—usually high in sugar and calories. Pastry rolls—usually high in sugar and calories. Bagel—usually high in sugar and calories.
Beverages Skim milk—many nutrients and not much fat. Lowfat 2% milk—many nutrients and not much fat.	Whole milk—many nutrients but high in fat content.	Cocoa—nutrients but high in sugar and fat, so very fattening. Chocolate milk—nutrients, but high in sugar and fat, so very fattening.
Fruits and Vegetables Vegetables—fresh, high in vitamins. Fruits—fresh/dried, high in vitamins.	Canned fruits/vegetables packed in water and natural juices—lose some nutrients in processing.	Canned fruits/vegetables packed in syrup—lose nutrients in processing and syrup has added sugar.
Extras		Butter—high in fats. Jelly—usually a lot of sugar.

Students may want to suggest foods to eat for breakfast that are not normally considered to be breakfast foods.

Searching for the Terrible Three

-10
high in fat
+10
-10
high in sodium
-10
extra high in sugar
+10
-20
· high in fat
· high in sodium
+10
-20
· high in sugar
high in fat

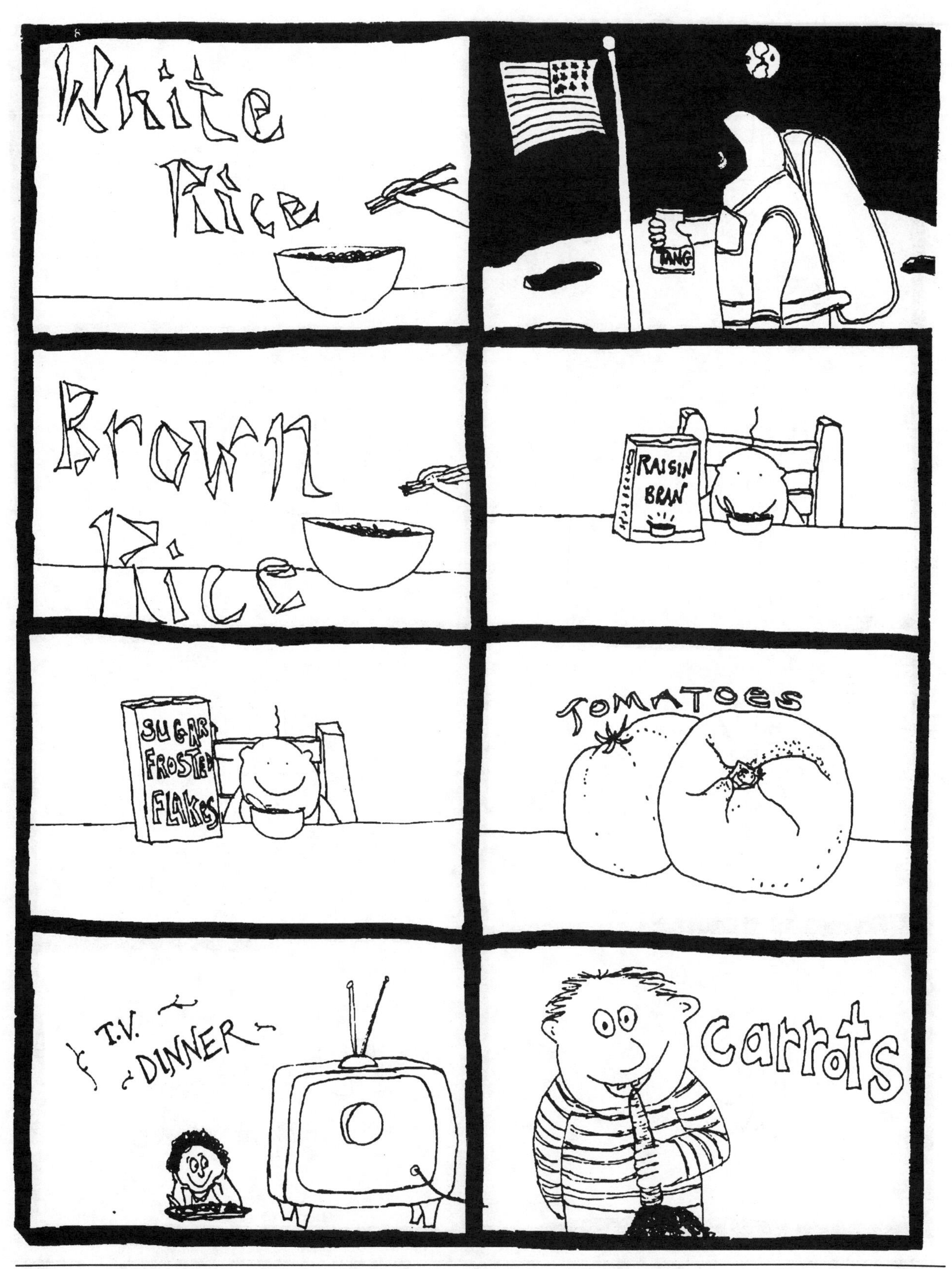
White Rice
TANG
Brown Rice
RAISIN BRAN
SUGAR FROSTED FLAKES
TOMATOES
T.V. DINNER
Carrots

-10
extra high in sugar
+0
+10
+10
+10
-10
high in sugar
+10
-10
high in sodium

eggs
cottage cheese
COTTAGE CHEESE
oatmeal
WHITE BREAD
SKIM MILK
hot dog
Soda Pop
Pop
cookie

+10	+0
-10 · high in sodium	+10
-20 · high in fat · high in sodium	+10
-10 · extra high in sugar	-10 · extra high in sugar

Peas
King
Vitamin
Cereal
fresh
fish
STEAK
Sugar
Frosted
Flakes
WHOLE
MILK
BACON
BAKED POTATO
ON
ON

-10 · high in sugar	+10
-10 · high in fat	+10
-10 · high in fat	-10 · high in sugar
+10	-20 · high in fat · high in sodium

Swiss Cheese
Lentil Soup
BROCCOLI
PEANUT BUTTER
YUM !?
TASTY POTATO CHIPS
MARGARINE
PUDDING
PEACHES packed in water

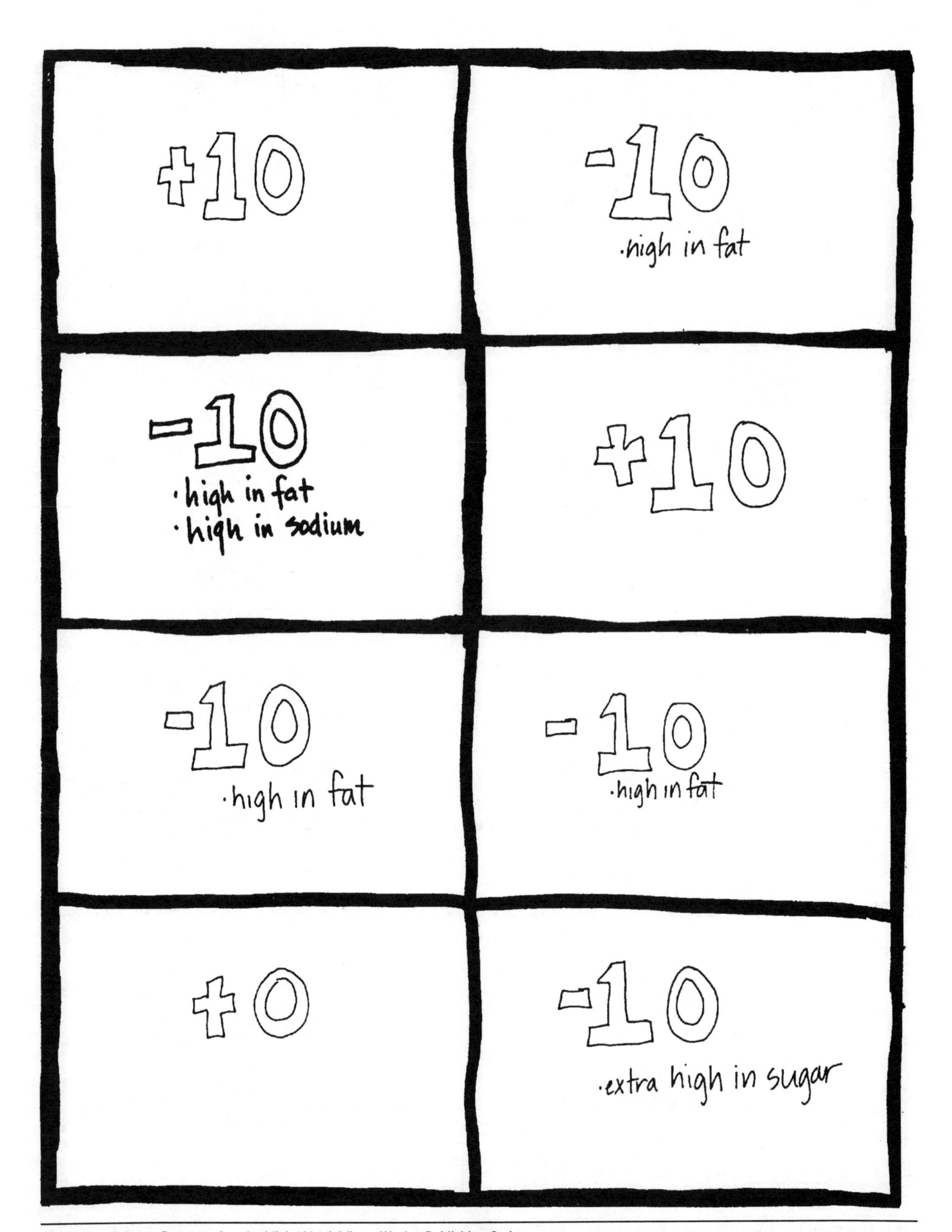
+10
-10
·high in fat
-10
·high in fat
·high in sodium
+10
-10
·high in fat
-10
·high in fat
+0
-10
·extra high in sugar

Lo-Fat
Yogurt
CANNED PEACHES IN SYRUP
CANNED PEACHES IN SYRUP
SWEETENED
APPLE
SAUCE
orange
juice
CHOCOLATE
Milk
SALAD
SMITH-JONES
BOLOGNA
SLICES
SOYBEANS

-10
·high in sugar
+10
+10
-10
·high in sugar
+10
-20
·extra high in sugar
+10
-20
·high in sodium
·high in fat

Meal Planning Chart

	Milk Group	Protein Group	Vegetable/Fruit Group	Grain Group	Extra Group
Breakfast					
Lunch					
Dinner					
Snack					

Remember: We need 3 servings from the Milk Group; 4 servings from the Vegetable/Fruit Group; 2 servings from the Protein Group; 4 servings from the Grain Group. We do not need foods from the Extra Group.

The Nutrient Chart

	NUTRIENT	FUNCTION	FOOD GROUP
Cats	carbohydrates	energy	grains fruits and vegetables
Wait	water	maintenance: carries nutrients in blood—maintains temperature	all
For	fats	energy/maintenance carries some vitamins	all
Mice	minerals	maintenance: regulate and maintain body functions	fruits and vegetables
Very	vitamins	maintenance: regulate and maintain body functions	fruits and vegetables
Patiently	protein	growth; tissue building and repair	proteins

Adapted from *Peanut Butter and Pickles,* Humboldt County Office of Education.

NAMES DATE

Buy Me! Buy Me!

What you bought	inexpensive	easy to fix	advertised on TV	has high nutritional value	has nothung artificial in it	my parents buy it	want to try something new	catchy packaging prize inside	friend said to try it	on sale	my family likes it	other
Totals												

Building a Coldframe for Your Life Lab

Coldframes extend your growing season into the colder months of winter and give you a headstart on seedlings before spring has officially arrived.

You can build many different types of coldframes, depending on your needs and budget. Following is a design for a simple coldframe. Using math and art skills, students can help to design, plan, and build. Try to recycle materials when possible. You can use excess lumber and plastic film. Plan to have coldframes situated facing south so that they receive at least six hours of direct sunlight during the spring and summer. Most of all, coldframes should be in an accessible area, where students can reach, water, observe, and work within them with ease.

You can grow most anything inside a coldframe, although it is best to stay away from plants that tend to spread out when growing, such as peas or melons. Try growing leafy greens for fresh salads in the wintertime, and start seedlings for early spring crops. Grow tomatoes and peppers in late summer for a fall crop.

Some Points to Remember: Since gardening in coldframes is intensive and requires nutrient-rich soil, you might want to add some compost to the soil to help enrich it.

Temperatures in a coldframe can get quite hot if it is not ventilated properly. Be sure to have a thermometer inside so that temperatures can be regulated. Ventilate when necessary by opening vents or lifting sides of plastic, opening windows, and so on. The best temperatures for growing most plants can differ from regular outdoor gardening. Cooler season plants grow well with coldframe temperatures between 40° to 70° F and warm weather plants at 60° to 80° F.

To increase heat retention during very cold weather, you can place the coldframes 8"-10" in the ground or paint the insides black. Plastic jugs filled with water placed inside the coldframe also help to increase heat retention. On very cold nights, you can cover the coldframe with straw or even blankets.

Since coldframes help retain water, it is important to be sure that you don't overwater. Overwatering can cause mildew and mold to grow and ruin your crops. Water condensation inside the coldframe is a sign of overwatering. If the north side of the coldframe is raised 6" above the south side, rain water will runoff and heat retention will improve.

A Simple Coldframe Design

Materials:

Lumber (All wood preferably redwood or pressure-treated wood.)

Two 7' long 2 X 12s
Two 4' long 2 X 12s
One 7' long 2 X 10
Two 4' tapered 2 X 10s
Four 3 1/2' long 1 X 2s
Four 44" long 1 X 2s

4 hinges
4 L-brackets

Two 3 1/2' X 4' pieces polyethylene film or plastic

Tacks

Hammer

Foam insulation strips (weatherstripping)

Thermometer

Step 1:
Frame base of coldframe using 2 X 12 boards as per diagram. Add L-brackets in each inside corner for stability.

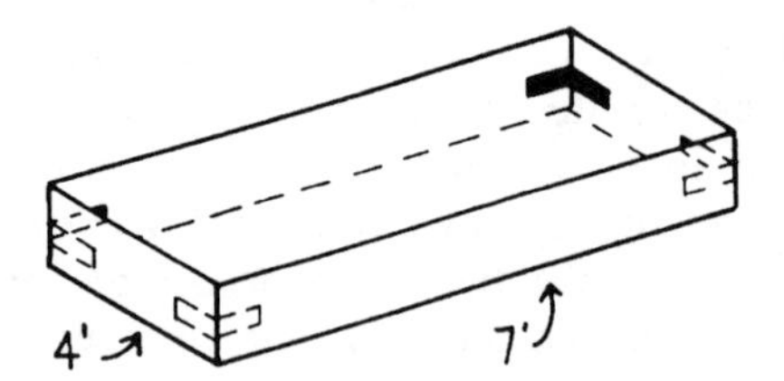

Step 2:
Add tapered boards and back piece (2 X 10s) to base to add height to coldframe.

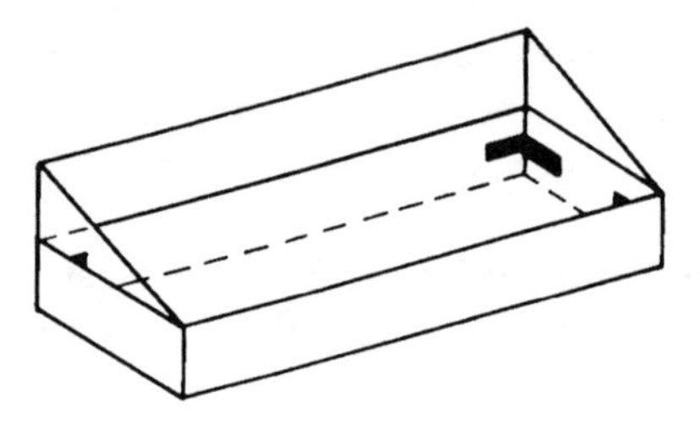

Step 3:
Frame covers as per diagram. Then tack plastic film to outside of each cover.

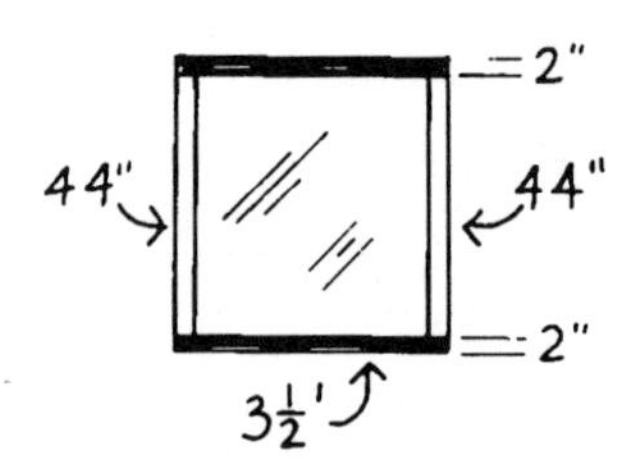

Step 4:
Using hinges, attach cover to top back side.
Note: Add foam insulation strips on frame below cover if necessary to add insulation and cut down on heat loss..

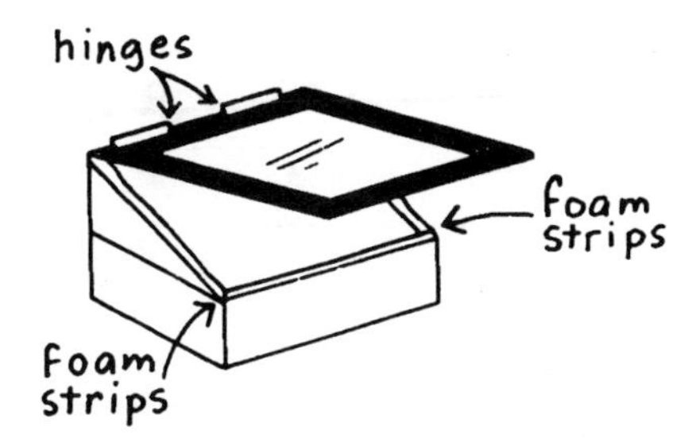

Step 5:
Add thermometer to inside back wall so that it may easily be read.
Note: Coldframe can be propped open with small sticks for ventilation. Soil can be mounded around edges for added insulation.

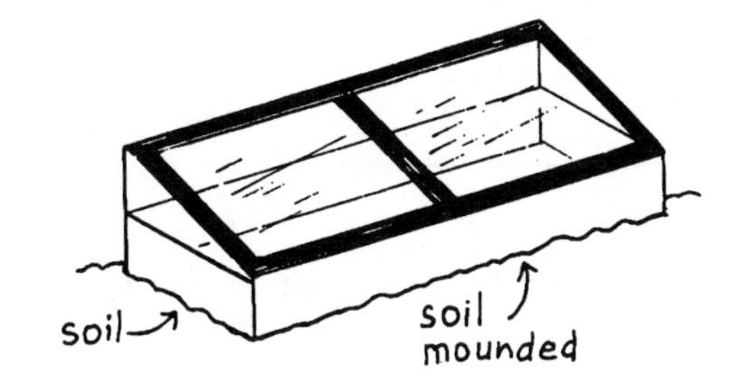

Plans for a Planter Box

48"L X 11.5"H X 20.75"W

Materials:

Lumber (All lumber should be redwood; otherwise it will rot.)

Two 4' long 1 X 12s
Two 20 3/4" long 1 X 12s
Two 4' long 1 X 10s
Five 19" long 2 X 4s

8d Galvanized nails

Bottom:
Place 2 1 X 10s next to each other with a slight space in between. Nail 2 X 4s to slats.

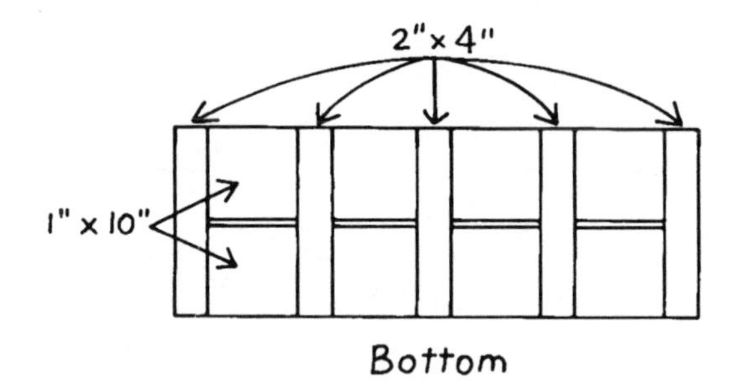

Side:
Nail through the side, both into the bottom 1 X 10s and into 2 X 4s.

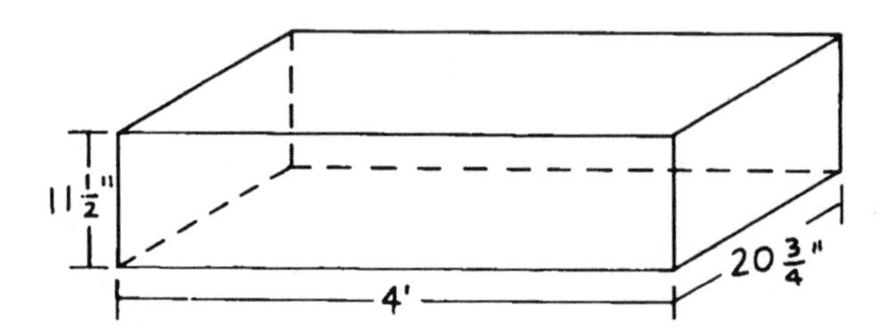

End:
The last piece goes over the whole end. It should measure 20 3/4" wide but you should check the measurement first. Nail through the end piece to existing sides and bottom.

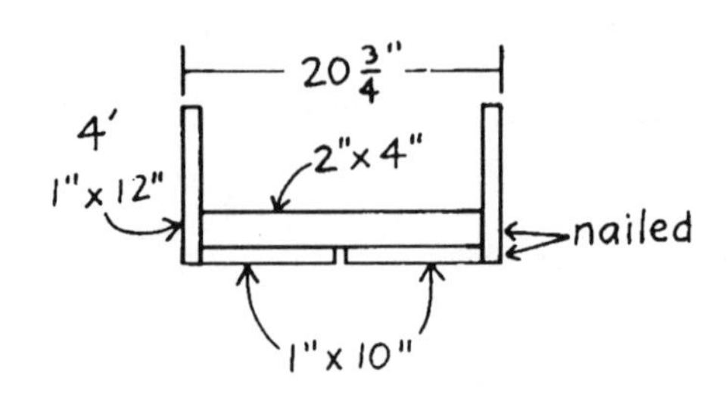

Flat-Making

Materials: (for one regular flat)

Two 1" X 4" X 11 1/2" redwood or fir boards (end pieces)

Five 1/4" X 4" X 18" redwood pieces of bender board (sides and bottom)

Hammer

Twenty 4 penny galvanized nails (Dull the tips of the nails by hammering them once. This will lessen the chance of the bender board splitting while assembling the flat.)

All the pieces will be nailed onto the two 1" X 4" X 11 1/2" end pieces. Follow the simple steps below to build one flat.

Nail bender board sides, two 1" X 4" X 11 1/2", to the two 1" X 4" X 11 1/2" end pieces. This forms the rectangular frame to nail all other pieces onto. Two nails at each end are suggested for adequate reinforcement.

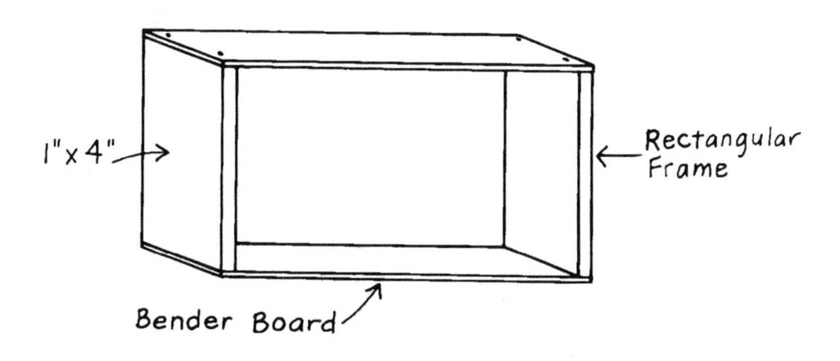

Align the remaining 3 pieces of bender board along the frame to form the bottom. Leave approximately 1/8" space between each board for drainage and air movement. Again, two nails at each end are suggested.

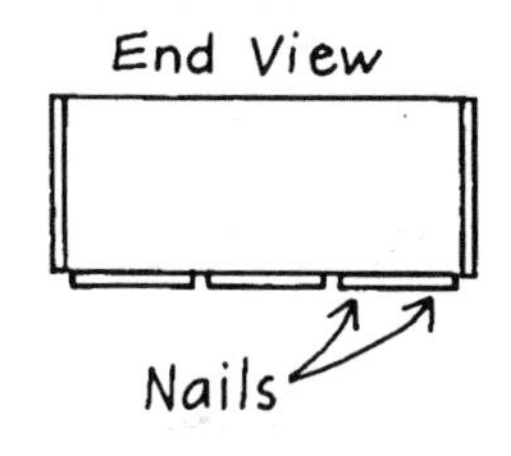

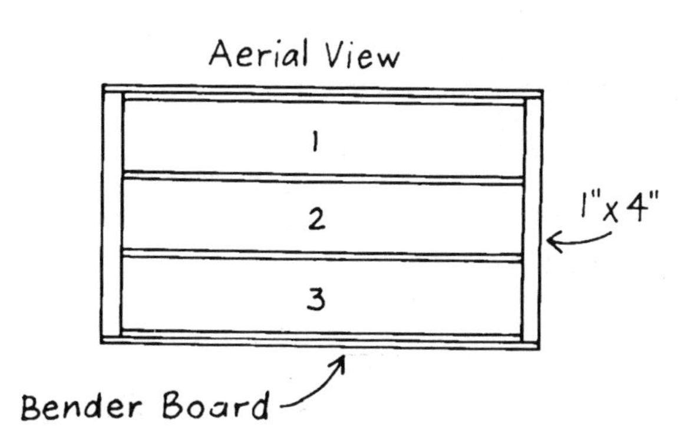

Making and Using Root View Boxes

Some root view box experiments

Show the movement of water through various kinds of soils.

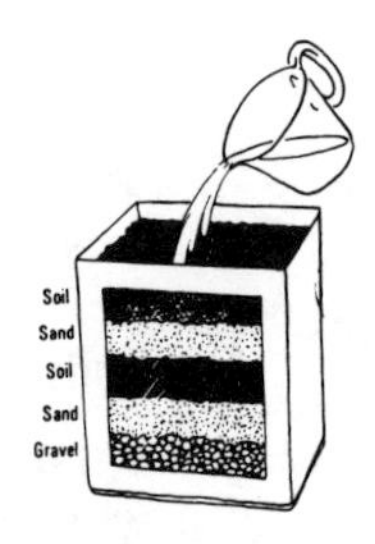

Show how water moves through soil layers; how it must build up a head to move out of one into another.

See how the roots and the top of a plant change their growth when the box is laid on its side after the plant has developed in the upright position.

Milk carton root view boxes

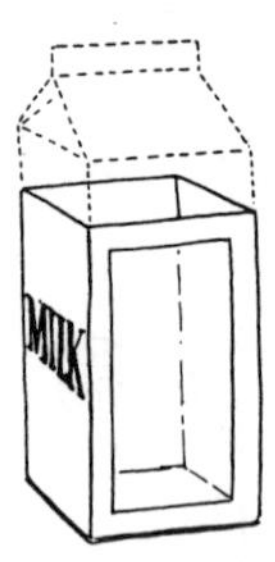

Cut the top (for a vertical box) or the side (for a horizontal box) from a half-gallon milk carton.

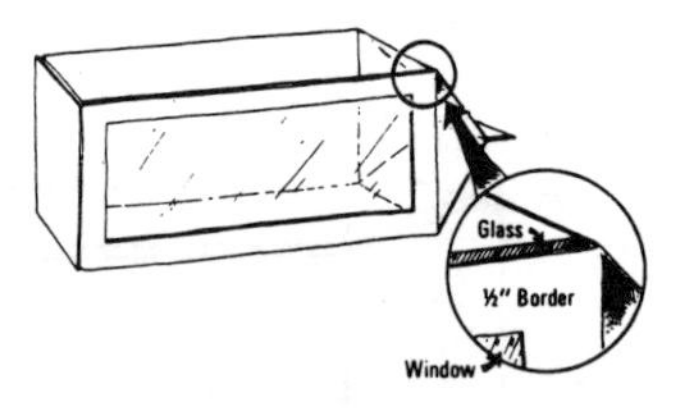

Cut out window area leaving about 1/2 inch of carton between corner and window. Cut glass to fit tightly into the corners of the carton. Waterproof glue or pruning paint may be used for a tighter seal.

Since roots tend to grow straight down, the window must be slanted (carton tipped) to keep roots growing against the glass and all of their action visible.

How to build a root view box

Nail wood pieces together and, before inserting window, spray interior with pruning paint or other water-proofing. Fill to within 1 inch of the top with soil mix and plant seeds 1/2 inch from window. Keep soil moist but not soggy.

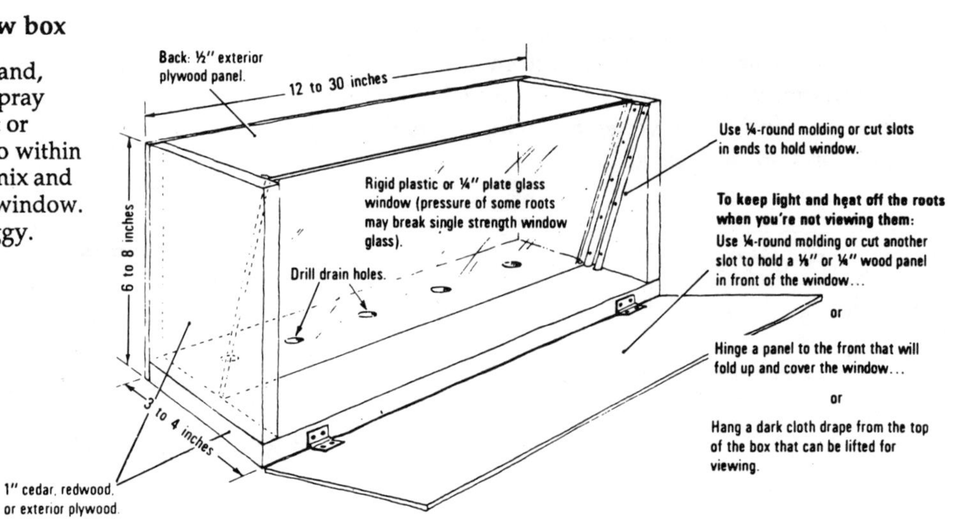

Adapted from *A Child's Garden*, Chevron Chemical Company.

Worm Box

Materials:

Two 5/8" X 35" X 12" CDX* plywood pieces (side pieces)

Two 5/8" X 23" X 12" CDX plywood pieces (end pieces)

One 5/8" X 24" X 36" CDX plywood pieces (top piece)

One 1/4" X 1/4" hardwood cloth, 24" X 36" (bottom of box)

12 metal hinges

30 heavy gauge staples

12 ardox nails**

Drill with 1/2" bit

Hammer

* CDX plywood is exterior grade, good on one side; #2 pine boards or scrap lumber can be substituted.

** Ardox nails have a spiral shape that increases their holding power.

The illustration below shows how to interlock the corners of the box for greater strength. Once the sides are nailed together with about three nails per side, secure the hardware cloth onto the bottom with heavy gauge staples. The hardware cloth on the bottom of the box provides good drainage and aeration. Attach the hinges to the 24" X 36" piece of plywood, and then secure the lid onto the box.

Place the box on pieces of wood or bricks so that it is up off the ground. This will also help to promote aeration and drainage.

Insect Collecting Net

Materials: (for one net)

One 52" piece of #8 iron wire

One 1" aluminum tubing, 4" long

One 7/8" X 3' wood dowel

One 48" X 30" piece of muslin

Needle and thread

1. Take wire and fold as shown.

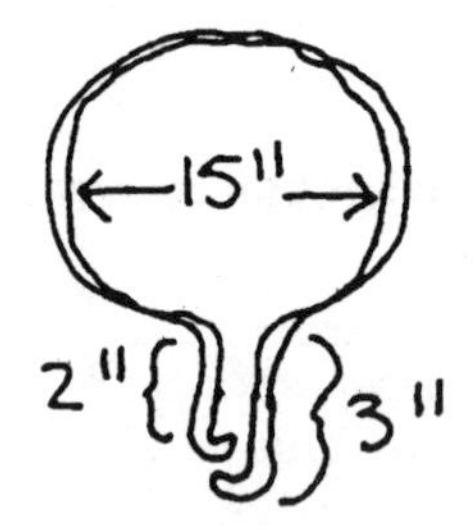

2. Take dowel and cut 2 grooves to hold the wire.

3. Slide piece of aluminum tubing up to the base of the wire. A small screw keeps the tubing from slipping.

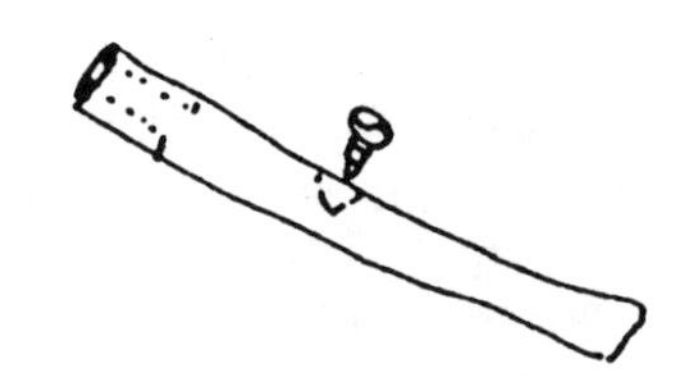

4. Fold the netting and sew as shown, leaving a 1 1/2" hem.

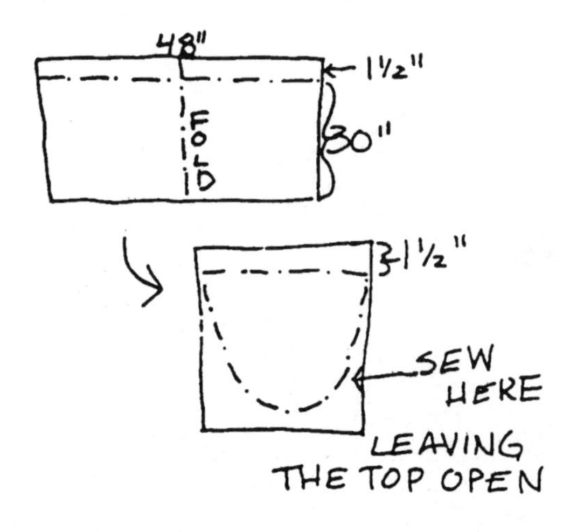

5. Hem the netting around the wire and you've got it!

Directions: Ski-Making

Materials:

Two 10 foot 2" X 4"s

84 feet of strong cord

24 eyehooks

Hammer

Hammer the eyehook into the 2 X 4.

Slip a 7' length of rope through the loop and make a strong knot.
Note: For younger students shorter boards with fewer ropes can be used.

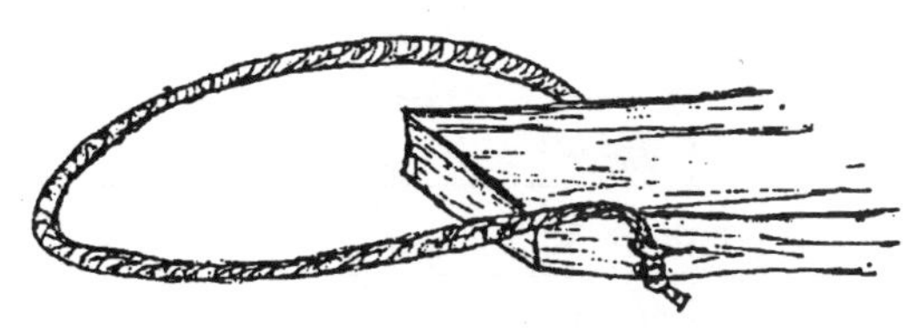

Do the same on the other side.

Now do the same with the other 2 X 4 and begin!

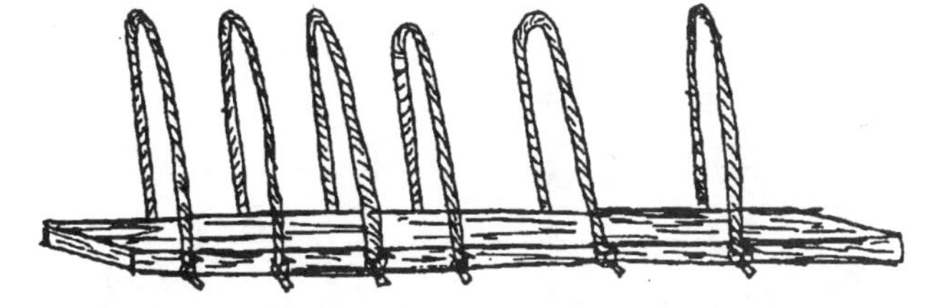

Make 6 of these rope handle along the board and you should end up with something like this!

Vegetable Planting Guide

Vegetable	Warm Weather	Cool Weather	Sow in Flat	Sow Direct	Spring Planting		Fall Planting
					Start seeds*	Set-out*	**
Bush Beans	•			•	3-4 before	1-2 after	12 before
Pole Beans	•			•	3-4 before	1-2 after	12 before
Beets	•			•		2-4 before	8-10 before
Broccoli		•	•		5-8 before	5-8 before	14-17 before
Brussel Sprouts		•	•		5-8 before	4-6 before	17 before
Cabbage		•	•		4-6 before	5 before	13-14 before
Carrots	•	•		•		2-4 before	13 before
Cauliflower		•	•		5-8 before	1-2 before	14 before
Celery	•		•		8-10 before	2-3 before	19 before
Chard	•	•		•		2-4 before	6 before
Corn	•			•	3-4 before	1-2 after	6 before
Cucumber	•			•	2-3 before	1-2 after	11 1/2 before
Eggplant	•		•		6-8 before	2-3 after	14 before
Garlic		•		•		6 before	Sept. w/mulch
Kale		•	•			5 bef-2 aft	6-8 before
Kohlrabi		•	•			5 bef-2 aft	10 before
Leeks		•	•		8-10 before	5 before	
Lettuce	•	•	•			2-4 b:3 aft	6-8 before
Onion		•	•			3 bef:2 aft	8 after spring
Parsley	•	•	•	•	4-6 before	1-2 after	
Peas		•		•		4-6 bef; 2-3 aft	12 before
Peppers	•		•			1-3 after	
Potatoes	•			•		4-6 before	
Pumpkin	•			•		after frost	
Radish	•	•		•		4-6 before	7 before
Spinach	•	•		•		3-6 before	6-8 before
Squash, Summer	•			•		1-4 after	10 before
Squash, Winter	•		•			2 after	13 before
Tomatoes	•					2-4 after	

*Weeks before or after last frost **Weeks before 1st frost

Vegetable	Days to Emerge	Days to Harvest	Spacing of Plants (inches)	Depth to Plant Seeds (inches)	Soil Temp. for Germination	Best Air Temp. for Growing
Bush Beans	4-10	50-60	6	1	60-85	60-80
Pole Beans	4-10	60-70	6-8	1	60-85	60-80
Beets	7-10	50-80	2-4	1/2	60-75	50-75
Broccoli	5-10	80-100	15-18	1/4	50-65	60-75
Brussel Sprouts	8-10	100-110	18	1/4	65-75	55-70
Cabbage	4-10	80-100	18	1/4	50-75	50-75
Carrots	10-17	50-75	2	1/4	55-75	45-75
Cauliflower	5-10	60-80	15-18	1/4	50-75	60-72
Celery	7-12	90-120	6	1/4	50-75	60-75
Chard	7-14	45-55	8	1	40-70	45-70
Corn	3-10	90-110	12-15	1	55-85	50-95
Cucumber	3-8	60	12-24	1	65-85	60-80
Eggplant	5-13	90	18	1/2	65-85	65-85
Garlic		180-200	4-6	1/2		
Kale		100-120	15	1/2	40-70	40-70
Kohlrabi	5-10	50-70	6-9	1/4	50-75	40-75
Leeks	7-14	130-160	4-6	1/2	below 70	60
Lettuce	2-10	60-80	10-12	1/4	45-70	55-70
Onion	4-12	85-200	4	1/4	50-80	60-85
Parsley	11-27	70-90	4	1/4		
Peas	6-15	60-80	4	1	40-75	55-75
Peppers	8-20	80-100	10-12	12	65-85	65-85
Potatoes	10-15	140-160	10-12	6	60-65	60-80
Pumpkin	7-10	110-130	36	1	65-85	50-90
Radish	3-10	25-40	1	1/4	40-85	45-75
Spinach	6-14	50-55	4-8	1/4	60-70	40-75
Squash, Summer	3-12	50-60	15-24	1	65-85	60-85
Squash, Winter	4-10	80-120	24-36	1/2 to 1	65-85	60-85
Tomatoes	6-14	80-100	18-24	1/4 to 1/2	65-85	65-85

Vegetable Planting Guide

Additional Comments

Bush Beans sensitive transplanting; pinch extra plants, don't pull them. Pick every 3 - 7 days.

Pole Beans sensitive transplanting; pinch extra plants, don't pull them. Pick every 3-7 days.

Beets Thin when young and cook tops as greens.

Broccoli Keep cool to get stocky plants, but don't go below 40'F. Transplant into beds up to first true leaves. Harvest main head when buds begin to loosen. Side heads will form after first head is cut.

Brussel Sprouts Keep cool to get stocky plants, but don't go below 40'F. Harvest sprouts when they are 1 1/2" wide. Pick lower ones first.40'F. Harvest sprouts when they are 1 1/2" wide. Pick lower ones first.

Cabbage Harvest when head is formed. Keep cool to get stocky plants but don't go below 40'F.

Carrots Thin when plants are small; harvest any size.

Cauliflower Tie outer leaves around head to protect from sun. Likes temperature between 57'-68'F.

Celery Must go below 60'F at night for seeds to germinate. Requires a lot of nutrients and water.

Chard Cut leaves close to ground when 8-10" high. Harvest outer leaves first.

Corn Sensitive to transplanting, pinch extra plants. Plant in blocks, harvest when kernals are milky.

Cucumber Somewhat sensitive to transplanting; pinch extra plants. Mound soil into hills;plant 3 seeds per hill.

Eggplant Grows well in hot weather, harden off carefully.

Garlic Harvest with digging fork when tops start to die.

Kale Keep cool to get stocky plants, but not below 40'F. Cut outer leaves closer to stem when10" or longer.

Kohlrabi Keep cool to get stocky plants, but not below 40'F. Harvest when enlarged stem is 3" in diameter.

Leeks Keep cool to get stocky plants, but not below 40'F. Plant out when 4" high.

Lettuce Keep cool to get stocky plants, but not below 40'F. Head lettuce likes repotting. Plant successively every two weeks. Will go to seed in high temperatures. Harvest outer leaves of leaf lettuce vs. head.

Onion Harvest with digging fork when tips start to die back.

Parsley Soak seeds overnight before planting to speed germination, be patient! Cut outer leaves near stem.

Peas Sensitive to transplanting, pinch extra plants, don't pull them. Harvest frequently.

Peppers Sensitive to cold, harden off gradually. Green peppers turn red when ripe.

Potatoes Very tender; cannot tolerate frost. Dig up with digging fork after tops have flowered.

Pumpkin Sensitive to transplanting; pinch, don't pull plants. Plant in hills, 3-4 plants per hill, 6-8 ft. apart.

Radish Plant every 10 days. Will get woody when over-mature.

Spinach Keep cool for stocky plants. Plant every 2 weeks. Will go to seed in hot weather.

Squash, Summer Sensitive to transplanting; pinch extra plants, don't pull them. Harvest freuqently

Squash, Winter Sensitive to transplanting; pinch extra plants, don't pull them. Can store through the winter.

Tomatoes Prefers warm days and cool nights. Taste the difference!

Companion Planting Guide

Vegetable	Plant with	Don't plant with
Beans	Potatoes, carrots, cucumbers cauliflower, cabbage, summer savory, most other vegetables and herbs	Onion, garlic, gladiolus
Beans, Bush	Potatoes, cucumbers, corn, celery, summer savory, sunflowers, strawberries	Onions
Beans, Pole	Corn, summer savory	Onions, beets, kohlrabi, sunflower
Beets	Onions, kohlrabi	Pole beans
Cabbage Family (cabbage, cauliflower, kale, kohlrabi, broccoli)	Aromatic plants, potatoes, celery, dill, chamomile, sage, peppermint, rosemary, beets, onions, thyme, lavender	Strawberries, tomatoes, pole beans
Carrots	Peas, leaf lettuce, chives, onions, leek, rosemary, sage, tomatoes	Dill
Celery	Leek, tomatoes, bush beans, cucumbers, pumpkin, squash	
Corn	Potatoes, peas, beans, cucumbers, squash, pumpkin	
Cucumbers	Beans, corn, peas, radishes, sunflowers	Potatoes, aromatic herbs
Eggplant	Beans	
Leek	Onions, celery, carrots	
Lettuce	Carrots and radishes (lettuce, carrots, and radishes make a strong team grown together), strawberries, cucumbers	
Onion/Garlic	Beets, strawberries, tomato, lettuce, summer savory, chamomile, beans (protects against ants)	Peas
Parsley	Tomatoes, asparagus	
Peas	Carrots, turnips, radishes, cucumbers, corn, beans, most vegetables, herbs (adds Nitrogen to soil)	Onions, garlic, gladioulus, potatoes

Vegetable	Plant with	Don't plant with
Potato	Beans, corn, cabbage, horseradish (should be planted at corners of patch), marigold, eggplant (as a lure for Colorado potato beetle)	Pumpkin, squash, cucumber, sunflower, tomato, raspberry
Pumpkin	Corn	Potato
Radish	Peas, nasturtium, lettuce, cucumbers	
Soybeans	Grows with anything; helps everything	
Spinach	Strawberries	
Squash	Nasturtium, corn	
Strawberry	Bush beans	
Sunflower	Cucumbers	Potato
Tomatoes	Chives, onion, parsley, asparagus, marigold, nasturtiums, carrots, limas	Kohlrabi, potato, fennel, cabbage
Turnip	Peas	

Herbs	Companions and Effects
Basil	Companion to tomatoes; dislikes rue intensely; improves growth and flavor; repels mosquitos and flies.
Beebalm	Companions to tomatoes; improves growth and flavor.
Borage	Companion to tomatoes, squash, and strawberries; deters tomato worm; improves flavor and growth.
Caraway	Plant here and there; loosens soil.
Catnip	Plant in borders; deters flea beetle.
Chamomile	Companion to cabbages and onions; improves growth and flavor.
Chervil	Radishes; improves growth and flavor.
Chives	Companion to carrots; improves growth and flavor; plant around base of fruit trees to discourage insects climbing trunks.
Dill	Companion to cabbage; dislikes carrots; improves growth and health of cabbage.
Fennel	Plant away from garden; most plants dislike it.
Garlic	Plant near roses and raspberries; deters Japanese beetle; improves growth and health; plant liberally throughout garden to deter pests.
Horseradish	Plant at corners of potato patch to deter potato bug.

Herbs	Companions and Effects
Hyssop	Companion to cabbage and grapes; deters cabbage moth; keep away from radishes.
Lamb's Quarters	This edible weed should be allowed to grow in moderate amounts in the garden, especially in the corn.
Lemon Balm	Sprinkle throughout garden.
Marigolds	The workhorse of the pest deterrents; plant throughout garden especially with tomatoes; it discourages Mexican bean beetles, nematodes, and other insects.
Mint	Companion to cabbage and tomatoes; improves health and flavor; deters white cabbage moth.
Marjoram	Plant here and there in garden; improves flavor.
Nasturtium	Companion to tomatoes and cucumbers.
Petunia	Protects beans; beneficial throughout garden.
Purslane	This edible weed makes good ground cover in the corn.
Pigweed	One of the best weeds for pumping nutrients from the subsoil, it is especially beneficial to potatoes, onions, and corn; keep weeds thinned.
Rosemary	Companion to cabbage, bean, carrots, and sage; deters cabbage moth, bean beetles, and carrot fly.
Rue	Keep it far away from sweet basil; plant near roses and raspberries; deters Japanese beetle.
Sage	Plant with rosemary, cabbage, carrots, beans, and peas; keep away from cucumbers; deters cabbage moth and carrot fly.
Summer Savory	Plant with beans and onions, improves growth and flavor; deters bean beetles.
Tansy	Plant under fruit trees; companion to roses and raspberries; deters flying insects, Japanese beetles, striped cucumber beetles, squash bugs, and ants.
Tarragon	Good throughout the garden.
Thyme	Plant here and there in the garden; it deters cabbage worm.
Yarrow	Plant along borders, paths, near aromatic herbs; enhances essential oil production.

Adapted from *Organic Gardening and Farming*, February 1972, pp. 32-33, 54, and *The Encyclopedia of Organic Gardening*, Rodale Press, Inc., 1978, pp. 233-235.

Scope and Sequence

We suggest this Scope and Sequence for schools where more than one grade level is involved in Life Lab. This Scope and Sequence has been field tested with grades 2-6. We emphasize different units at different grade levels as well as repeat activities that present core concepts or that can be interpreted at a more sophisticated level in the upper grades. Most of all, we encourage you to use the plan as a starting point for designing a grade level sequence that works well for your school. Let us know what you devise and we'll share your ideas with other Life Labs.

Section	Grade 2	Grade 3
Let's Work Together/ Problem Solving and Communication	10-4 Good Buddy I Just Love Applause	10-4 Good Buddy I Just Love Applause Lighthouse Count Off Knots
We Are All Scientists/ Awareness and Discovery	Sharp Eyes Candid Camera Ear-Ye, Ear-Ye Everyone Needs A Rock See No Evil, Hear No Evil	Candid Camera Ear-Ye, Ear-Ye Six of One, Half Dozen of the Other See No Evil, Hear No Evil
The Living Earth/Soil	Tools and Us Dig Me and Dig Me Again	Space Travelers Sensual Soil Living in the Soil Tools and Us To Dig or Not To Dig Dig Me and Dig Me Again Soil Doctors What Good is Compost? The Matchmaker
Growing	Bioburgers Seedy Character Let's Get to the Root of This Seed Power Lotus Seeds Room To Live Seed To Earth Stem, Root, Leaf, or Fruit?	

Grade 4	Grade 5	Grade 6
10-4 Good Buddy I Just Love Applause Lighthouse Tell It Like It Is Knots Skis	Lighthouse The Connected Circle Knots Skis Sinking Ship	Tell It Like It Is Knots Skis Who Am I?
Candid Camera The Unnature Trail Big Ears Only The Nose Knows See No Evil, Hear No Evil Mystery Powders Little Munchkins Burma Shave	The Unnature Trail Only the Nose Knows Six of One, Half Dozen of the Other Mystery Powders On Location	Little Munchkins Burma Shave On Location
Tools and Us Dig Me and Dig Me Again Soil Doctors	Space Travelers Sensual Soil The Nitty-Gritty Water, Water Everywhere Tools and Us Dig Me and Dig Me Again Soil Doctors A Soil Prescription What Good Is Compost? What's To Worry? Splash Day at the Races	Tools and Us Dig Me and Dig Me Again Soil Doctors
Bioburgers Seedy Character Let's Get to the Root of This ZIP Code Seeds Adapt-a-Seed It's Getting Stuffy in Here Growing, Growing, Gone Glass Seed Sandwich Run Root Run Let's Get a Handle on This Which Way Did It Grow? Sugar Factories Plant Food Magic	ZIP Code Seeds	Let's Get to the Root of This ZIP Code Seeds Sugar Factories Sipping Through a Straw Plant Sweat Magical Mystery Tour Plants Need Light Too Plant Food Magic Star Food

Section	Grade 2	Grade 3
Living Laboratory/ Outdoor Gardening	It's As Simple As One, Two, Three... Four So What? Sow Seeds! Transplanting	It's As Simple As One, Two, Three... Four So What? Sow Seeds! Transplanting
Cycles and Changes	The Cycle Hunt Compost Bags Let's Make a Compost Cake	Adopt-a-Tree Me and The Seasons Collector's Corner
Interdependence		I Eat the Sun Lunch Bag Ecology Part One The Hungry Bear
Garden Ecology		Magic Spots
Garden Creatures		Earth, Planet of Insects The Great and Powerful Earthworm The Butterfly Flutter By
Climate	Degrees Count Temperature Hunt	Temperature Hunt The Station Creation Keeping Track
Food Choices	Keep Me Running I Eat My Peas with Honey The Good, the Bad, and the Ugly	Keep Me Running How Do We Make a Horse into Jell-O? Are You A Natural? Hear Us, Hear Us The Good, the Bad, and the Ugly On Your Mark, Get Set, Breakfast! Button Up for Breakfast

Grade 4	Grade 5	Grade 6
It's As Simple As One, Two, Three... Four So What? Sow Seeds! Transplanting Water We Doing?	It's As Simple As One, Two, Three... Four So What? Sow Seeds! Inch By Inch, Row By Row Transplanting What's in a Name? Weeding, Writing, Arithmetic	It's As Simple As One, Two, Three... Four So What? Sow Seeds! Inch By Inch, Row By Row Transplanting What's in a Name?
The Power of the Circle Roundabout What a Deal Bring in the Clean-Up Crew Compost Bags Let's Make a Compost Cake Mother Earth	What a Deal Dr. Jekyl and Mr. Hyde You Look Different... Compost Bags Let's Make a Compost Cake	Compost Bags Let's Make a Compost Cake The Cycle of Recycle A Human Paper Factory
Eat the Earth Lunch Bag Ecology Part One Lunch Bag Ecology Part Two	The Hungry Bear We're Just Babes in the Woods	The Hungry Bear You Are What You Eat Lunch Bag Ecology Part Two DDT Chew The Day They Parachuted Cats Into Borneo Caught in the Web of Life
Magic Spots I Need My Space	Flower Power Part One Flower Power Part Two Magic Spots Natural Defense Under the Big Top	Flower Power Part One Magic Spots Garden Puzzle Companion Planting Under The Big Top Plant Architects
Earth, Planet of Insects Who Lives Here? Following the Thread	Earth, Planet of Insects Insect Anatomy Buggy Diner Ladybug, Ladybug	Earth, Planet of Insects Slimy Characters on Trial My Friend Tummy-Foot
I'm the Hottest A Ravishing Radish Party We've Got Solar Power!	The Station Creation Keeping Track Wind, Rain, Sleet, or Hail	A Shoebox of Sunshine What's the Angle? Glow Globe A Journey to Different Lands
Keep Me Running When I Was Little The Roots of Food The Fabulous Familiar Four Searching for the Terrible Three Dear Diary Food Planners	Keep Me Running	

Section	Grade 5	Grade 6
Nutrients *(most appropriate for Grades 5 and 6—adapt for lower grades)*	Invisible Gold The Cat's in the Carbohydrates The Sugar Is in the Cat... Fat Cats Burn Out Water You Made Of? The Body's Helping Hands Patient Proteins to Grow On We Complement Each Other Nutrition Charades	
Consumerism *(most appropriate for Grades 5 and 6—adapt for lower grades)*		Supermarket Snoop Yes, It's Got No Bananas The $1,000,000 Orange This Little Lettuce Went to Market The New, *Improved* Madison Avenue Diet Try It, You'll Like It Buy Me! Buy Me! Feast or Famine Market Garden

Materials

This list does not refer to ordinary items such as paper, pencils, and so on, or to the blackline masters in this book (see each activity for blackline master requirements).

Unit 1: Let's Work Together (Problem Solving and Communication)

Classroom Materials

squares of wide masking tape for each student, numbered sequentially

blindfold

Garden Materials

hammer

Special Materials

84 ft. of strong cord

Two 10-foot 2x4s

24 eye hooks

Unit 2: We Are All Scientists (Awareness And Discovery)

Classroom Materials

15-20 objects (see The Unnature Trail for description)

Life Lab journals

hand lenses

1 blindfold per student

3x5 cards

various objects from around the classroom (see Ear-Ye, Ear-Ye)

one garden map outline per student

5 different pairs of noisemakers

1 noisemaker unlike any other

Garden Materials

1 rock per student

Special Materials

string

old sheet

cloth

ammonia

various fragrances for Only the Nose Knows

one opaque container (film container) for each fragrance

sugar

flour

one egg carton per group of three

salt

11 paper bags with ear holes cut out

6 containers such as pie tins

powdered milk

baking soda

cement (plaster of Paris)

Unit 3: The Living Earth (Soil)

Classroom Materials

hand lenses

strong tape

Life Lab journals

large white sheets of construction paper

grease pencil

string

Garden Materials

4 containers of soil: clay, sand, compost, garden soil

bucket/gallon pot

compost

topsoil

sand

flats

seeds

soil test kit

garden tools (see Introduction)

seedlings

2 or 3 varieties of mature legume plants, at least one plant per student—planted in garden at least 6 weeks before activity

a selection of seeds from the legume family: bell beans, fava beans, red clover, alfalfa, peas, purple vetch

legume cover crop to plant, min. 2 seeds per student

garden plant markers and grease pencil

2 minimum-maximum thermometers mounted on sticks, or soil thermometers

mulching material: straw, leaves

sand or pebbles to anchor empty milk cartons

one watering can with sprinkler head per group

ready-to-use compost

seedlings, approximately 20 of the same variety: broccoli, lettuce

Special Materials

one glass jar with lid per group of 5

4 lamp chimneys

screen or cheesecloth

quart jars (4)

measuring cups

legume inoculant (optional)

microscope (optional)

1/2 gal. milk cartons

several copies of local yellow pages

4 lunch bags

clock or watch

2 spoons

fork

5 shoe boxes

5 blocks

strong tape

Classroom Materials	Garden Materials	Special Materials
	4 dry soil samples of very different soils: sand, garden soil, compost, clay	
	tool cleaning materials: wire brushes, sand, motor oil	
	2 experimental garden beds, planted with the same crops	
	sod	
	planter box (see appendix)	

Unit 4: Growing

Classroom Materials	Garden Materials	Special Materials
tray	variety of seeds	toothpicks
Life Lab journals	2 pkg pea seeds	balloons
labels	beans	paper bags
tweezers	lima beans	tacks
eyedropper	small flat with soil	cork
grease pencils	climbing pea seeds (1 pkg)	feathers
paper towels	small potted plants	metal springs
	potting soil	wire
	radish seeds (3 pkg)	4 small bottles with cork stoppers
	5 different soils	2 glass jars
	vermiculite	rubbing alcohol
	seeds: pumpkin, bean, corn, pea	2 pieces of glass (8"x10")
	3 identical plant containers for each group	piece of wood 4"x1"
	several 4" pots	fresh celery stalks with leaves
	soil	poster or sketch of hamburger
	mature plants (15)	magnifying glasses
	bean seeds (1 pkg)	bottle of bromothymol blue
	seeds with different root systems	two sprigs of aquarium plant (Elodea)
	2 garden beds or one long one	test tubes or bottles with stoppers and pipettes or straws

Classroom Materials

Garden Materials

seedlings to transplant: broccoli, tomatoes

vegetables for snack

one soaked pinto bean per student

12 each of four seeds: lima bean, corn, pumpkin, radish

one plant with large leaves

seedling (marigold, pea, bean)

Special Materials

5 jars

alcohol burner

pins

2 beakers

burner stand

iodine

aluminum foil (recycled from home)

seed catalogues

list of recommended vegetables and flowers from County Extension

knife

clear glasses or cups

4 clear plastic containers with holes for drainage

milk carton with clear plastic or glass front

small sticks

red food coloring

measuring cups

plastic ties

root view boxes (see appendix)

fresh spices

cutting board

dip for snack

large plastic bags with ties or rubberbands (15-20)

Unit 5: Living Laboratories (Outdoor Gardening)

Classroom Materials

labels

grease pencils

graph paper

newspaper

Garden Materials

one flat per group of five

seeds

soil mix ingredients

flat mix

Special Materials

map or list of last year's garden

seed packets and catalogues

magnetic compass

Classroom Materials

class chart

string

Life Lab journals

Garden Materials

2 dug and fertilized beds

different varieties of one particular vegetable

bean or pea seeds (1 pkg)

plant stakes

water sprayers or sprinkling cans

crop seed for one bed: bush beans, radishes

potting soil

garden tools

seedlings ready to be transplanted from a flat

Special Materials

1/2 pint milk cartons

measuring cups or graduated cylinders

squares of paper labeled A-E for each student

Unit 6: Cycles and Changes

Classroom Materials

Life Lab journals

4 large sheets of butcher paper

season labels for each student (see The Cycle Hunt)

brads (1 per child)

corner of classroom with display board and table

paper for decorative wall or table background

hand lenses

meter stick

Garden Materials

soil

sand

hose with fan sprayer

water access with hose and fan sprayer

compost materials

pebbles, rocks, gravel

small plants: ferns or houseplant seedlings

humus, moss (optional)

materials gathered from garden or schoolyard characteristic of season

grass clippings

markers (sticks, stones)

Special Materials

clear containers with lids

clear plastic bags

petri dishes with covers

pieces of fruit, vegetable, bread

various decomposing and nondecomposing materials

1 large plastic bag with closing tie

3 grapes

plastic cup

lettuce leaves

2 nails

white bread (one slice)

toilet paper

whole wheat bread (one slice)

sifting screen

walnut shell

apple

Classroom Materials

Garden Materials

Special Materials

old bicycle

microscope (optional)

plastic tube

sheets of used white writing paper

pan

bucket or plastic tub

blender

fine mesh screen to fit inside pan

piece of cloth or old bedsheet

sponge

large glass container (fishbowl, aquarium, or large pickle jar)

untreated charcoal

plastic wrap or glass

spray bottle

materials for collage (See Roundabout)

enough objects or pictures of objects representing each of the four seasons so each student can find one

plastic cup

soil thermometer

Unit 7: Interdependence

Classroom Materials

string

students' lunches

8 sweaters or jackets

Garden Materials

2 stakes

Special Materials

apples (a few)

knife

21 paper plates

12 colored flags or markers

reference books on plants and animals (optional)

tape measure (to mark 300' path)

Unit 8: Garden Ecology

Classroom Materials

Life Lab journals

drawing boards

insect reference books (optional)

pictures of corn, bean, squash plants

grease pencil

hand lenses (optional)

meter stick

Garden Materials

flowers

radish seeds (4 pkg)

garden markers

seeds or seedlings

3 small plots in garden (2' x 2')

Special Materials

four 1 gal. plant containers

leaves of eucalyptus, bay, pine, tomato

cardboard tubes

bug boxes (optional)

coat hangers or sticks for stems

Unit 9: Garden Creatures

Classroom Materials

Life Lab journals

plant and insect reference guides

hand lenses

Garden Materials

variety of seedlings (flowers and weeds)

soil mix

lettuce leaves

garden snails (1 per pair of students)

compost, soil, leaves, and straw to layer in each worm box

Special Materials

2 terrariums with tight mesh covers

one clear plastic tray per 2 students (1 per pair)

fingernail polish

black cloth or paper

2 clear plastic containers or 2 root view boxes

5"x7" file cards with box

10 petri dishes with covers

ladybugs (see Resources)

red worms (1 container)

one bug box per group of three students

one clear jar with lid per group of five students

jar lid filled with water

one small jar lid or 35 mm film cannister lid per pair of students

3 T. of margarine or butter

a few apples

Classroom Materials

Garden Materials

Special Materials

one balance scale (with 1,2,5, and 10g masses)

newspaper

Unit 10: Climate

Classroom Materials

large piece of poster board

Life Lab journals

large bulletin board space for graph

grease pencil

paint (white, black, and colors)

paintbrushes

encyclopedia or reference books

several pints of black tempera paint

globe

magazines to use to cut out pictures (optional)

Garden Materials

white sand

dark soil

five flats

potting soil to fill 5 flats

radish seeds for 5 flats

garden labels

one potted pea or bean plant per pair of students

Special Materials

Air thermometers

70cm white elastic strip, one inch wide

70cm colored elastic strip, one inch wide

hammer

nails

Velcro

exchange class (see Prep. in lesson)

weather map—national temperature chart from newspaper

40cm square board to set experiment on

three baby food jars per pair of students

clip-on lamp

24 glow-in-the-dark stickers

strong light (flashlight)

8 large paper cups

5 soil/compost thermometers

one shoe box for each group of four or two students

insulating materials (newspaper, puffed cereal, packing materials)

flashlight

one thermometer capable of measuring at least 140°F per group

aluminum foil

Classroom Materials

Garden Materials

Special Materials

cellophane

cardboard

one 11"x13" clear plastic food bag per pair of students

letter envelopes

postage stamps

Unit 11: Food Choices

Classroom Materials

Life Lab journals

pictures of food

pictures of plants and animals that are sources of food

buckets of water

posterboard or cardboard

large sheets of newsprint

Garden Materials

Special Materials

samples of: wheat plant, wheat berries, whole grain flour, white flour

hand flour grinder

sample cereal cartons

ingredients high in sugar, fat, salt

safety pins or button making machine

food pictures of basic food groups, plus extra foods

ingredients for breadmaking

Breakfast Food Cards (available from Dairy Council, see p. 480)

Velcro

Unit 12: Nutrients

Classroom Materials

3" squares of brown paper

Life Lab journals

Garden Materials

Special Materials

small brown bags

pictures of food

nutrient labels

2 clear cups

cornstarch

5 eyedroppers

iodine

Classroom Materials	Garden Materials	Special Materials
		food samples:

flour water	orange
bread	sugar water
apple	zucchini
potato	oil
milk	margarine
egg	peanuts
tuna fish	vegetable
fruit juice	cookies
tofu	soda

small cups

Tes-tape

metal pie plate

candle

jar

1 walnut

1 sugar cube

1 iron tablet

1 2"x3" piece of aluminum foil

matches

salt

cooking oil

cracker

tongs

cotton

rubbing alcohol

4 paper plates

food samples (see The Body's Helping Hands)

5 small jars

sodium hydroxide solution

copper sulfate solution

international cookbooks

recipe ingredients and cooking materials (for We Complement Each Other)

slice of bread

Classroom Materials	Garden Materials	Special Materials
		rusty nails in a plastic bag
		3"x5" blank cards

Unit 13: Consumerism

Classroom Materials	Garden Materials	Special Materials
Life Lab journals	seeds for chosen crops (beans, squash, peas, tomatoes)	variety of packages with ingredient labels
8"x11" sheets of oak tag		one can banana pudding
		one orange with price tag
		1/2 a fresh orange per student
		hand juicer
		bowl
		pitcher
		one coffee stirrer or juice bar stick per student
		one small paper cup per student
		freezer space
		knife
		cutting board
		measuring cup
		seasonal list of locally grown fruits and vegetables (available from County Agricultural Extension)
		one grocery store newspaper ad per group of 4 students
		whole food ads from newspapers and magazines
		food samples of food to be advertised by class
		Chemical Cuisine Chart ordered from Let's Get Growing (optional)
		peanuts with shells

English/Spanish Vocabulary List

KEY:
(a) = adjective
(n) = noun
(v) = verb
[ ] = words between brackets are not a correct translation, but they are frequently misused in the same sense of the correct translation.
Translated by Francisco Javier Espinoza

English	Spanish
abdomen	abdomen, vientre, [panza]
adaptation	adaptación, ajuste
advertisement	anuncio
advertising	publicado (n), anunciando (v)
agriculture	agricultura
air	aire
air pressure	presión de aire
amino-acids	aminoácidos
animals	animales
antennae	antenas
atmosphere	atmósfera
barometer	barómetro
"beds"	"camas", camas agrícolas
calcium	calcio
calories	calorías
carbohydrate	carbohidrato, hidrato de carbono
carbon dioxide	dióxido de carbono
carnivore	carnívoro
change	cambio
chlorophyll	clorofila
circle	círculo, [rueda]
clay	arcilla, barro
climate	clima
cloud	nube
cold	frío
communication	comunicación
community	comunidad
compaction	compactación, apisonamiento
companion	acompanante, compañero or compañera
compost	composta
compromise	compromiso
consumer	consumidor
"control" (for experiment)	testigo
convection	convección
cooperation	cooperación
consumerism	el consumo
crop	cultivo
cycle	ciclo
deciduous	decíduo, cáduco
decision	decisión
decomposers	detritívoros, descomponedores
decomposition	descomposición, degradación
degrees	grados
depend	depender (v)
depth	profundidad, hondura
development	desarrollo
drain	drenaje
earthworm	lombriz de tierra
ecosystem	ecosistema
embryo	embrión
encouragement	apoyo, estimulación, animar
energy	energía
environment	medio ambiente
evaporation	evaporación
exchange	intercambio
exoskeleton	exoesqueleto
farmer	agricultor, campesino, [ranchero]
fat	gordo (a), grasa (n)
filament	filamento, [hilo]
"(seedling tray) flat"	cajita para sembrar
flower	flor
fog	niebla, neblina

food	comida, alimento
food chain	cadena, alimenticia
fuel	combustible
fungi	hongos
garbage	basura, desperdicio
garden	jardín
gases	gases
geotropism	geotropismo
germination	germinación
grain	grano
gravity	gravedad
greenhouse	invernadero
group	grupo
growth	crecimiento
habitat	hábitat
hail	saludar (v), vitorear (v), granizo(n)
harvest	cosecha (n), cosechar (v)
hear	oir (v)
heat	calentar (v), calor (n)
herbivore	herbívoro
hot	caliente
humans	humanos
humidity	humedad
humus	humus, [tierra negra]
hydrotropism	hidrotropismo
ingredients	ingredientes
insect	insecto
insulation	aislamiento, térmico o eléctrico
interrelationship	correlación, relación mutua, interrelación
iron	hierro, fierro
jaw	mandíbula, quijada
leaf	hoja
maintenance	mantenimiento
manure	guano, boñiga, [caca de ganado o gallinas]
meat-eater	carnívoro
microclimate	microclima
milk	leche
minerals	minerales
mold	moho (pronounce mo-o)
molecules	moléculeas
mouth	boca
nitrogen	nitrógeno
nutrients	nutrientes
nutrition	nutrición
ovary	ovario
oxygen	oxígeno ["aire"]
pests	plagas
petal	pétalo
phosphorus	fósforo
photosynthesis	fotosíntesis
phototropism	fototropismo
pistil	pistilo
planting	siembra, plantar, sembrando
plants	plantas
poison	veneno, ponzoña
pollen	polen
pollination	polinización
potassium	potasio
predator	predador
prey	presa
process	proceso
producers	productores
protein	proteína
psychrometer	psicrómetro
radiation	radiación
rain	lluvia
rain gauge	pluviómetro
rake	rastrillo
recycle	reciclar, reusar
reflection	reflexión
resources	recursos
root	raíz
safety	seguridad
salt	sal
sand	arena
seasons	estaciones del año
seed coat	cubierta de semilla, cáscara de semilla
seedling	plántula, [plantita]
seeds	semillas
senses	sentidos
sepal	sépalo
shade	sombra
shovel	pala
sight	sentido de la vista, vistazo, visión
silt	limo
sleet	granizo
smell	olfato (n), oler (v)
soil	suelo, tierra
solar	solar
solar collector	colector solar
spacing	espaciar, espaciamiento
spading fork	trinche, rastrillo
stamen	estambre
starch	almidón
style	estilo

sugar	azúcar
support	apoyo (n), apoyar (v)
systems	sistemas
taste	sentido de gusto, saborear (v) [probar (v)]
temperature	temperatura
texture	textura
thermometer	termómetro
thigmotropism	tigmotropismo
thorax	tórax, torso
tool	herramienta
topsoil	capa superior del suelo
touch	tacto (n), tocar (v), toque (n)
transpiration	transpiración, sudar
transplant	transplantar (v), transplante (n), injerto (n)
trophic levels	niveles tróficos
variety	variedad
vegetable	verdura
vegetarian	vegetariano or vegetarian
Vitamin A	vitamina A
Vitamin C	vitamina C
vitamins	vitaminas
warm	tibio or tibia
weather	tiempo ["clima"]
weeds	malezas, hierbas
wheelbarrow	carretilla
wind	viento ["aire"]
wind meter	anemómetro
wind vane	veleta
wing	ala

Seed Company List*

West

Bountiful Gardens
5798 Ridgewood Rd.
Willits, CA 94590
Organically grown, open pollinated seed, cover crops, grains, misc. Biological pest controls.

Exotica Seed and Rare Fruit Nursery
P.O. Box 160
Vista, CA 92083
Tropical vegetable and fruit seed, nursery stock; catalog available.

Shepherd's Garden Seeds
7389 West Zayante Rd.
Felton, CA 95018
Also has non-hybrid seeds.

Native Seed Search
7389 West New York Avenue
Tuscon, AZ 85745

Redwood City Seed Company
P.O. Box 362
Redwood City, CA 94064
Unusual herb and vegetable seeds; also California native plant seeds.

J. L. Hudson, Seedsman
P.O. Box 1058-TB
Redwood City, CA 94064
Rare and unusual seeds, including culinary, medicinal and dye herbs, imported vegetables, ornamentals, wildflowers.

Northwest

Nichols Garden Nursery
1190 North Pacific Hwy.
Albany, OR 97321
Herb seeds and plants, rare and Oriental vegetable seeds; non-hybrid seeds.

Territorial Seed Company
P.O. Box 27
Lorane, OR 97451
Non-hybrid seeds.

Midwest

Archias Seed Store
P.O. Box 109
Sedalia, MO 65301
Open pollinated corn, cover crops, small fruit, general vegetable, one of the oldest seed houses.

Field (Henry) Seed and Nursery
Shenandoah, IA 51602
Full line of vegetable and flower seeds; green manure crop seeds; unusual varieties and rare vegetables.

Cross Seed Co., Inc.
HC 69, Box 2
Bunker Hill KS 67626
Grains, cover crops, beans, cole crops, organically grown, misc.

Seed Savers Exchange
c/o Whealy
Rt. 2
Princeton, MO 64673

* Seed companies are listed regionally for your convenience. However, most companies provide seeds that can be grown in a variety of climates.

Northeast

Rohrer's
P.O. Box 5
Smoketown, PA 17576
Cover crops, general vegetable seeds.

Pinetree Garden Seeds
New Gloucester, Maine 04260
Untreated vegetable, flower, and herb seeds, especially varieties suited to intensive gardening.

Johnny's Selected Seeds
Albion, ME 04910
Hardy, short-seasoned varieties; organic seed and crop research.

Chas. C. Hart Seed Co.
Main and Hart Streets
Wethersfield, CT 06109
Wide selection of vegetable and flower seeds, including old-fashioned and non-hybrid vegetables.

W. Atlee Burpee Co.
Warminster, PA 18974
Vegetables, flowers, shrubs, trees, garden aids.

Stokes Seeds, Inc.
Box 548, Main P.O.
Buffalo, NY 14240
Quality flower and vegetable seeds, good variety for northern growers.

Thompson & Morgan, Inc.
Box 24
401 Kennedy Blvd.
Somerdale, NJ 08083
Large selection of flower and vegetable seeds.

Southeast

George Park Seed Company
Box 31
Greenwood, SC 29646

Hastings: Seedman to the South
434 Marietta St., NW
P.O. Box 4274
Atlanta, GA 30302
Vegetables and flowers exclusively for southern gardens.

Additional Resources

Books

Bringing Home The Bacon: School Gardens & Home Careers In Urban Farming, John Smith, 1980., Rancho Vejar, Inc. 37 Mountain Drive, Santa Barbara, CA 93013. A look at a youth farm and garden project with ways to integrate curricular areas.

BugPlay, Marlene Hapai and Leon Burton, 1990, Addison-Wesley Publishing Co., Menlo Park, CA 94025. This teacher sourcebook provides activities with insects for young children. Interesting information, blackline master drawings, poems, and tape cassette of songs teach children about these fascinating creatures.

Build it Better Yourself, William Hylton, Rodale Press, 33 E. Minor Stret, Emmaus, PA 18049. A book on how to construct almost anything you may need for your garden.

Children's Gardens: A Field Guide For Teachers, Parents, and Volunteers, Bremner and Pursey, 1982. Common Ground Gardens Program, 2615 S. Grand Avenue, Suite 400, Los Angeles, CA 90007. This excellent guide includes all information needed for successful children's gardening, even if you've never gardened. It includes special projects and activities appropriate for children ages 2-10. Common Ground also publishes bilingual leaflets on basic gardening.

Food First Curriculum, Laurie Rubin, 1984, Institute for Food and Development Policy, 1885 Mission St., San Francisco, CA 94103. A multicultural curriculum about food, the American and international roots of hunger, and local action for this global problem. The 35 worksheet activities are intended for grade 6 but may be modified for grades 4-5.

How To Grow More Vegetables, John Jeavons, 1982. Ten Speed Press, P.O. Box 7123, Berkeley, CA 94707. One of the original guides to organic gardening. In clear language, Jeavons explains the Biodynamic/French Intensive Method and teaches basic technique. Alan Chadwick, the father of the Biodynamic/French Intensive Method, called it "a masterpiece."

Project Seasons, Shelbourne Farms Resources, Shelbourne, VT 05482. Educational materials intended to assist classroom teachers in the task of "teaching the stewardship of natural and agricultural resources." An interdisciplinary program for grades K-6 that emphasizes science, math, language arts, and social studies. An excellent curriculum for northern climates.

Sharing Nature With Children, Joseph Bharat Cornell, 1979. Ananda Publishers, 14618 Tyler Foote Rd., Nevada City, CA 95959. Lessons on observation, sensory exploration, interdependence, and ecology in the out-of-doors.

Sunset New Western Garden Book, 1979, Lane Publishng Co., Menlo Park, CA 94025. This thorough gardening text covers climate zones, basic plant selection and care, landscaping, and an encyclopedia of terms and plant names.

Winter Gardening in the Maritime Northwest, Binda Colebrook, 1984, Maritime Publications, P.O. Box 527, Everson, WA 98247. A winter gardening guide with techniques for extending garden production through October to May cool weather season. Contains detailed information on principles of cool weather vegetable production, appropriate vegetable varieties, sowing dates, and resources to contact for more information specific to your area.

Worms Eat My Garbage, Mary Appelhof, 1982, Flower Press, 10332 Shaver Rd., Kalamazoo, Michigan 49002. One-hundred pages of information and instructions for a proper vermicomposting system (or worm bin).

The Youth Gardening Book., Lynn Ocone/Eve Pranis, revised 1987. National Gardening Association, 180 Flynn Ave., Burlington, VT 05401. A comprehensive, how-to guide for teachers, youth leaders, and parents. Topics such as planning and organizing a garden program, designing and maintaining a site, and fundraising are discussed.

Magazines

Science and Children, Monthly magazine published for grades K-8 teachers by the National Science Teachers Association, 1742 Connecticut Ave., NW, Washington, DC 20009. High quality articles, reviews, news, posters, and inserts for the elementary science teacher.

Ranger Rick's *NatureScope*, National Wildlife Federation, 1412 16th Street, NW, Washington, DC 20036-2266. A creative education series dedicated to inspiring in children an understranding of the natural world while developing the skills they will need to make responsible decisions about the environment.

National Gardening (formerly *Gardens For All*), monthly magazine for gardeners by the National Gardening Association, 180 Flynn Ave., Burlington, VT 05401. (Subscriptions: NGA Member Subscription Service, Sepot Sq., Peterborough, NH 03458); subscription includes membership to NGA.) Basic gardening tips, creative gardening ideas, and addresses for seed exchanges (great for letter writing by elementary school children.)

Supplies

Let's Get Growing, General Feed & Seed Company, 1900-B Commercial Way, Santa Cruz, CA 95065. Telephone (408) 476-5344. A catalogue of supplies for the Life Lab curriculum and other school garden projects. Catalogue includes information about the Banana Slug String Band and tapes.

Carolina Biological Supply Company, Burlington, North Carolina 27215. Telephone (800) 334-5551. Source for general science supplies.

Rincon-Vitova Insectories, P.O. Box 95, Oak View, CA 93022. Telephone (805) 643-5407. A good supply of beneficial insects.

Harmony Farm Supplies, 4050 Ross Rd., Sebastopol, CA 95472. Telephone (707) 823-9125. A good supply of insects and other garden supplies.

National Dairy Council, 6300 North River Road, Rosemont, IL 60018. Telephone (312) 696-1020.

Banana Slug String Band, P.O. Box 717, Pescadero, CA 94060. Telephone (408) 423-7807. Songbooks, tapes, lyrics, and teacher friendly chords for songs used in the Life Lab curriculum and other ecology programs. (Band is available for live performances, too.)